T0139844

Green Energy and Technology

Climate change, environmental impact and the limited natural resources urge scientific research and novel technical solutions. The monograph series Green Energy and Technology serves as a publishing platform for scientific and technological approaches to "green"—i.e. environmentally friendly and sustainable—technologies. While a focus lies on energy and power supply, it also covers "green" solutions in industrial engineering and engineering design. Green Energy and Technology addresses researchers, advanced students, technical consultants as well as decision makers in industries and politics. Hence, the level of presentation spans from instructional to highly technical.

Indexed in Scopus.

More information about this series at http://www.springer.com/series/8059

Nicu Bizon

Optimization of the Fuel Cell Renewable Hybrid Power Systems

 Springer

Nicu Bizon
Faculty of Electronics,
Communication and Computers
University of Pitesti
Pitesti, Romania

ISSN 1865-3529 ISSN 1865-3537 (electronic)
Green Energy and Technology
ISBN 978-3-030-40243-3 ISBN 978-3-030-40241-9 (eBook)
https://doi.org/10.1007/978-3-030-40241-9

This Springer imprint is published by the registered company Springer Nature Switzerland AG
The registered company address is: Gewerbestrasse 11, 6330 Cham, Switzerland

Foreword

The conventional centralized energy systems are comprised by large power plants such as hydro plants, thermoelectric plants and Combined Heat and Power (CHP) plants. The aging power systems cause to several grid problems such as intermittency, power quality issues, and blackouts since a few decades. Therefore, the power infrastructure requires serious troubleshooting studies in a wide manner. The Distributed Generation (DG) and integration of large distributed power plants such as Photovoltaic (PV) plants, wind turbines, energy storage systems and fuel cells have led to intensive interest on DG power grid infrastructure. The widespread use of DG based decentralized grids and microgrids decrease significant power losses and facilitate the operation and maintenance of microgrids. Besides, increased use of DC loads is easily supplied by DC microgrids that eliminate the requirement for power inverters. It is noted that the elimination of DC–AC power conversion can prevent power losses of entire system with 7% up to 15%, which is a remarkable ratio for a microgrid. In this context, the microgrid and distributed generation infrastructures have been widespread thanks to hybrid power plants which are comprised by integration of Renewable Energy Sources (RESs).

The hybrid power systems are comprised by integration of conventional and various power sources in a single plane which generates microgrid concept that are operated either in island mode in remote areas or normal mode defining grid-integrated operation. The most widely used hybrid plants are installed with integration of PV and wind power systems. However, the intermittent structure of solar and wind resources require additional precaution and ubiquitous sources such as batteries, fuel-cells, and energy storage back-ups for improving reliability and feasibility of power plant. Although the regulations, technical improvements, and market share aspects surround the next generation power system, it is driven by a number of enhancements. The regulatory approaches are more robust among others, but the technical improvements are also rather important. The prominent technical and regulatory enhancements for power systems can be listed as follows; cost reductions due to use of RESs, developments in information and communication technologies and data management technologies, growing concerns on energy security, reliability, and resiliency, gradually changing consumer profile, environmental concerns, changing revenue and investment predictions.

These aspects of hybrid power systems have been comprehensively presented and introduced by the author of this book. The modeling of energy storage systems, fuel-cell based power plants, and PV plants have been drawn, and control methods such as global extremum seeking algorithms, power optimization strategies, and fuel economy maximization are presented in an efficient and successful way in terms of theory and application approaches.

This book can be very essential for undergraduate education in classroom, and can be suggested for graduate researchers, professionals, and engineers. It will be an important resource for in hybrid energy systems area.

Prof. Dr. Ersan Kabalci
Nevsehir Haci Bektas Veli University
Nevsehir, Turkey

Preface

This book presents and discusses innovative solutions in the field of Hybrid Power Systems (HPS) and covers the modeling, design, performance and challenges related to a HPS based on the Renewable Energy Sources (RES) and a Proton Exchange Membrane Fuel Cell (PEMFC) system. It demonstrates how the PEMFC system under power-following control may compensate the variable power from RESs and loads with less support from the Energy Storage Systems (ESS). The batteries' stack will compensate the energy deviation of the power flow balance on the DC bus and the power storage device (such as the ultracapacitors' stack) will compensate the power pulsed on the DC bus. Thus, the batteries' stack will be operated in charge-sustained mode with many advantages compared with the charge-discharge mode of the battery that usually appears in other energy management strategies that use the ESS to compensate the variable power from the RESs and loads. Different hybrid ESS topologies such as passive, semi-active, and active hybrid ESS are analyzed. The performance of the active hybrid ESS topology is analyzed under power pulses on the DC bus. A brief review of the state-of-the-art energy management strategies for RES/FC HPS is performed. Potential combinations of control of the two energy generation systems (the RES and the FC system) are discussed and analyzed.

The highlights of this book are as follows: (1) presents innovative and performance hybrid power systems architectures and the challenges of RES integrating; (2) analyzes the RES/FC HPS architectures as clean and safe solutions to generate energy; (3) proposes the power-following control of the FC system to compensate the power flow balance on the DC bus under variable power of RESs and loads; (4) shows the advantages of the battery operating in sustained charging mode due to the power generated by the FC system or consumed by the electrolyzer when the RES power is lower or higher than the load demand in order to compensate the power flow balance without support from the ESS; (5) proposes the optimization of the FC system's operation through one or two optimization loops; (6) analyzes the performance of the optimization functions that may increase the FC net power or improve the fuel economy; (7) proposes a new performance index to evaluate the Global Maximum Power Point Tracking (GMPPT) algorithms used for

Photovoltaic (PV) under Partially Shaded Condition (PSCs); (8) analyzes the performance of the GMPPT algorithm based on the Global Extremum Seeking (GES) technique; (9) describes the benefits of the RES/FC HPS architectures under power-following control.

The structure of the chapters that responds to the aforementioned highlights is as follows.

Chapter 1 presents the main research directions identified in the literature to be of great interest in the Hybrid Power Systems (HPS), which will be detailed in the next chapters of this book.

Chapter 2 presents the Hybrid Power System (HPS) architecture as an optimal combination of Renewable Energy Sources (RESs) with a backup energy source to sustain the dynamic load demand. The Proton Exchange Membrane Fuel Cell (PEMFC) system is proposed as a non-polluting backup energy source, instead of a diesel generator. The models used in simulation for the PEMFC system, battery, ultracapacitors, wind turbine, photovoltaic panel, power converters and load demand were presented. Also, the performance indicators for efficient use of a RES/FC HPS have been presented.

Chapter 3 proposes the Load-Following (LFW) strategy for a FC HPS without RES support to sustain the power flow balance on the DC bus with the battery operating in charge-sustaining mode. The excess of energy during light load stage can supply an electrolyzer or can be sold if FC HPS is connected to the network. Optimization of the FC HPS operation may be performed using well-known optimization algorithms such as the Global Maximum Power Point Tracking (GMPPT) algorithms, Global Maximum Efficiency Point Tracking (GMEPT) algorithms, or fuel economy algorithms. The performance of an advanced fuel economy strategy using fueling regulators switching is analyzed. Also, a new mitigation strategy of the power pulses based on the anti-pulse control of the Hybrid Storage System (HSS) is analyzed.

Chapter 4 analyzes the asymptotic Perturbed-based Extremum Seeking Control (aPESC) schemes for global search using performance indicators such as searching speed, searching accuracy, tracking efficiency and searching resolution. Also, the performance indicators such as fuel economy and electrical energy efficiency have been used for the PEMFC system.

Chapter 5 analyzes the improvements in net power of the PEMFC systems based on the Load-Following (LFW) control, and one or two optimization loops. Thus, seven FC net power maximization strategies have been compared to a reference commercial strategy using performance indicators such as the electrical energy efficiency, the fuel consumption efficiency, and the fuel economy.

Chapter 6 presents a systematic analysis of fuel saving strategies for PEMFC systems based on the Load-Following (LFW) control, and one or two optimization loops. The FC power may be controlled via the fuel and air regulators, and the controller of the FC boost converter. So, three control inputs are available to operate the PEMFC system for best fuel economy. From these inputs, one will be used to implement the LFW control mode, and the other two or only one will be used to optimize the fuel economy. Thus, seven fuel economy maximization strategies have been compared to a reference commercial strategy.

Chapter 7 highlights the performance of the Maximum Power Point Tracking (MPPT) algorithms compared to the Global MPPT algorithms to harvest the available energy of a photovoltaic system that are partially shaded. An evaluation index based on seven criteria is proposed for objective evaluation of the MPPT algorithms is proposed.

Chapter 8 highlights the performance of the Power-Following (PFW) control to mitigate the power pulses on the DC bus. The PFW strategy is proposed for a FC/RES HPS as a variant of the LFW control based on the difference between the load and available renewable energy instead of the load. The FC/RES HPS under PFW control operates the battery in charge-sustained mode. The most efficient energy strategies and best fuel economy strategies proposed in Chaps. 5 and 6 are analyzed using an electrolyzer instead of a dump load to ensure the charge-sustained mode for the battery.

Therefore, the proposed book tries to clarify the new approaches for RES/FC HPS by presenting the design of the advanced energy management strategies and detailed commentary of the obtained results. Furthermore, the book puts forward some practical energy management strategies for FC vehicles. Thus, the book will be helpful for the future research in the field of FC/RES HPS and FC vehicles, researches which will be made by young researchers and practitioners working in these very current and challenging fields of electrical engineering.

Pitesti, Romania   Nicu Bizon

Contents

Abbreviations

η_{sys}	FC system efficiency
A-ECMS	Adaptive Equivalent Consumption Minimization Strategy
ACS	Ant Colony Systems
AirFr	Air Flow rate
ANN	Artificial Neural Network
aPESC	Asymptotic PESC
aPESCH1	Asymptotic PESC based on FFT to compute the H1 magnitude
aPESCLy	Asymptotic PESC based on Lyapunov switching function
aPESCs	Asymptotic PESC based on scalar ESC scheme
AV	Average
BESS	Battery Energy Storage Systems
BMS	Battery Management System
BPF	Band-Pass Filter
CCL	Charge current limit
CS	Charge-Sustaining
DCL	Discharge current limit
DE	Differential Evolution
DOD	Depth of Discharge
EA	Evolutionary Algorithms
ECMS	Equivalent Consumption Minimization Strategy
ELZ	Electrolyzer
EMO	Energy Management and Optimization
EMS	Energy Management Strategy
EMU	Energy Management Unit
EPR	Equivalent Parallel Resistors
ES	Extremum Seeking
ESC	Extremum Seeking Control
ESR	Equivalent Series Resistors
ESS	Energy Storage System
EUDC	Extra-Urban Driving Cycle

FC	Fuel Cell
FCHPS	Fuel cell hybrid power system
FES	Flywheel energy systems
FFT	Fast Fourier Transform
FLC	Fuzzy Logic Controller
$Fuel_{eff}$	Fuel consumption efficiency
$Fuel_T$	Total Fuel consumption
FuelFr	Fuel Flow rate
GA	Genetic Algorithm
GaPESC	Global aPESC scheme based on one BPF
GaPESCbpf	Global aPESC scheme based on two BPFs
GaPESCd	Global aPESC scheme based on derivative operator
GaPESCH1	Global aPESC scheme based on FFT to compute the H1 magnitude
GES	Global extremum seeking
GM	Global maximum
GMPP	Global Maximum Power Point
GMPPT	Global Maximum Power Point Tracking
H1	First Harmonic
HC	Hill Climbing
HEV	Hybrid electric vehicle
HIL	Hardware in a loop
HPF	High Pass Filter
HPS	Hybrid Power System
HSS	Hybrid Storage System
IC	Incremental Conductance
ICDA	Initialization and Control of the Dither Amplitude
LFW	Load-following
LiPo	Li-ion polymer
LMPP	Local Maximum Power Point
LPF	Low Pass Filter
MEP	Maximum Efficiency Point
MEPT	Maximum Efficiency Point Tracking
MPP	Maximum Power Point
MPPT	Maximum Power Point Tracking
MV	Mean Value
NEDC	New European Driving Cycle
OER	Oxygen excess ratio
OMC	Operating Mode Control
P_{FCnet}	FC net power
PEM	Polymer Electrolyte Membrane
PEMFC	Proton Exchange Membrane Fuel Cell
PESC	Perturbed-based Extremum Seeking Control
PFW	Power-following
PHC	Percent of the Hit Count
PHS	Pumped hydroelectric storage systems

PI	Proportional-Integral
PID	Proportional-Integral-Derivative
PSC	Partially Shaded Condition
PSO	Particle Swarm Optimization
P & O	Perturb and Observe
PV	Photovoltaic
RCC	Ripple Correlation Control
RES	Renewable Energy Source
RFC	Renewable Fuel Cells
RT	Real-time
RTO	Real-time Optimization
Scalar PESC	PESCs
sFF	Static Feed-Forward
SIDO	Single-Input Double-Outputs
SISO	Single-Input Single-Output
SOC	State-Of-Charge
SoH	State of health
SMES	Superconducting Magnetic Energy Storage
UCs	Ultracapacitors
UIC	Uniform Irradiance Conditions
VI	Virtual instruments
VWP	Variable-weather-parameter
WT	Wind Turbines
ZEB	Zero Energy Building

Chapter 1
Introduction

1.1 Sustainable Energy System Development Using Renewable Energy Sources and Hydrogen Infrastructure

Hydrogen is not only a source of energy, but it is also a carrier vector for the energy produced from different sources. The flexibility of hydrogen production will have to be combined and adapted to current energy distribution and usage systems. If new energy sources will be discovered and developed, these can also be used to produce hydrogen without affecting the energy system at which consumers are linked. The idea of using hydrogen as an energy carrier, in energy storage and transport, has been studied by many researchers around the world [1], and the interest in developing the hydrogen infrastructure is growing steadily (see Fig. 1.1).

In nature, different combinations of hydrogen are found in large quantities, but obtaining them as molecular hydrogen is an energy-consuming process. Depending on the used energy resource and the by-products resulting, hydrogen may or may not fall into the category of sustainable fuels. The concept of sustainable hydrogen defines the hydrogen in whose life cycle raw materials and renewable energy sources (RESs) are clean (not-polluting). Thus, the production of sustainable hydrogen is based on the electrolysis of water using an electrolyzer supplied from renewable energy sources [2], and the interest in this challenging subject grows steadily (see Fig. 1.2).

1.2 Proton Exchange Membrane Fuel Cell System

Following the process of generating energy from the proton exchange membrane fuel cell (PEMFC) system, from hydrogen and oxygen results only water, which may be used as it is (being clean and drinkable) or the water will reenter the natural circuit [3]. In this context, the new concept of hydrogen-based technology, which includes

© The Editor(s) (if applicable) and The Author(s), under exclusive license to Springer Nature Switzerland AG 2020
N. Bizon, *Optimization of the Fuel Cell Renewable Hybrid Power Systems*,
Green Energy and Technology, https://doi.org/10.1007/978-3-030-40241-9_1

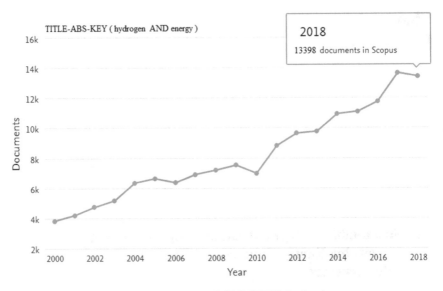

Fig. 1.1 Papers published in hydrogen energy field (SCOPUS database)

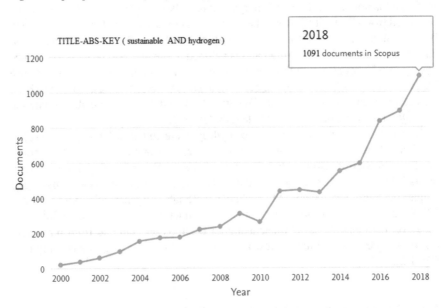

Fig. 1.2 Papers published in the production of sustainable hydrogen (SCOPUS database)

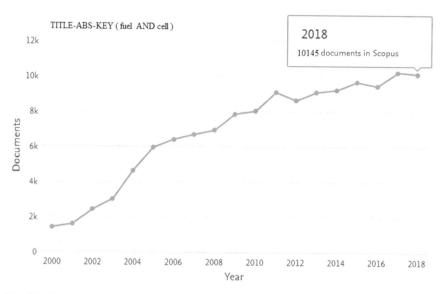

Fig. 1.3 Papers published in fuel cell subject (SCOPUS database)

the production, storage and use of hydrogen as an energy carrier, has become in the field of sustainable energy a fundamental element for the next decades of European policy and not only [4]. Interest in FC systems (including PEMFC types) has grown steadily over the past two decades (see Fig. 1.3).

Hydrogen cannot be found in nature in a pure state and must be extracted from natural gas, ethanol, methanol, biogas, and other different chemical compounds using the renewable energy. Since hydrogen can be produced relatively easy using water electrolysis, this will become an energy hub in the next decades, as is the electricity system today [5]. It is well known that the water electrolysis may be obtained using the following methods [6]:

(a) Biological process: Solar energy is captured based on the photosynthesis process.
(b) Photo-thermal technique: a process in which solar energy (using concentrators) or nuclear energy is harnessed to reach the high temperatures required for the decomposition of water into oxygen and hydrogen.
(c) Photo-electrochemical technique: Following the sunlight illumination of an electrode of the electrochemical cell, the conditions for the decomposition of water into hydrogen and oxygen can be created.

A generally agreed solution for obtaining sustainable hydrogen is the use of the solar energy as a primary energy source, in two ways [7]:

– Directly, the hydrogen being obtained using the solar energy for the electrolysis of water.

– Indirectly, the hydrogen being generated and stored as a result of FC/RES hybrid
   power system (HPS) operation;

1.3 Hybrid Power Systems

Hybrid power systems (HPSs) provide AC energy in remote locations without access
to the grid (electrical power system), but can be grid-connected into a micro- or
nano-grid as well, minimizing the power loss transfer over relatively long distances
[8]. Recent research addresses various aspects of HPS design and optimization, and
researchers' interest in this topic is validated by the number of publications in recent
years (see Fig. 1.4).

A hybrid power system includes additional power sources such as RESs (wind
turbines (WT), photovoltaic (PV) systems, biomass and hydroelectric generators,
etc.); energy storage systems (ESSs) and backup energy sources (FC system, diesel
generators, etc.); and different AC and DC loads (including an electrolyzer) [9]. The
HPSs supplying DC loads have been used for decades in telecommunications stations
(situated in isolated locations) and other low power applications (usually less than
5 kW) [10]. The RESs may be the key solution for sustainable energy production
(see Fig. 1.5).

The power flow balance between the power generated and consumed on the common bus may be ensured using different energy management strategies based on
appropriate control of the power flow of the ESS or/and the backup energy source

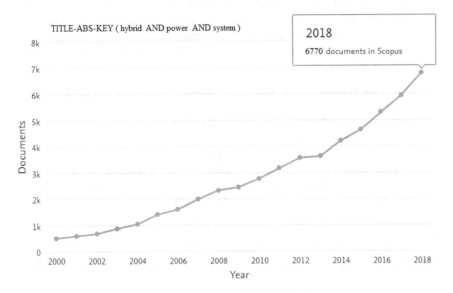

Fig. 1.4 Papers published in hybrid power systems (SCOPUS database)

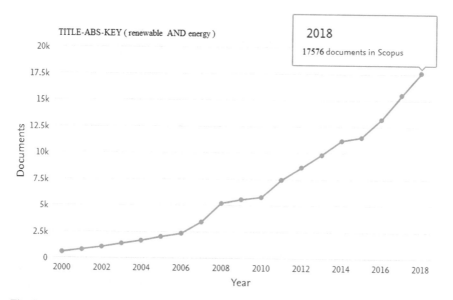

Fig. 1.5 Papers published in renewable energy sources (SCOPUS database)

[11, 12], so more than six thousand HPS optimization strategies have been proposed in the last twenty years (see Fig. 1.6). In this book, efficient energy management

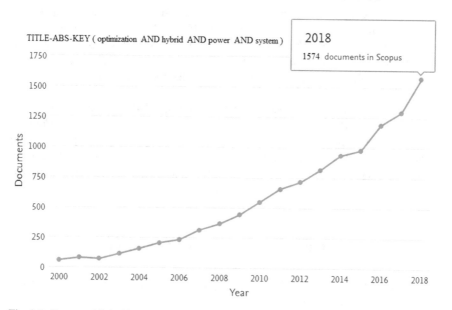

Fig. 1.6 Papers published in HPS optimization strategies (SCOPUS database)

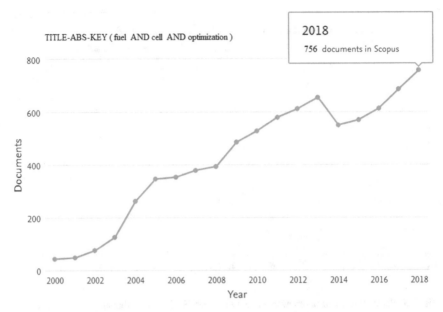

Fig. 1.7 Papers published in FC optimization strategies (SCOPUS database)

strategies based on the new control and optimization loops for the PEMFC power generated on the DC bus will be presented.

The PEMFC system may be used as a backup source due to the clear advantages compared to other FC technologies [9, 10], such as a high energy density, low emissions, and low-temperature operation [13, 14]. PEMFC is a competitive variant compared to the diesel generator, offering low costs and easy maintenance, besides the ecological functioning [15, 16], and the optimization of the FC system is intensively researched in Fig. 1.7.

So, based on the power generated by the controlled FC system, the RES/FC HPS integrating RES, FC, and ESS will efficiently manage the non-synchronized power flows of the dynamic loads and variable renewable power [17–22]. This power balance is difficult to obtain in RES HPS even if more RESs are integrated via the controlled power converters [23–31]. New topologies of energy-efficient power converters have recently been proposed (see Fig. 1.8).

1.4   Research Directions in Hybrid Power Systems

The research directions in hybrid power systems identified based on the recently published studies are as follows:

1.   The maximum power point (MPP) tracking control must be implemented for each RES in order to harvest all power available; thus, many advanced global

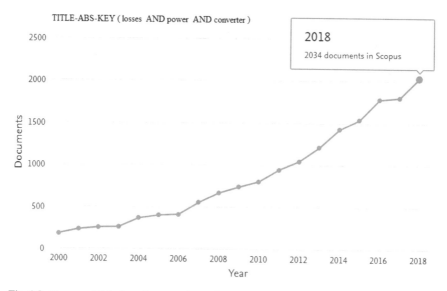

Fig. 1.8 Papers published on the power losses in the power converters (SCOPUS database)

MPP tracking algorithms have been proposed for PV array [32–43] and WT farms [4, 44] (see Fig. 1.9);

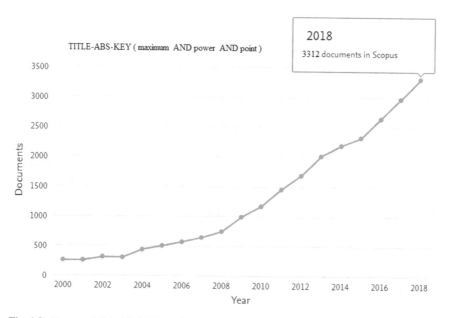

Fig. 1.9 Papers published in MPP tracking control (SCOPUS database)

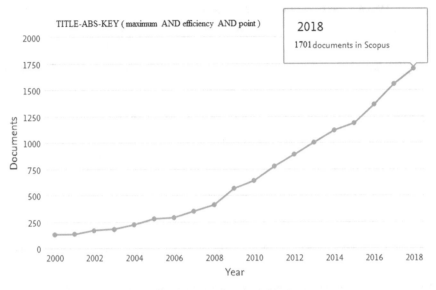

Fig. 1.10 Papers published in MEP tracking control (SCOPUS database)

2. The maximum efficiency point (MEP) tracking control must be implemented for the PEMFC system in order to efficiently generate the requested power on the DC bus based on the load-following and power-following control techniques [18, 22, 45–47] or other control strategies [48–51] (see Fig. 1.10);

3. The fuel economy strategies must be implemented for the FC vehicles, so many optimization strategies have been proposed [18–21, 47, 52–59] (see Fig. 1.11);

4. The ESS must be hybridized with power storage devices such as ultracapacitors and superconducting magnetic energy storage (SMES) in order to dynamically compensate the power pulses on the DC bus [60, 61] (see Fig. 1.12);

5. The number of charge–discharge cycle of the battery must be minimized in order to increase the battery's lifetime; for this, the load-following and power-following control techniques [18, 22, 45–47] have been proposed to ensure the charge-sustained mode for the battery from the RES/FC HPS [62], but other battery's management strategies and new high-performance technologies for the batteries have also been developed in the last two decades (Fig. 1.13);

6. The power losses of the power converters must be minimized using advanced topologies in order to have an energy efficiency higher than 95% [63, 64].

7. The RES power must be accurately predicted in order to accurately design the HPS (the maximum FC power, the battery's capacity, etc.) [65–67]; thus, about four hundred RES power prediction techniques have been proposed in the year 2018 (see Fig. 1.14).

8. The safe operation of the RES/FC HPS must be ensured by appropriate protection measures that are required especially for the safe operation of PEMFC systems [68–70], where hydrogen is needed (see Fig. 1.15);

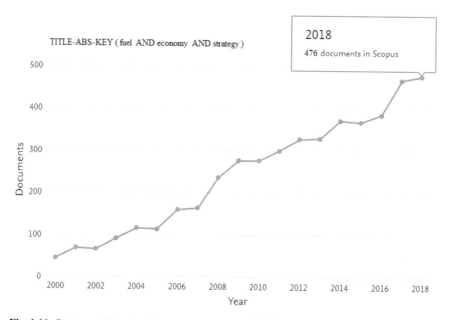

Fig. 1.11 Papers published in fuel economy strategies (SCOPUS database)

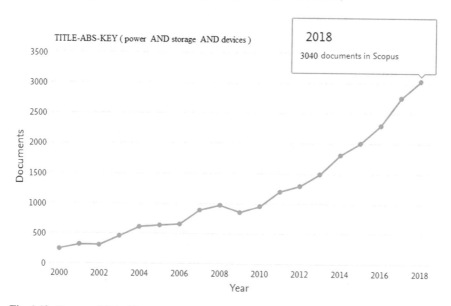

Fig. 1.12 Papers published in power storage devices (SCOPUS database)

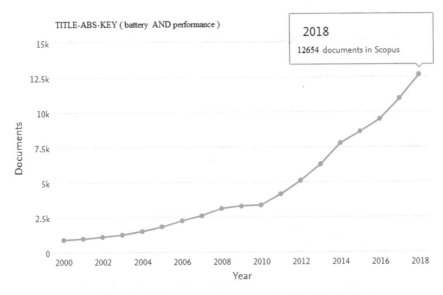

Fig. 1.13 Papers published on the battery's performance topic (SCOPUS database)

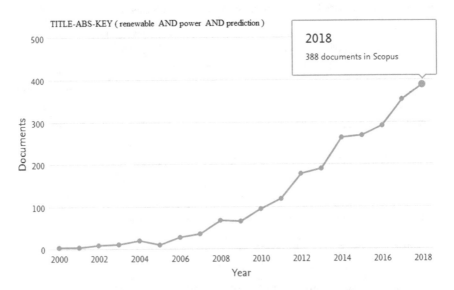

Fig. 1.14 Papers published on the RES power prediction techniques topic (SCOPUS database)

9. The safe production, transportation, and storage of the hydrogen technology is in the attention of the researchers (see Fig. 1.16) in order to develop lower-cost technologies to produce hydrogen based on fuel reformers [71, 72];

10. The clean transport for goods and persons must be implemented using electric vehicle and FC vehicles (see Fig. 1.17), which have an extended range operation

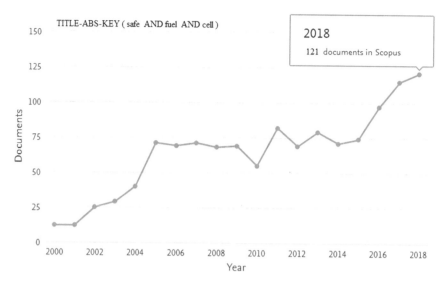

Fig. 1.15 Papers published on the safe operation of the PEMFC systems (SCOPUS database)

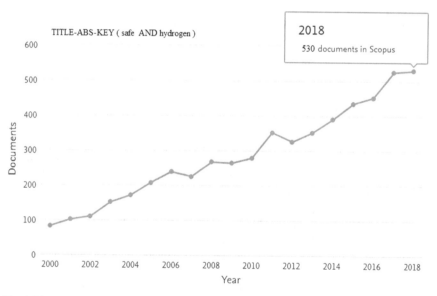

Fig. 1.16 Papers published on the safe production, transportation, and storage of the hydrogen (SCOPUS database)

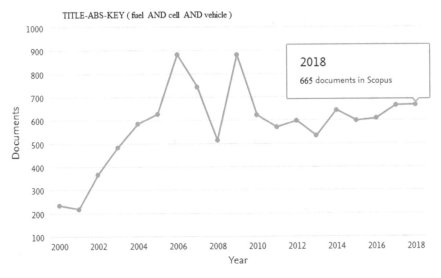

Fig. 1.17 Papers published on the FC vehicles topic (SCOPUS database)

compared to the electric vehicles due to the use of an onboard FC system [53–57, 73–80].

1.5   Conclusion

The topic of the hybrid power system based on the renewable energy sources (RES) and energy of hydrogen was, is, and will be necessary and challenging because the energy sustainable development that becomes mandatory be implemented due to climatic changes and new environmental regulations.

The proton exchange membrane fuel cell (PEMFC) system may be used as a backup energy source to sustain the power flow balance on the DC bus of the hybrid power system. The differences between the power generated on the DC bus (the PEMFC power and RES power) and the load demand may be compensated by the hybrid energy storage systems (ESS) or by the PEMFC system using a power-following control for one fueling regulator and optimization loops for the other fueling regulators and/or the boost DC-DC power converter. In the last case, the battery from the ESS will operate in charge-sustained mode with multiple advantages related to the lifetime, size, capacity, and cost. The maximum power point (MPP) tracking control must be implemented for each RES to harvest all available energy, and a maximum efficiency point (MEP) or fuel economy strategy may be implemented for the PEMFC system to optimize the whole hybrid power system.

References

1. European Commission (2019) Community research, introducing hydrogen as an energy carrier. http://ec.europa.eu/research/energy/pdf/hydrogen_22002_en.pdf. Accessed on 1 Aug 2019
2. Gielen D, Boshell F, Saygin D, Bazilian MD, Wagner N, Gorini R (2019) The role of renewable energy in the global energy transformation. Energy Strateg Rev 24:38–50
3. Wang F-C, Yi-Shao Hsiao Y-S, Yi-Zhe Yang Y-Z (2018) The optimization of hybrid power systems with renewable energy and hydrogen generation. Energies 11(8):1948. https://doi.org/10.3390/en11081948
4. Hesselink LXW, Chappin EJL (2019) Adoption of energy efficient technologies by households—barriers, policies and agent-based modelling studies. Renew Sustain Energy Rev 99:29–41
5. Lawan Bukar AL, Wei Tan CW (2019) A review on stand-alone photovoltaic-wind energy system with fuel cell: system optimization and energy management strategy. J Clean Prod 2019. https://doi.org/10.1016/j.jclepro.2019.02.228 (in press)
6. European Commission (2019) Community reserch, hydrogen production, overview of technology options. http://www1.eere.energy.gov/hydrogenandfuelcells/pdfs/h2_tech_roadmap.pdf. Accessed on 1 Aug 2019
7. Bizon N, Tabatabaei NM, Blaabjerg F, Kurt E (2017) Energy harvesting and energy efficiency: technology, methods and applications. Springer. Available on http://www.springer.com/us/book/9783319498744
8. Tabatabaei NM, Kabalci E, Bizon N (2019) Microgrid architectures, control and protection methods. Springer, London, UK. Available on https://www.springer.com/in/book/9783030237226
9. Larminie J, Dicks A (2000) Fuel cell system explained. John Wiley & Sons, Chichester
10. Vielstich W, Gasteiger H, Lamm A (2003) Handbook of fuel cells-fundamentals, technology, applications. Wiley, New York
11. Tabatabaei NM, Bizon N, Aghbolaghi AJ, Blaabjerg F (2017) Fundamentals and contemporary issues of reactive power control in AC power systems. Springer, London, UK. http://www.springer.com/gp/book/9783319511177
12. Bizon N, Tabatabaei NM, Shayeghi H (2013) Analysis, control and optimal operations in hybrid power systems—advanced techniques and applications for linear and nonlinear systems. Available on http://dx.doi.org/10.1007/978-1-4471-5538-6. Springer
13. Gou B, Na WK, Diong B (2010) Fuel cells: modeling, control, and applications. CRC Press, Boca Raton, p 6
14. Pukrushpan JT, Stefanopoulou AG, Peng H (2004) Control of fuel cell power systems: principles, modeling, analysis, and feedback design. Springer Verlag, London
15. Lior N (2010) Sustainable energy development: the present (2009) situation and possible paths to the future. Energy 35:3976–3994
16. International Energy Agency (IEA) (2019) Clean energy investment trends 2019. https://webstore.iea.org/clean-energy-investment-trends-2019. Accessed on 1 Aug 2019
17. Bizon N, Thounthong P (2018) Real-time strategies to optimize the fueling of the fuel cell hybrid power source: a review of issues, challenges and a new approach. Renew Sustain Energy Rev 91:1089–1102. https://doi.org/10.1016/j.rser.2018.04.045
18. Bizon N, Mazare AG, Ionescu LM, Thounthong P, Kurt E, Oproescu O, Serban G, Lita I (2019) Better fuel economy by optimizing airflow of the fuel cell hybrid power systems using the load-following control based on the fuel flow. Energies 12(14):2792–2810. https://doi.org/10.3390/en12142792
19. Efficient fuel economy strategies for the fuel cell hybrid power systems under variable renewable/load power profile. Appl Energ 251:113400. https://doi.org/10.1016/j.apenergy.2019.113400 (2019)

20. Bizon N, Hoarcă CI (2019) Hydrogen saving through optimized control of both fueling flows of the fuel cell hybrid power system under a variable load demand and an unknown renewable power profile. Energ Convers Manage 184:1–14. https://doi.org/10.1016/j.enconman.2019.01.024
21. Bizon N, Lopez-Guede JM, Kurt E, Thounthong P, Mazare AG, Ionescu LM, Iana G (2019) Hydrogen economy of the fuel cell hybrid power system optimized by air flow control to mitigate the effect of the uncertainty about available renewable power and load dynamics. Energ Convers Manage 179:152–165. https://doi.org/10.1016/j.enconman.2018.10.058
22. Bizon N, Oproescu M (2018) Experimental comparison of three real-time optimization strategies applied to renewable/FC-based hybrid power systems based on load-following control. Energies 11(12):3537–3569. https://doi.org/10.3390/en11123537
23. Nikolova S, Causevski A, Al-Salaymeh A (2013) Optimal operation of conventional power plants in power system with integrated renewable energy sources. Energy Convers Manage 65:697–703
24. Tascikaraoglu A, Boynuegri AR, Uzunoglu M (2014) A demand side management strategy based on forecasting of residential renewable sources: a smart home system in Turkey. Energ Build 80:309–320
25. Tascikaraoglu A, Erdinc O, Uzunoglu M, Karakas A (2014) An adaptive load dispatching and forecasting strategy for a virtual power plant including renewable energy conversion units. Appl Energ 119:445–453
26. Ahmed NA, Miyatake M, Al-Othman AK (2008) Power fluctuations suppression of stand-alone hybrid generation combining solar photovoltaic/wind turbine and fuel cell systems. Energy Convers Manage 49:2711–2719
27. García-Trivino P et al (2014) Long-term optimization based on PSO of a grid-connected renewable energy/battery/hydrogen hybrid system. Int J Hydrogen Energ 9(21):10805–10816
28. Tabatabaei NM, Ravadanegh SN, Bizon N (2018) Power systems resiliency: modeling, analysis and practice. Springer, London, UK. https://www.springer.com/in/book/9783319944418
29. Wang Y, Sun Z, Chen Z (2019) Rule-based energy management strategy of a lithium-ion battery, supercapacitor and PEM fuel cell system. Energy Procedia 158:2555–2560
30. Das V, Padmanaban S, Venkitusamy K, Selvamuthukumaran R, Siano P (2017) Recent advances and challenges of fuel cell based power system architectures and control—a review. Renew Sustain Energy Rev 73:10–18
31. Ettihir K, Boulon L, Agbossou K (2016) Optimization-based energy management strategy for a fuel cell/battery hybrid power system. Appl Energ 163:142–153
32. Bizon N (2016) Global maximum power point tracking (GMPPT) of photovoltaic array using the extremum seeking control (ESC): a review and a new GMPPT ESC scheme. Renew Sustain Energy Rev 57:524–539. https://doi.org/10.1016/j.rser.2015.12.221
33. Bizon N (2016) Global extremum seeking control of the power generated by a photovoltaic array under partially shaded conditions. Energy Convers Manage 109:71–85. https://doi.org/10.1016/j.enconman.2015.11.046
34. Bizon N (2016) Global maximum power point tracking based on new extremum seeking control scheme. Prog Photovoltaics Res Appl 24(5):600–622. http://onlinelibrary.wiley.com/doi/10.1002/pip.2700/full WOS:000373624100002
35. Bizon N (2017) Searching of the extreme points on photovoltaic patterns using a new asymptotic perturbed extremum seeking control scheme. Energ Convers Manage 144:286–302
36. Belhachat F, Larbes C (2019) Comprehensive review on global maximum power point tracking techniques for PV systems subjected to partial shading conditions. Sol Energy 183:476–500
37. Eltamaly AM, Farh HMH, Othman MF (2018) A novel evaluation index for the photovoltaic maximum power point tracker techniques. Renew Sustain Energy Rev 174:940–956
38. Rezka H, Fathy H, Abdelaziz AY (2017) A comparison of different global MPPT techniques based on meta-heuristic algorithms for photovoltaic system subjected to partial shading conditions. Renew Sustain Energy Rev 74:377–386
39. Jiang LL, Srivatsan R, Maskell DL (2018) Computational intelligence techniques for maximum power point tracking in PV systems: a review 85:14–45

40. Belhachat F, Larbes C (2018) A review of global maximum power point tracking techniques of photovoltaic system under partial shading conditions. Renew Sustain Energy Rev 92:513–553

41. Danandeh MA, Mousavi GSM (2018) Comparative and comprehensive review of maximum power point tracking methods for PV cells. Renew Sustain Energy Rev 82:2743–2767

42. Batarseh MG, Za'ter ME (2018) Hybrid maximum power point tracking techniques: a comparative survey, suggested classification and uninvestigated combinations. Sol Energy 169:535–555

43. Ahmad R, Murtaza AF, Sher AH (2019) Power tracking techniques for efficient operation of photovoltaic array in solar applications—a review. Renew Sustain Energy Rev 101:82–102

44. Bizon N (2018) Optimal operation of fuel cell/wind turbine hybrid power system under turbulent wind and variable load. Appl Energ 212:196–209

45. Bizon N (2018) Real-time optimization strategy for fuel cell hybrid power sources with load-following control of the fuel or air flow. Energ Convers Manage 157:13–27

46. Bizon N (2014) Load-following mode control of a standalone renewable/fuel cell hybrid power source. Energ Convers Manage 77:763–772

47. Bizon N (2019) Real-time optimization strategies of FC hybrid power systems based on load-following control: a new strategy, and a comparative study of topologies and fuel economy obtained. Appl Energ 241C:444–460

48. Daud WRW, Rosli RE, Majlan EH, Hamid SAA, Mohamed R, Husaini T (2017) PEM fuel cell system control: a review. Renew Energ 113:620–638

49. Wang F-C, Lin K-M (2019) Impacts of load profiles on the optimization of power management of a green building employing fuel cells. Energies 12(1):57. https://doi.org/10.3390/en12010057

50. Nejad HC, Farshad M, Gholamalizadeh E, Askarian B, Akbarimajd A (2019) A novel intelligent-based method to control the output voltage of proton exchange membrane fuel cell. Energ Convers Manage 185:455–464

51. Ziogou C, Papadopoulou S, Pistikopoulos E et al (2017) Model-based predictive control of integrated fuel cell systems—from design to implementation. Adv Energy Syst Eng: 387–430

52. Bizon N, Thounthong P (2018) Fuel economy using the global optimization of the fuel cell hybrid power systems. Energ Convers Manage 173:665–678

53. Ahmadi S, Bathaee SMT, Hosseinpour AH (2018) Improving fuel economy and performance of a fuel-cell hybrid electric vehicle (fuel-cell, battery, and ultra-capacitor) using optimized energy management strategy. Energy Convers Manage 160:74–84

54. Kaya K, Hames Y (2019) Two new control strategies: for hydrogen fuel saving and extend the life cycle in the hydrogen fuel cell vehicles. Int J Hydrogen Energy. https://doi.org/10.1016/j.ijhydene.2018.12.111 (in press)

55. Sulaiman N, Hannan MA, Mohamed A, Ker PJ, Majlan EJ, Daud WRW (2018) Optimization of energy management system for fuel-cell hybrid electric vehicles: Issues and recommendations. Appl Energ 228:2061–2079

56. Fernández RÁ, Caraballo SC, Cilleruelo FB, Lozano JA (2018) Fuel optimization strategy for hydrogen fuel cell range extender vehicles applying genetic algorithms. Renew Sustain Energ Rev 81(1):655–668

57. Yuan J, Yang L, Chen Q (2018) Intelligent energy management strategy based on hierarchical approximate global optimization for plug-in fuel cell hybrid electric vehicles. Int J Hydrogen Energ 43(16):8063–8078

58. Chena S, Kumar A, Wong WC, Chiu M-S, Wang X (2019) Hydrogen value chain and fuel cells within hybrid renewable energy systems: Advanced operation and control strategies. Appl Energ 233–234:321–337

59. Haseli Y (2018) Maximum conversion efficiency of hydrogen fuel cells. Int J Hydrogen Energ 43(18):9015–9021

60. Bizon N (2019) Hybrid power sources (HPSs) for space applications: analysis of PEMFC/Battery/SMES HPS under unknown load containing pulses. Renew Sustain Energy Rev 105:14–37. https://doi.org/10.1016/j.rser.2019.01.044

61. Bizon N (2018) Effective mitigation of the load pulses by controlling the battery/SMES hybrid energy storage system. Appl Energ 229:459–473
62. Arenas LF, Ponce de León C, Walsh FC (2019) Redox flow batteries for energy storage: their promise, achievements and challenges. Curr Opin Electrochem 16:117–126
63. Sajedi S, Farrell M, Basu M (2019) DC side and AC side cascaded multilevel inverter topologies: a comparative study due to variation in design features. Int J Elec Power 113:56–70
64. Han W, Corradini L (2019) Wide-range ZVS control technique for bidirectional dual-bridge series-resonant DC-DC converters. IEEE T Power Electr 34(10):10256–10269
65. Koike M, Ishizaki T, Ramdani N, Imura JI (2020) Optimal scheduling of storage batteries and power generators based on interval prediction of photovoltaics—monotonicity analysis for state of charge. Control Syst Lett 4(1):49–54
66. Dong N, Chang J-F, Wu A-G, Gao Z-K (2020) A novel convolutional neural network framework based solar irradiance prediction method. Int J Elec Power 114:105411. https://doi.org/10.1016/j.ijepes.2019.105411
67. Piotrowski P, Baczyński D, Kopyt M, Szafranek K, Helt P, Gulczyński T (2019) Analysis of forecasted meteorological data (NWP) for efficient spatial forecasting of wind power generation. Electr Pow Syst Res 175:105891. https://doi.org/10.1016/j.epsr.2019.105891
68. Valente A, Iribarren D, Dufour J (2019) End of life of fuel cells and hydrogen products: From technologies to strategies. Int J Hydrogen Energ 44(38):20965–20977
69. Darvish Falehi A, Rafiee M (2019) Optimal control of novel fuel cell-based DVR using ANFISC-MOSSA to increase FRT capability of DFIG-wind turbine. Soft Comput 23(15):6633–6655
70. Restrepo C, Ramos-Paja CA, Giral R, Calvente J, Romero A (2012) Fuel cell emulator for oxygen excess ratio estimation on power electronics applications. Comp Elec Eng 38:926–937
71. Mousavi SMA, Piavis W, Turn S (2019) Reforming of biogas using a non-thermal, gliding-arc, plasma in reverse vortex flow and fate of hydrogen sulfide contaminants. FUEL PROCESS TECHNOL: 378–391
72. Herdem MS, Sinaki MY, Farhad S, Hamdullahpur F (2019) An overview of the methanol reforming process: comparison of fuels, catalysts, reformers, and systems. Int J Energy Res 43(10):5076–5105
73. Sorrentino M, Cirillo V, Nappi L (2019) Development of flexible procedures for co-optimizing design and control of fuel cell hybrid vehicles. Energ Convers Manage 185:537–551
74. Pan ZF, An L, Wen CY (2019) Recent advances in fuel cells based propulsion systems for unmanned aerial vehicles. Appl Energ 240:473–485
75. Yue M, Jemei S, Gouriveau R, Zerhouni N (2019) Review on health-conscious energy management strategies for fuel cell hybrid electric vehicles: degradation models and strategies. Int J Hydrogen Energy 44(13):6844–6861
76. Chen H (2019) The reactant starvation of the proton exchange membrane fuel cells for vehicular applications: a review. Energ Convers Manage 182:282–298
77. Ahmadi P, Torabi SH, Afsaneh H, Sadegheih Y, Ganjehsarabi H, Ashjaee M (2019) The effects of driving patterns and PEM fuel cell degradation on the lifecycle assessment of hydrogen fuel cell vehicles. Int J Hydrogen Energy. https://doi.org/10.1016/j.ijhydene.2019.01.165 (in press)
78. Han J, Yu S (2018) Ram air compensation analysis of fuel cell vehicle cooling system under driving modes. Appl Therm Eng 142:530–542
79. Zhang H, Li X, Liu X, Yan J (2019) Enhancing fuel cell durability for fuel cell plug-in hybrid electric vehicles through strategic power management. Appl Energ 241:483–490
80. Li Q, Wang T-H, Dai C-H, Chen W-R, Ma L (2018) Power management strategy based on adaptive droop control for a fuel cell-battery-supercapacitor hybrid tramway. IEEE Trans Veh Technol 67(7):5658–5670

Chapter 2
Hybrid Power Systems

2.1 Introduction

In general, a hybrid system means a combination of different subsystems, such as nature through controlled physical processes to meet an imposed goal. Consequently, hybridization is the operation of adding sub-sites to the base system to best meet the required goal. Optimization objectives are specifically defined for each hybrid system [1, 2].

The hybrid systems discussed in this book are specific to the energy field, where the hybridization concept describes a combined system of energy sources and energy storage systems [3, 4]. Thus, a definition of hybrid power systems (HPSs) that is commonly used in the literature highlights that this concept represents a combination of different energy generation systems to optimally produce energy according to load demand [5, 6].

By combining renewable technologies with other clean technologies (such as based on wind, solar, etc. [7] and fuel cell (FC) systems [8], respectively), the power supply system for a home [9] or office building [10] is obvious an HPS that must generate energy efficiently in both stand-alone and grid-connected modes [11, 12]. Also, the drive train of the hybrid electric vehicles and FC-based vehicles are powered by a hybrid power system [13–15].

The differences between the energy generated and load demand are compensated in power flow balance via a battery//ultracapacitors-based energy storage system (ESS) [16] or using a backup energy source such as a diesel genset or FC system [17] in order to ensure maximum reliability and security of supply [18] and Paris climate agreement goals [19].

© The Editor(s) (if applicable) and The Author(s), under exclusive license to Springer Nature Switzerland AG 2020
N. Bizon, *Optimization of the Fuel Cell Renewable Hybrid Power Systems,*
Green Energy and Technology, https://doi.org/10.1007/978-3-030-40241-9_2

The Paris climate agreement sets the main objectives to protect the environment, stopping further deterioration based on distributed generation that use the RESs available locally for energy production. The renewable energy generation technology based on solar and wind energy is green, but it is worth mentioning that its availability is variable due to weather, seasonal changes, and environment conditions (such as solar irradiance and wind speed) [20, 21]. Thus, the hybridization of the RESs available locally must be efficiently performed [22, 23] for a sustainable energy development [19, 24]. Distributed generation based on RES HPSs can be the saving solution from a global environmental disaster through innovative and efficient control of power and load demand variability while simultaneously reducing pollutant emissions and deployment costs [25, 26].

Electricity consumption in the residential sector is known to be more than 35% of total energy consumption [9, 22], so optimizing energy use in homes and office buildings has become mandatory [10, 27, 28]. The FC system and the FC vehicle can function as backup power sources, replacing the pollutant variant of a diesel generator with a non-polluting energy source that generates the necessary energy in the absence of renewable energy under a load tracking control [29, 30]. This FC power generation must avoid reported issues related to the fueling starvation and the stress of the membrane if a polymer electrolyte membrane fuel cell (PEMFC) will be used as backup power source [12, 31, 32], as it is the case of the HPSs analyzed in this chapter. PEMFC system can be safely operated based on the load-based control of the fueling regulators [33–36], which will supplementarily use 100 A/s slope limiters at their inputs [37].

When the energy from renewable sources is higher than the load demand, the surplus energy can be sold [38] or feed an electrolyzer to get green and cheap hydrogen, which can be stored under pressure in PEMFC fuel tanks [39].

The load-based strategy can control efficiently a solid oxide FC system [40], a mixed heat and power system [41], or other type of distributed generation system, conventional or not [6, 20].

Compared to other FC technologies, hydrogen-based FC systems have many advantages mentioned in [42, 43] and, moreover, hydrogen is available in nature or can be simply and environmentally produced by reforming natural gas, ethanol, methanol, biogas, or, as mentioned before, by electrolysis of water [44, 45]. In addition, if PEMFC heat will be recovered, a mixed solution for power and heating of a zero-energy building (ZEB) could increase PEMFC efficiency up to 70% [46–49]. Also, it is worth mentioning that besides the efficient and environmentally friendly operation of RES/ FC HPS [50, 51], the maintenance costs are much lower for a PEMFC system compared to the variant of a diesel generator as backup energy source [52, 53].

Therefore, the variants of the integration of RES, FC, and ESS into a HPS can be obtained on the basis of a common DC or AC bus, but also using an AD/DC hybrid bus topology with an AC-DC bidirectional converter for balancing the power flow on both buses (see Fig. 2.1) due to the different types of power generated by PV (DC) panels, wind turbines (AC) and FC (DC) systems and the hybrid bus-specific connection.

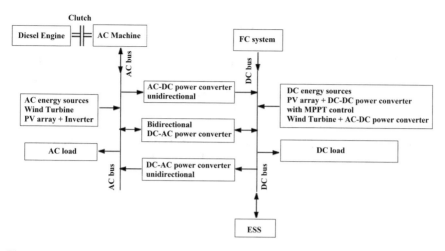

Fig. 2.1 Hybrid power systems with an AD/DC hybrid bus topology

Using a battery pack to store energy with the size as low as possible or even without battery in ESS, the energy management strategies (EMSs) must always ensure load demand without large imbalances in the power flow balance between DC and AC buses. The objectives of the EMSs and performance indicators will be presented in this chapter and detailed in the next chapters of this book.

Thus, the structure of this chapter is as follows. Section 2.1 defines the hybrid power systems as a combination of different energy generation systems to produce optimal energy according to load demand, introduces the topology based on common DC or AC bus, or AD/DC hybrid bus, and highlights the advantages of using a backup energy to mitigate the variability of the renewable energy and load demand. The hybrid power system architecture used in this book is presented in Sect. 2.2. The energy storage system topologies and the most used technologies for energy and power storage devices are presented in Sect. 2.3. The battery and ultracapacitors models are also presented in Sect. 2.3. The simplified and detailed models of the PEMFC system are presented in Sect. 2.4. The models of the wind turbine and photovoltaic panel are presented in Sect. 2.5. Section 2.6 presents two models for the load demand. The models of the power converters are presented in Sect. 2.7. The performance indicators of the PEMFC system and renewable energy sources are presented in Sect. 2.8. Sections 2.9 and 2.10 present the main objectives and directions of research for the hybrid power system, and conclusions of this chapter, respectively.

2.2 Hybrid Power System Architecture

The architecture of a renewable/fuel cell hybrid power system (RES/FC HPS) with common DC bus topology is presented in Fig. 2.2. The subsystems of the RES/FC HPS are as follows: renewable energy sources (RESs), proton exchange membrane fuel cell (PEMFC) system, energy storage system (ESS) using a semi-active hybrid topology based on the batteries and ultracapacitors, the power converters used as interfaces for RESs, PEMFC, and ultracapacitors, and the DC load that represents the equivalent load demand on the DC bus for DC loads and AC loads via the inverter.

The schematic of the HPS architecture (which has been detailed in Fig. 2.2) is shown in Fig. 2.3. The power flow balance on the DC bus is (2.1):

$$C_{DC} v_{dc} \frac{dv_{dc}}{dt} = p_{FC} + p_{ESS} - p_{DC} \tag{2.1}$$

where p_{FC} the FC net power generated by the FC system.

The average value of (2.1) during a load cycle will be given by (2.2):

$$0 \cong P_{FC} + P_{ESS} - P_{DC} \tag{2.2}$$

If the power exchanged by the ESS with the DC bus ($p_{ESS} = p_{ESS-P} + p_{ESS-E}$) is zero during a load cycle ($P_{ESS} = 0$), then see (2.2) and also Fig. 2.2:

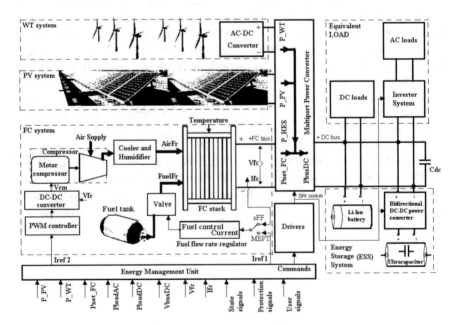

Fig. 2.2 Renewable/fuel cell hybrid power system with common DC bus topology [12]

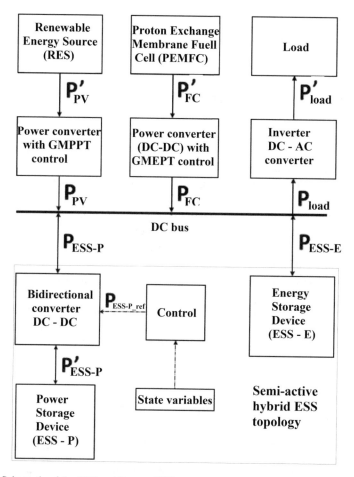

Fig. 2.3 Schematic of the HPS architecture [10]

$$P_{FCnet} = P'_{FC} = \frac{P_{FC}}{\eta_{boost}} = \frac{P_{DC}}{\eta_{boost}} \cong \frac{P_{load} - P_{RES}}{\eta_{boost}} \tag{2.3}$$

So, the reference $I_{ref(PFW)}$ of the DC power-based control will be given by (2.4):

$$I_{ref(PFW)} = I_{FC} = \frac{P_{DC}}{V_{FC} \cdot \eta_{boost}} \tag{2.4}$$

The C_{DC} is the filtering capacitor of the DC voltage (v_{dc}), and η_{boost} is the efficiency of the boost DC-DC power converter.

The power-based control (2.4) and the DC voltage regulation based on the voltage error, $e_v = V_{DC} - V_{DC(ref)}$, will be presented in Chap. 3. This chapter is focused on the models used for all HPS subsystems. Thus, modeling of the energy storage devices (ESS-E) and power storage devices (ESS-P) will be presented in the next section.

2.3 Energy Storage System (ESS)

The energy storage devices that are commonly used in ESS in HPSs are batteries, regenerative FC systems, and flywheels [16]. ESS hybridization to meet dynamic power requirements is achieved with power storage devices [54, 55]. The most used devices are ultracapacitors (UCs), superconducting magnetic energy storage (SMES), and high-speed flywheels [56–58].

Energy storage system (ESS) is generally overstated to maintain the stability of the hybrid energy source in the presence of power demand variations resulting from the variability of power flows generated by renewable energy sources and the dynamics of the required load profiles.

Over the past decade, storage technologies have become diverse, but the most used are battery energy storage systems (BESS) due to the proposed advanced technologies [59, 60] in addition to pumped hydroelectric storage systems (PHS), hydrogen fuel cells (PEMFC systems), flywheel energy systems (FES), superconducting magnetic energy storage (SMES) and other potential energy storage systems (see Fig. 2.4) [61, 62].

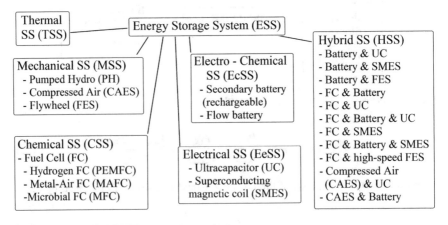

Fig. 2.4 Classes of ESS [57]

2.3.1 Hybrid ESS Topologies

The energy of the battery from the storage systems can be used to offset the energy flow balances on the DC bus, but frequent charging and discharging, sometimes in deep discharge cycles that are evaluated by depth of discharge (DOD), reduce battery life. Consequently, BESS hybridization with power storage devices is required to produce viable energy storage solutions in intelligent power grids that integrate renewable energy sources [63], but also into other stationary power supply and transport systems [64–67]. The problems of operation, maintenance, and potential environmental hazards can be mitigated or avoided by appropriate control [68–71].

Generally, BESS hybridization uses an active or semi-active topology due to their flexibility and performance compared to passive topology (see Fig. 2.5).

In general, the ESS1 is a battery bank, so the ESS2 must be a power storage device (UC, SMES, or high-speed FES) to dynamically offset the balance of power flows on the DC bus.

2.3.2 Modeling the Semi-active Hybrid ESS Topology

The model of the semi-active ESS topology used in simulations is presented in Fig. 2.6. The subsystems of the semi-active ESS are as follows: 100 Ah Li-ion battery, 100 F ultracapacitors' stack, and the buck-boost DC-DC power converter and its controller to regulate the DC voltage at $V_{DC(ref)} = 200$ V.

The energy storage device and the power storage device technologies that can be used to hybridize the BESS will be briefly presented in the following two sections.

2.3.3 Energy Storage Devices Technologies

The Li-ion battery is still used in ESS even if its energy density (which is of about 150 Wh/kg) is four times lower than that of the regenerative FC systems [26, 72].

2.3.3.1 Batteries

Choosing a type of battery from different classes of battery technologies (such as acid lead batteries, nickel-metal batteries, nickel-cadmium hydride batteries, sodium batteries, and lithium-ion batteries) is done by taking into account the cost criteria, required maintenance, minimum energy density, lifetime imposed, etc. [73, 74].

Battery capacity is specifically estimated using (2.5) [75]:

Fig. 2.5 ESS topologies
[71]

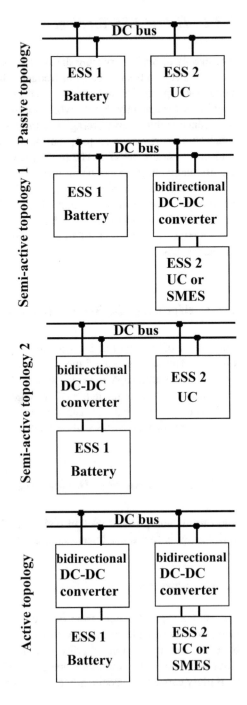

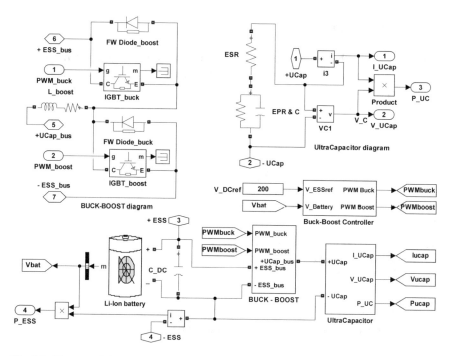

Fig. 2.6 Diagram of the semi-active ESS topology [10]

$$C_{\text{Bat}} = \frac{\Delta P_{\text{load}} \cdot T_{\text{cycle(Bat)}} \cdot n_{\text{cycles}}}{\Delta V_{\text{Bat}}} \tag{2.5}$$

where ΔV_{bat} is the voltage drop under an energy pulse $\Delta P_{\text{load}} \cdot T_{\text{cycle(Bat)}}$ repeated in each load cycle, and n_{cycle} is the number of load cycles.

The technological objectives for the next decade based on different cathode chemistry and anodic nanomaterials are as follows: energy density and lifetime of more than 1000 kWh/kg and 15 years, operation in the increased temperature range (from -150 °C up to 450 °C) and a DOD less than 30% [26, 74, 76].

However, the degradation of battery performance parameters will depend to load cycle and the strategy implemented for HPS energy management [77–79]. It is worth mentioning that if the objective of the energy management strategy (EMS) is to mitigate the load pulses, then an active ESS topology must be used. The pulse mitigation will be performed by appropriate control of the bidirectional converter of the ultra-capacitors, and the loop for DC voltage regulation will be moved to battery side [57, 80].

In the simulations presented in this book, a generic model for lithium-ion battery taken from MATLAB–Simulink® 2013 is used [81]. This is briefly explained in the next section

2.3.3.2 Modeling the Battery

The signal S sets the battery operation in charge mode ($S = 1$) or discharge ($S = 0$). Switching modes is done depending on the battery current filtered by a low pass filter, $i_{LPFbatt}$. The battery voltage value, E_{Batt}, will be given by the charge (f_c) or discharge function (f_d), given by (2.6):

$$f_d(t) = E_w - K_r \frac{Q}{Q - q} \cdot i_{LPFbatt} - K_c \frac{Q}{Q - q} \cdot q + E_{exp}$$

$$f_c(t) = E_w - K_r \frac{Q}{q + Q/10} \cdot i_{LPFbatt} - K_c \frac{Q}{Q - q} \cdot q + E_{exp} \qquad (2.6)$$

where:

E_w	is the working voltage of the battery (V);
K_c	Constant polarization V/(Ah);
K_r	Polarization resistance (Ω);
$i_{LPFbatt}$	Filtered battery current (A);
Q	Battery capacity (Ah).

The battery charging state (SoC) is calculated with (2.7):

$$SOC_{batt} = 100 \cdot \left(1 - \int_0^t i_{batt} dt / Q \right) \qquad (2.7)$$

Battery voltage is given by (2.8):

$$V_{batt} = E_{batt} - r_{batt} \cdot i_{batt} \qquad (2.8)$$

where i_{LPF} is the battery current, and r_{batt} is internal battery resistance.

The equivalent circuit parameters can be changed to represent a particular type of battery specifying nominal voltage, charging capacity, and discharge characteristic parameters (see Fig. 2.7).

For example, a 160 V/38 Ah battery has the following parameters: $A = 15$ V, $B = 0.375$ (Ah)$^{-1}$, $K_c = 0.03$ V/(Ah), $K_r = 0.03$ Ω, $E_w = 163$ V, and $Q = 43$ Ah. The result of the simulation is shown in Fig. 2.8.

2.3.3.3 Regenerative Fuel Cells

As mentioned, the current energy density of renewable fuel cells (RFCs) is four times higher compared to Li-ion-based batteries [82, 83], but may become 10 times higher in the coming years, so RFCs will be used intensely for storage applications [83].

Technological objectives for the next years are as follows: specific energy greater than 1500 Wh/kg, 70% efficiency and minimum 10,000 h lifetime [26].

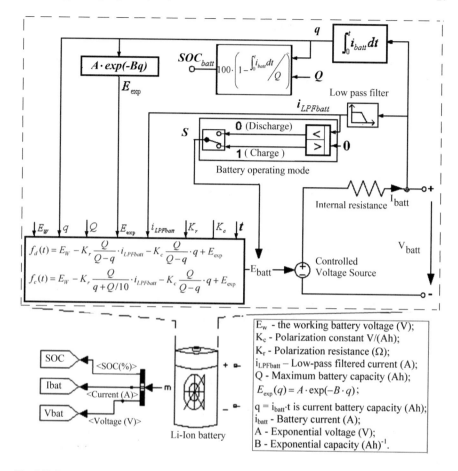

$$f_d(t) = E_W - K_r \frac{Q}{Q-q} \cdot i_{LPFbatt} - K_c \frac{Q}{Q-q} \cdot q + E_{exp}$$

$$f_c(t) = E_W - K_r \frac{Q}{q+Q/10} \cdot i_{LPFbatt} - K_c \frac{Q}{Q-q} \cdot q + E_{exp}$$

E_w - the working battery voltage (V);
K_c - Polarization constant V/(Ah);
K_r - Polarization resistance (Ω);
$i_{LPFbatt}$ – Low-pass filtered current (A);
Q - Maximum battery capacity (Ah);
$E_{exp}(q) = A \cdot \exp(-B \cdot q)$;
$q = i_{batt} \cdot t$ is current battery capacity (Ah);
i_{batt} - Battery current (A);
A - Exponential voltage (V);
B - Exponential capacity (Ah)$^{-1}$.

Fig. 2.7 Battery model [10]

Fig. 2.8 Battery discharging characteristic

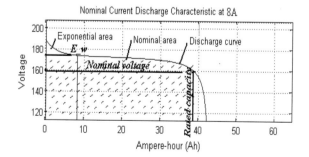

2.3.4 Power Storage Devices Technologies

As mentioned before, the ESS hybridization can be made using UCs, SMESs, and high-speed flywheel [16, 84].

2.3.4.1 Capacitors and Ultracapacitors

Capacitors (C_{DC}) and ultracapacitors (C_{UC}) are used for filtering and mitigation of power pulses.

C_{DC} capacity is specifically estimated using (2.9) [75]:

$$C_{DC} \cong \frac{2 \cdot \Delta P_{load} \cdot T_{cycle(Cds)} \cdot n_{cycles}}{V_{DC}^2 \cdot \left[\left(1 + \frac{\Delta V_{DC}}{V_{DC}}\right)^2 - \left(1 - \frac{\Delta V_{DC}}{V_{DC}}\right)^2 \right]} \tag{2.9}$$

where ΔV_{DC} is the voltage drop on the DC bus under an energy pulse $\Delta P_{load} \cdot T_{cycle(Bat)}$ repeated in each load cycle, and n_{cycle} is the number of load cycles.

Also, C_{UC} capacity is specifically estimated using (2.9) [75]:

$$C_{UC} \cong \frac{2 \cdot \Delta P_{load} \cdot T_{cyle(UC)} \cdot n_{cycles}}{V_{UC}^2 \cdot \left[\left(1 + \frac{\Delta V_{UC}}{V_{UC}}\right)^2 - \left(1 - \frac{\Delta V_{UC}}{V_{UC}}\right)^2 \right]} \tag{2.10}$$

where ΔV_{UC} is the ultracapacitors voltage drop.

The technological objectives for the next decade based on nanomaterials are as follows: a specific energy and power of about 100 Wh/kg and 10,000 W/kg and increased temperature range (from $-60\ ^\circ$C up to 300 $^\circ$C) [26].

2.3.4.2 Modeling the Ultracapacitors

A first-order model is used for capacitors and ultracapacitors (Fig. 2.9), where R_s and R_p are the series and parallel resistance, respectively.

The capacity of the ultracapacitor is evaluated on the basis of Stern's mathematical model, which combines the Helmholtz and Gouy-Chapman models using (2.11) [85]:

$$C_{UC} = \left[\frac{1}{C_H} + \frac{1}{C_{GC}} \right]^{-1} \tag{2.11a}$$

$$C_H = \frac{N_c \varepsilon_r \varepsilon_0 A_i}{d} \tag{2.11b}$$

$$C_{GC} = \frac{F Q_c}{2 N_c R T} \sin h \left(\frac{Q_c}{N_c^2 A_i \sqrt{8 R T \varepsilon_r \varepsilon_0 c}} \right) \tag{2.11c}$$

Fig. 2.9 Ultracapacitor
model

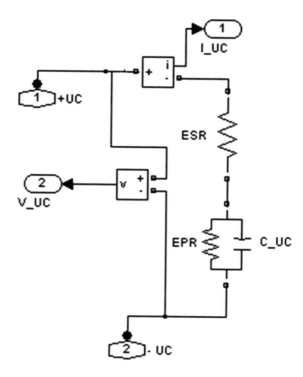

where C_H și C_{GC} are the capacities of the ultracapacitors defined by Helmholtz and Gouy-Chapman, N_c is the number of layers of the electrode, $\varepsilon_r\varepsilon_0$ is the electrical permeability of the material (F/m), A_i is the common surface between electrode and electrolyte (m^2), d is the distance between the layers (m), Q_c is the electrical charge (C), and c is the molar concentration (mol m^{-3}).

2.3.4.3   Superconducting Magnetic Energy Storage (SMES) Devices

In addition to using in microgrids and smart grids (ensuring the stability of the electricity system, black start, and uninterruptible power supply when is the case) [86, 87], SMES devices have recently attracted researchers' attention for use in space applications [88, 89] and military equipment [90, 91] (see Fig. 2.10).

2.3.4.4   High-Speed Flywheels

The high-speed flywheel can store energy from a few kWh to MWh. The mass of the ESS can be reduced using a flywheel to replace the momentum wheels and the batteries [92]. Flywheel energy systems using the drop control are able to mitigate power pulses [57, 71, 80].

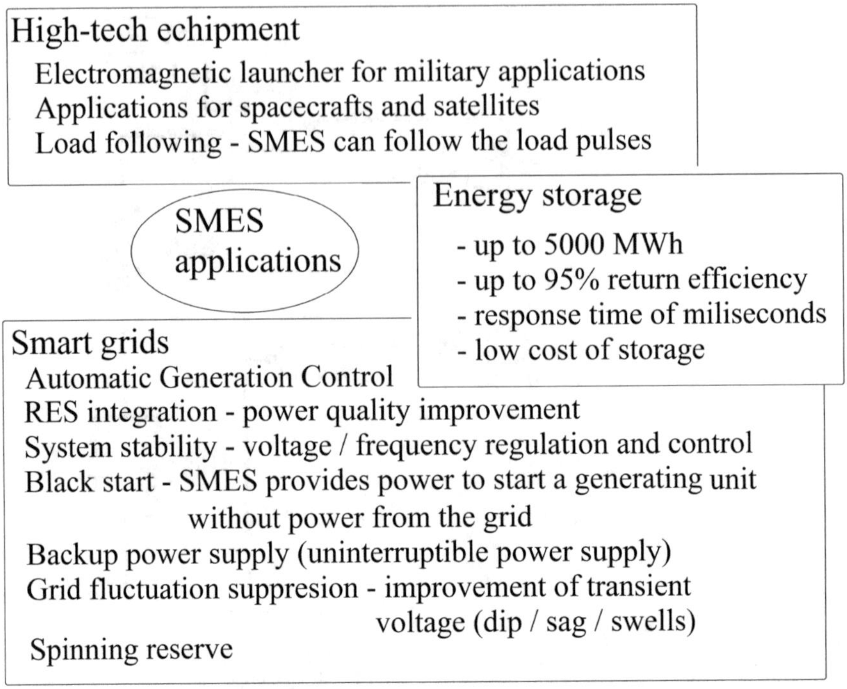

High-tech echipment

Electromagnetic launcher for military applications
Applications for spacecrafts and satellites
Load following - SMES can follow the load pulses

SMES
applications

Energy storage

- up to 5000 MWh
- up to 95% return efficiency
- response time of miliseconds
- low cost of storage

Smart grids
Automatic Generation Control
RES integration - power quality improvement
System stability - voltage / frequency regulation and control
Black start - SMES provides power to start a generating unit
                        without power from the grid
Backup power supply (uninterruptible power supply)
Grid fluctuation suppresion - improvement of transient
                              voltage (dip / sag / swells)
Spinning reserve

Fig. 2.10 SMES applications [57]

The technological objectives for the next decade based on carbon nanofiber and carbon composite rotors are as follows: 2.700 Wh/kg specific energy and 20 years lifetime [26, 93, 94].

The energy sources (PEMFC and RESs) will mainly ensure the load demand, so a dynamic model must be considered for them in simulation. Modeling of PEMFC, RESs and their power flows, and DC load will be presented in the next sections.

2.4 Proton Exchange Membrane Fuel Cell (PEMFC) System

PEMFC systems have been increasingly used in the past decade for mobile applications [20, 40, 41] and stationary [94]. The technological objectives for the next decade are as follows: specific energy of about 100 W/kg, efficiency greater than 50 or 70% (if thermal energy is recovered) and a lifetime of more than 5000 h [95–97].

2.4.1 Modeling the PEMFC System

The modeling of the PEMFC system is difficult if phenomena such as stack temperature variation, uneven humidification, mass transport influence, channel pressure loss, and so on are included.

In the simulations presented in this book, a PEMFC model taken from MATLAB–Simulink® 2013 is used [81], which is an electrical model based on the aforementioned assumptions. This has two variants (simple and improved model), which will be briefly explained in the next two subsections.

2.4.1.1 Simplified Model of the PEMFC System

Simplified model of the PEMFC system neglects the voltage due to the effect of the electric load concentration at the PEM level, the fuel cell operating voltage being given by (2.12):

$$V_{FC} = V_{Nernst} - V_{act} - V_{ohmic}$$
$$V_{act} = N_C \cdot A \cdot \ln\left(\frac{i_{FC}}{i_0}\right) \cdot \frac{1}{sT_{FC}/3 + 1}$$
$$V_{ohmic} = r_{ohmic} \cdot i_{FC} \tag{2.12}$$

where V_{Nernst} is the thermodynamic voltage given by the Nernst relationship (2.13), V_{act} and V_{ohmic} represent activation losses and resistive ohmic losses, N_C is the number of cells in series, A_T is the Tafel slope (V), i_0 is the internal current at the PEM level (A), r_{ohmic} is the internal resistance of the FC system (Ω), T_{FC} is the response time of the FC system, and V_{FC} și i_{FC} represent voltage and current of the FC system (see Fig. 2.11).

Nernst voltage can be given by approximation variants (2.13):

$$V_{Nernst1} \cong E = 1.299 + (\theta_{FC} - 25)\frac{-44.43}{2F} + \frac{R \cdot (273 + \theta_{FC})}{2F}\ln\left(P_{H_2} \cdot P_{O_2}^{\frac{1}{2}}\right)$$
$$V_{Nernst2} \cong E_n = E - K_u\left(f_{O_2} - f_{O_2 nom}\right) \tag{2.13}$$

where $(\theta_{FC} + 273) = T$ is the temperature of the FC system (K), P_{H_2} and P_{O_2} are the partial pressures at the anode and cathode (atm), f_{O2nom} is the nominal oxygen consumption (%), K_u is the voltage drop adjustment parameter, and F and R are the Faraday constant (A s/mol) and the constant of the ideal gases (J/mol K), respectively.

Density of internal current (i_0) and Tafel slope (A) is calculated with (2.14):

$$i_0 = \frac{2 \cdot F \cdot k\left(P_{H2} + P_{O_2}\right)}{R \cdot h} \cdot \exp\left(\frac{-\Delta G}{R \cdot T}\right)$$

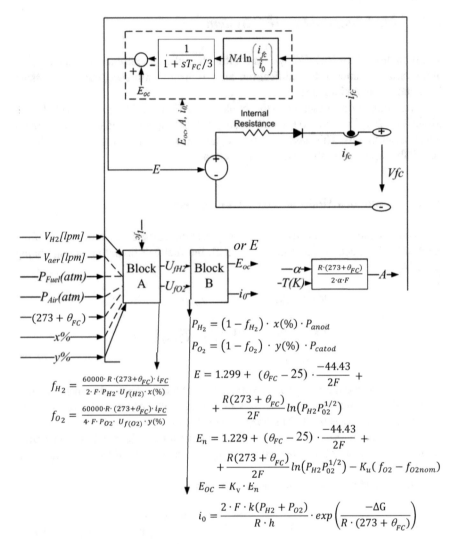

Fig. 2.11 Simplified model of the PEMFC system

$$A_T = \frac{R \cdot T}{2 \cdot \alpha \cdot F} \tag{2.14}$$

where ΔG is the enthalpy (J), α is the transfer coefficient of the electrical charge, and k and h are the Boltzmann (J/K) and Planck (J s) constant.

The partial pressures are calculated using (2.15):

$$P_{H_2} = \left(1 - f_{H_2}\right) \cdot x(\%) \cdot P_{\text{anod}}$$
$$P_{O_2} = \left(1 - f_{O_2}\right) \cdot y(\%) \cdot P_{\text{catod}} \tag{2.15}$$

where f_{H_2} și f_{O_2} are coefficients of use for hydrogen and oxygen (2.16), P_{anod} and P_{catod} are the supply pressure from the anode and the cathode (atm), and x and y are the percentage concentrations of hydrogen and oxygen in the fuel.

$$f_{H_2} = \frac{60{,}000 \cdot R \cdot (273 + \theta_{FC}) \cdot i_{FC}}{2 \cdot F \cdot P_{H_2} \cdot U_{f(H_2)} \cdot x(\%)}$$

$$f_{O_2} = \frac{60{,}000 \cdot R \cdot (273 + \theta_{FC}) \cdot i_{FC}}{4 \cdot F \cdot P_{O_2} \cdot U_{f(O_2)} \cdot y(\%)} \tag{2.16}$$

where V_{H_2} și V_{aer} are the values of hydrogen and airflows (liters/min or lpm), and i_{FC} is the FC current.

The unloaded FC system has the FC voltage given by (2.17):

$$E_{OC} = K_v \cdot E_{Nernst} \tag{2.17}$$

where K_v is a parameter for adjusting the Nernst voltage.

2.4.1.2   Improved Model of the PEMFC System

Improved model of the PEMFC system is synthetically presented in Fig. 2.12, where the relationships implemented by each subsystem are mentioned and explained below.

The time variation (t) of PEMFC (T) depends on the FC current (I_{FC}) of the pile (IFC) and can be calculated with (2.18):

$$\theta_{FC} + 273 = T = T_0 + (T_0 - T_{rt} + T_{ic} \cdot I_{FC}) \cdot \left(1 - \exp\left(\frac{t \cdot I_{FC}}{T_{it}}\right)\right) \tag{2.18}$$

where parameters T_0, T_{rt}, T_{ic}, and T_{it} are experimentally determined
Dynamics of partial pressures (P_{H2} and P_{O2}) is given by (2.19):

$$\frac{d}{dt}P_{H2} = \frac{R \cdot (\theta_{op} + 273)}{V_{anode}} q_{H2} = \frac{R \cdot (\theta_{op} + 273)}{V_{anode}} (Q_{H2}^{in} - Q_{H2}^{out} - Q_{H2}^{r}) \Rightarrow$$

$$\frac{d}{dt}P_{H2} = \frac{R \cdot (\theta_{op} + 273)}{V_{anode}} \left(Q_{H2}^{in} - k_{H2} \cdot P_{H2} - 2\frac{N_C}{4FU_{f(H2)}}I_{FC}\right) \tag{2.19}$$

where:

- R—is the universal constant of the ideal gas (atm · kmol K)$^{-1}$);
- F—Faraday constant (C/kmol);
- V_{anode}—The volume at the anode (m^3);
- $(\theta_{FC} + 273) = T$—Operating temperature of the PEMFC system;
- $Q_{H_2}^{in}$—Input hydrogen flow (kmol s^{-1} or l/min);
- $Q_{H_2}^{out}$—Output hydrogen flow;

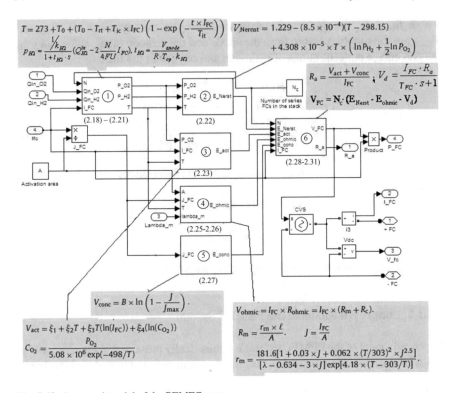

Fig. 2.12 Improved model of the PEMFC system

- $Q_{H_2}^r$—The hydrogen flow from the reaction;
- k_{H_2}—Hydrogen molar constant [kmol · (atm s)$^{-1}$];
- U_{fH_2}—Hydrogen utilization factor (%);
- N_C—Number of FC cells in series.

Applying the transformed Laplace to both members of the equation above is obtained (2.20)

$$p_{H2} = \frac{1/k_{H2}}{1 + t_{H2} \cdot s}\left(Q_{H2}^{in} - 2\frac{N_C}{4FU_{f(H2)}}I_{FC}\right), \quad t_{H2} = \frac{V_{anode}}{R \cdot (\theta_{FC} + 273) \cdot k_{H2}} \quad (2.20)$$

where t_{H_2} is the time constant of hydrogen.

Similarly, partial oxygen pressure is obtained:

$$p_{O2} = \frac{1/k_{O2}}{1 + t_{O2} \cdot s}\left(Q_{O2}^{in} - \frac{N_C}{4FU_{f(H2)}}I_{FC}\right), \quad t_{O2} = \frac{V_{catode}}{R \cdot (\theta_{FC} + 273) \cdot k_{O2}} \quad (2.21)$$

where:

- t_{O_2}—is the time constant of oxygen (s);

- $V_{cathode}$—The volume at the cathode (m^3);
- $Q_{H_2}^{in}$—Input oxygen flow (kmol s^{-1} sau l/min);
- k_{O_2}—Oxygen utilization factor [kmol (atm s)$^{-1}$].

Relationships (2.18)–(2.21) are used to implement subsystem 1. Partial hydrogen and oxygen pressures are inputs for the PEMFC model.

The Nernst voltage (2.22) is implemented in subsystem 2:

$$V_{Nernst} = 1.229 - 8.5 \times 10^{-4} \cdot (\theta_{FC} - 25) + 4.308 \times 10^{-5} \cdot (\theta_{FC} + 273) \cdot (\ln P_{H_2} + 1/2 \ln P_{O_2}) \tag{2.22}$$

The activation voltage given by (2.23) is implemented in subsystem 3:

$$V_{act} = e_1 + e_2 \cdot (\theta_{op} + 273) + e_3 \cdot (\theta_{op} + 273) \cdot \ln(I_{FC}) + e_4 \cdot \ln(C_{O_2}) \tag{2.23}$$

where the parameters e_i, $i = 1$–4, are determined experimentally, and C_{O_2} represents the concentration of oxygen dissolved in the cathodic catalyst:

$$C_{O_2} = \frac{P_{O_2}}{5.08 \times 10^{-6} \cdot \exp(-498/\theta_{FC} + 273)} \tag{2.24}$$

The ohmic voltage can be calculated with (2.25) and is implemented in subsystem 4:

$$V_{ohmic} = I_{FC} R_{ohmic}; \; J_{FC}/A; \; R_{ohmic} = R_m + R_c; \; R_m = r_m \cdot t_m / A \tag{2.25}$$

where:

- J_{FC}—is the density of the current FC (A/cm^2);
- R_{ohmic}—The ohmic resistance of the FC system (Ω);
- R_m—Equivalent resistance of the membrane (Ω);
- R_c—Resistance between membrane and electrodes (Ω);
- t_m—Thickness of the membrane (cm);
- A—Activation surface (cm^2);
- r_m—The resistivity of the membrane (Ω cm^2 m^{-1}) given by Nafion's relationship (2.26):

$$r_m = \frac{181.6 \cdot \left[1 + 0.03 \cdot J_{FC} + 0.062 \cdot (J_{FC})^{2.5} \cdot (\theta_{FC} + 273/303)^2\right]}{[\lambda_m - 0.0634 - 3 \cdot J_{FC}] \cdot \exp[4.18 \cdot ((\theta_{FC} + 273) - 303/\theta_{FC} + 273)]} \tag{2.26}$$

where λ_m represents the water content of the membrane and is an input variable for the PEMFC model.

Concentration voltage can be calculated with (2.27) and is implemented in subsystem 5:

$$V_{\text{conc}} = B \cdot \ln(1 - J_{\text{FC}}/J_{\text{max}}) \tag{2.27}$$

where B is a modeling constant (V), and J_{max} is the maximum current density.

The output voltage of the PEMFC system can be calculated with (2.28) and is implemented in subsystem 6:

$$V_{\text{FC}} = N_C(V_{\text{Nernst}} - V_{\text{ohmic}} - V_{\text{d}}) \tag{2.28}$$

where $V_{\text{d}} = V_{\text{act}} + V_{\text{conc}}$.

The differential equation (2.29) represents the dynamics of the PEMFC system (expressing the current FC by the currents trough the resistance R_{a} in parallel to the capacitor C_{FC}:

$$I_{\text{FC}} = \frac{V_{\text{d}}}{R_{\text{a}}} + C_{\text{FC}} \times \frac{\text{d}}{\text{d}t} V_{\text{d}} \tag{2.29}$$

where R_{a} (Ω) is:

$$R_{\text{a}} = \frac{V_{\text{d}}}{I_{\text{FC}}} = \frac{V_{\text{act}} + V_{\text{conc}}}{I_{\text{FC}}} \tag{2.30}$$

By applying the Laplace transform to both members of Eq. (2.29) is obtained (2.31):

$$V_{\text{d}} = \frac{I_{\text{FC}} \cdot R_{\text{a}}}{T_{\text{FC}} \cdot s + 1}, \quad T_{\text{FC}} = C_{\text{FC}} \cdot R_{\text{a}} \tag{2.31}$$

where T_{FC} represents the time constant of the PEMFC system.

Relationships (2.28)–(2.31) are used to implement subsystem 6.

All subsystems are coupled according to the interdependence relationships above, resulting in the dynamic PEMFC model in Fig. 2.12.

2.4.2 Modeling the Fueling Regulators

In this study, the flow fuel rate (FuelFr) and airflow rate (AirFr) given by (2.32) will be used as search variables of the optimum for a given optimization function:

$$\text{FuelFr} = \frac{60{,}000 \cdot R \cdot (273 + \theta_{\text{FC}}) \cdot N_C \cdot I_{\text{ref(H}_2)}}{2F \cdot (101{,}325 \cdot P_{f(\text{H}_2)}) \cdot (U_{f(\text{H}_2)}/100) \cdot (x_{\text{H}_2}/100)} \tag{2.32a}$$

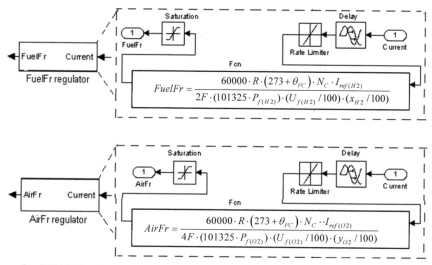

$R = 8.3145$ J/(mol K); $F = 96485$ As/mol;
N_C - number of cells in series (65);
θ_{FC} - operating temperature (65° Celsius)
$U_{f(H2)}$, $U_{f(O2)}$- nominal utilization of hydrogen (99.56%) and oxygen (59.3%);
$P_{f(H2)}$, $P_{f(O2)}$- pressure of the fuel (1.5 bar) and air (1 bar);
x_{H2}, y_{O2}- composition of fuel (99.95%) and oxidant (21%);
$I_{ref(H2)}$, $I_{ref(O2)}$- reference currents.

Fig. 2.13 Fueling regulators [12]

$$AirFr = \frac{60,000 \cdot R \cdot (273 + \theta_{FC}) \cdot N_C \cdot I_{ref(O_2)}}{4F \cdot (101,325 \cdot P_{f(O_2)}) \cdot (U_{f(O_2)}/100) \cdot (y_{O_2}/100)} \qquad (2.32b)$$

where $I_{ref(H_2)}$ and $I_{ref(O_2)}$ are the reference currents set by the energy management strategy (EMS) for the fueling regulators (see Fig. 2.13). A slope limiter is used to avoid high variations of the fueling flow rates due to rapid changes in the reference currents, preventing the fuel starvation.

2.4.3   Modeling the Compressor

The response time of the compressor is less than the PEMFC time constant, so it could be neglected in PEMFC performance analysis. In the simulations presented in this book, the compressor dynamic is described by a second-order system [98].

$$G_{d(cm)} = \frac{\omega_{cm}^2}{s^2 + 2\xi_{cm}\omega_{cm} \cdot s + \omega_{cm}^2} \qquad (2.33)$$

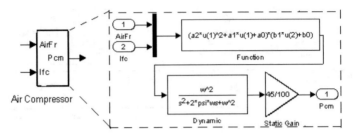

Fig. 2.14 Diagram of the air compressor [12]

where $t_{cm} = 2\pi/\omega_{cm}$ and ξ_{cm} are the time constant and amortization parameter.

The static model of the air compressor is nonlinear, so the consumed power (P_{cm}) is given by (2.34) [99].

$$P_{cm} = I_{cm} \cdot V_{cm} = k_{cm} \cdot \left(a_2 \cdot \text{AirFr}^2 + a_1 \cdot \text{AirFr} + a_0\right) \cdot (b_1 \cdot I_{FC} + b_0) \quad (2.34)$$

where $a_0 = 0.6$, $a_1 = 0.04$, $a_2 = -0.00003231$, $b_0 = 0.9987$ and $b_1 = 46.02$, and V_{mc} and I_{mc} are the voltage and current of the air compressor.

Parameter k_{cm} sets the consumed power, P_{cm}. For example, if $k_{cm} = 0.45$, then the compressor power is about 15% from the FC nominal power. The air compressor model is presented in Fig. 2.14.

The air compressor is the main consumer of the FC system, so, neglecting other losses, the FC net power, P_{FCnet}, can be approximated using (2.35):

$$P_{FCnet} = V_{FC} \cdot I_{FCnet} \cong P_{FC} - P_{cm} = V_{FC} \cdot I_{FC} - V_{cm} \cdot I_{cm} \quad (2.35)$$

The oxygen excess ratio (OER) depends on the FC current and can be controlled by the air compressor to vary within the admissible range (usually around 2, but always higher than 1) [98, 99]:

$$\text{OER} \equiv \lambda_{O_2} = \frac{a_3 \cdot I_{FC}^3 + a_2 \cdot I_{FC}^2 + a_1 \cdot I_{FC} + a_0}{b_1 \cdot I_{FC} + b_0} \quad (2.36)$$

where $a_0 = 402.4$, $a_1 = 402.4$, $a_2 = -0.8387$, $a_3 = 0.027$, $b_0 = 1$, and $b_1 = 61.4$.

Except the dynamical models for air compressor and power converters, the models for PEMFC system and battery presented above, respectively, the models for RESs and DC load that will be presented below, all are taken from the SimPowerSystems library of the MATLAB–Simulink® 2013.

2.5 Modeling the Renewable Energy Sources

2.5.1 Modeling the Wind Turbine

The wind turbine (WT) model presented in Fig. 2.15 and used in this book is slightly modified to the model included in the MATLAB–Simulink® toolbox so that both characteristics of power (P_m) and torque (T_m) can be drawn versus wind speed (see the bottom of Fig. 2) [100]. The WT model is described by (2.37) [101]:

$$P_m = c_p(\lambda, \beta) \frac{\rho A_{WT}}{2} v_{wind}^3 \tag{2.37a}$$

$$c_p(\lambda, \beta) = c_1 \cdot \left(\frac{c_2}{\lambda_i} - c_3\beta - c_4 \right) \cdot \exp\left(-\frac{c_5}{\lambda_i} \right) + c_6\lambda \tag{2.37b}$$

$$\frac{1}{\lambda_i} = \frac{1}{\lambda + 0.08\beta} - \frac{0.035}{\beta^3 + 1}, \tag{2.37c}$$

$$T_m = \frac{P_m}{\omega_m} \tag{2.37d}$$

$$J \frac{d\omega_m}{dt} = T_e - T_m - B \cdot \omega_m \tag{2.37e}$$

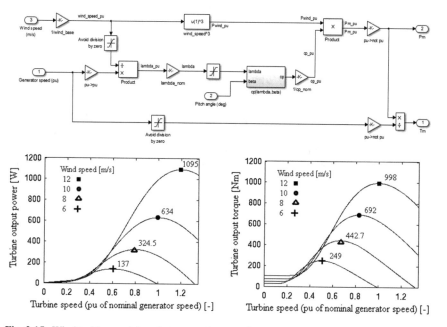

Fig. 2.15 Wind turbine model, and power and torque characteristics

where:

T_e	The electromagnetic torque (N m);
J	Moment of inertia (kg m^2);
B	Viscous friction coefficient (N m s/rad);
c_p	WT performance coefficient;
ρ	Air density (kg/m^3);
$A_{WT} = \pi R^2$	WT swept area (m^2);
R	WT blade radius (m);
β	Blade pitch angle (degree);
λ	Tip speed ratio (TST) defined by (2.38):

$$\lambda = \frac{R \omega_m}{v_w} \tag{2.38}$$

The WT power using the pu values is:

$$P_{m(pu)} = k_p \cdot c_{p(pu)}(\lambda, \beta) \cdot v_{wind(pu)}^3 \tag{2.39}$$

The simulation has been performed using the default values for the WT model: 1500 W power, a reference of 1.5/0.9 MVA for the electrical generator power and 12 m/s for the wind speed in order to evaluate their normalized values (pu values), and the coefficients as follows: $k_p = 0.73$, $c_1 = 0.5176$, $c_2 = 116$, $c_3 = 0.4$, $c_4 = 5$, $c_5 = 21$, and $c_6 = 0.0068$.

2.5.2 Modeling the Photovoltaic Panel

The photovoltaic (PV) panel model presented in Fig. 2.16 and used in this book is the PV model included in the MATLAB–Simulink® toolbox based on a silicon PV cell of polycrystalline and amorphous type.

The *one-diode* model is described by the following relationships (2.40) [3, 102, 103]:

$$I_O = I_{PV} - I_D - I_{sh}$$
$$I_D = I_R(e^{\frac{V_D}{\eta V_T}} - 1)$$
$$I_{sh} = \frac{V_O + I_O R_S}{R_{sh}} \tag{2.40}$$

where:

I_D and V_D are the current and voltage of the diode;
I_O and V_O are the output current and voltage of the FC cell;

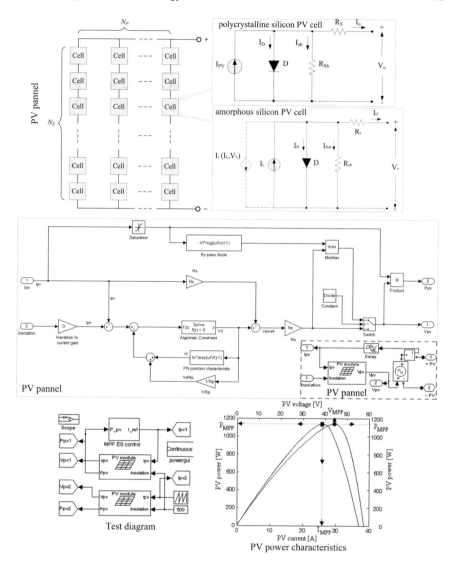

Fig. 2.16 Models of the PV cell and PV panel

I_R	Reverse blocking current;
V_T	Thermal voltage;
η	The ideality factor of the diode;
I_{sh}	Current of shunt resistor. (8.3)

By simple manipulations of (2.40), the relationship between the output current (I_O) and the output voltage (V_O) is obtained as follows:

$$I_O = I_{PV} - I_R \left(e^{\frac{V_O + I_O R_s}{\eta_I V_T}} - 1 \right) - \frac{V_O + I_O R_S}{R_{sh}} \qquad (2.41)$$

The model of the amorphous silicon PV cell is also presented in Fig. 2.16. In addition, this has a current source controlled by the photo-current and PV voltage.

A PV module contains Np strings of PV cells, where a string has Ns PV cells in series (see in Fig. 2.16). The current of the PV module (I_A) is given by (2.42):

$$I_A = N_P I_{PV} - N_P I_O \cdot \left(\exp \left(\frac{q \left(V_A + I_A \frac{N_S}{N_P} R_s \right)}{\eta_I N_s V_T} \right) - 1 \right) - \frac{V_A + I_A \frac{N_S}{N_P} R_S}{\frac{N_S}{N_P} R_{sh}}$$

$$(2.42)$$

The PV power characteristics and the diagram to obtain these characteristics are presented in Fig. 2.16 as well.

2.5.3 Modeling of the Renewable Power Profile

The real profiles of the irradiance and wind speed for various weather worldwide conditions are available in many databases. In simulations presented in this book, the renewable energy flow combines the PV and WT profiles with a random power, using weighting parameters to simulate the level of variability (Fig. 2.17).

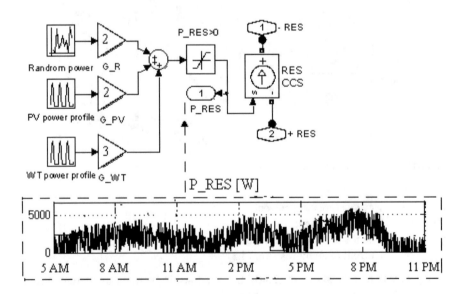

Fig. 2.17 Modeling of the renewable power profile [12]

Also, other real load demand profiles for various homes and residential buildings around the world are available in various databases.

2.6 Modeling the Load Demand on the DC Bus

The load demand variability may be modeled by adding random power to variable profile, using a weighting parameter to simulate this level of variability (Fig. 2.18).

Modeling of the load demand that contains power pulses and low-frequency ripples is presented in Fig. 2.19. This model can be useful to test the PEMFC HPS used in FC vehicles, where large power pulses can appear during a real route, and, as it is known, the inverter ripple normally appears on the DC bus.

The inverter ripple from the DC bus is back-propagated to the PEMFC via the power converters [104]. So, because the passive filtering is costly and uses bulky components, the active filtering is recommended [105]. An interesting anti-ripple generation technique using a bi-buck power converter topology is used to mitigate in real time the low-frequency ripples [106].

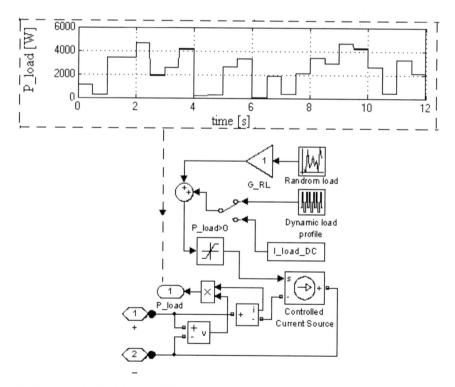

Fig. 2.18 Variable load demand [12]

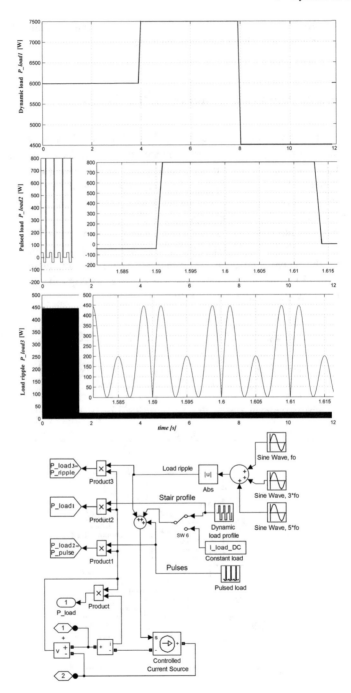

Fig. 2.19 Modeling load demand containing power pulses and low-frequency ripples [71]

The modeling of the power converters has been, is, and will be a subject of interest [107, 108]. In this book, the well-known models of the boost DC-DC power converter and buck-boost bidirectional DC-DC power converter will be used in simulation, so these will be briefly presented in the next section.

2.7 Modeling the Power Converters

The buck-boost bidirectional DC-DC power converter (see Fig. 2.6) and the boost DC-DC power converter (see Fig. 2.20) interface the ultracapacitors' stack and the PEMFC with the DC bus, respectively. The power devices (such as the insulated-gate bipolar transistor (IGBT) and freewheeling (FW) diode) and passive components have been selected from the SimPowerSystem® toolbox.

For example, the boost DC-DC power converter (see Fig. 2.20) operates using (2.43):

$$v_{FC} = (r_L + R_{DS(on)}) \cdot i_{FC} + L\frac{di_{FC}}{dt} + V_{DS(on)} \qquad (2.43a)$$

$$v_{FC} = (r_L + R_{D(on)}) \cdot i_{FC} + L\frac{di_{FC}}{dt} + V_{D(on)} + v_{DC} \qquad (2.43b)$$

where the on-state parameters for the resistance and voltage of the IGBT and FW diode are $R_{DS(on)}$ and $V_{DS(on)}$, and $R_{D(on)}$ and $V_{D(on)}$, respectively.

The relationships (2.43a) and (2.43b) describe the power converter operation during the conduction and blocking phase of the IGBT (when the FW diode is in on-state), respectively.

Both power converters have been designed to operate in continuous current mode (CCM) [107, 108].

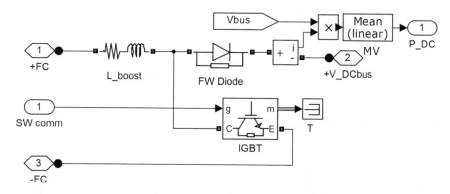

Fig. 2.20 Simulation diagram of the boost DC-DC power converter

2.8   The Performance Indicators Hybrid Power System

The hybrid power system generates power based on the WT and PV RESs that must be operated close to the maximum power point (MPP) and global MPP, respectively, in order to harvest all the available energy. The PEMFC system generates power to sustain load demand during lack of RES power, and this must be operated close to the maximum efficiency point (MEP) or to minimize the fuel consumption. The performance indicators used to assess how close to the optimal operating point the energy sources were operated will be presented below.

2.8.1   Performance Indicators of the PEMFC System

The performance indicators of the FC system are as follows: the FC net power (P_{FCnet}) given by (2.35), the fuel consumption efficiency (Fuel$_{eff}$) that measure the FC net power generated for consumed fuel (2.44), the FC system efficiency (η_{sys}) that evaluate the percent of the FC net power in the total FC power generated (2.45), and the total fuel consumption (Fuel$_T$) that is estimated during a load demand cycle (2.46):

$$\text{Fuel}_{eff} = \frac{P_{FCnet}}{\text{FuelFr}} \tag{2.44}$$

$$\eta_{sys} = \frac{P_{FCnet}}{P_{FC}} \tag{2.45}$$

$$\text{Fuel}_T = \int \text{FuelFr}(t)\mathrm{d}t \tag{2.46}$$

The fuel economy represents the difference in total fuel consumptions (ΔFuel_T) estimated for an efficient EMS compared to a reference strategy.

2.8.2   Performance Indicators of the Renewable Energy Sources

2.8.2.1   Wind Turbine

The power and torque characteristics of a wind turbine have a unique maximum (see Fig. 2.15) called MPP and MTP, which must be found accurately in real time.

The tracking accuracy (T_{acc}) represents how close this maximum can be found during a stationary regime of constant wind speed:

$$T_{acc} = \frac{y_{max}}{y^*_{max}} \cdot 100[\%] \tag{2.47}$$

where y_{max} and y^*_{max} are the maximum value and the value found.

The searching speed for value found is another performance indicator that measures the performance during the transitory regimes using the tracking efficiency (T_{eff}):

$$T_{eff} = \frac{\int_0^t y dt}{\int_0^t y^* dt} \cdot 100[\%] \tag{2.48}$$

where y is the WT power, and y^* is the value found of this power using an MPP tracking algorithm.

2.8.2.2 Photovoltaic Panel

The power characteristics of a photovoltaic panel have a unique maximum only under uniform and constant irradiance (see Fig. 2.16) called MPP. Generally, the power characteristics of a photovoltaic panel under partial shading conditions exhibit many maximums, so the global maximum (called GMPP) must be found accurately in real time.

So, besides the tracking speed, tracking accuracy (T_{acc}) and tracking efficiency (T_{eff}) defined for global search as (2.47') and (2.48'), the search resolution (S_R), which defines how close GMPP and local maximum power point (LMPP) can be on a multimodal pattern (2.29), and the hit count, which represents the success rate to find the GMPP during repetitive tests, will be used as performance indicators:

$$T_{acc} = \frac{y_{GMPP}}{y^*_{GMPP}} \cdot 100[\%] \tag{2.47'}$$

$$T_{eff} = \frac{\int_0^t y dt}{\int_0^t y^* dt} \cdot 100[\%] \tag{2.48'}$$

$$S_R = \frac{\min_i |y_{GMPP} - y_{LMPPi}|}{y_{GMPP}} \cdot 100[\%] \tag{2.49}$$

where y_{GMPP}, y_{LMPP}, and y^*_{GMPP} are the GMPP value, the LMPP and the founded value, respectively, and y and y^* are the PV power and the value of power tracked by a GMPP algorithm.

The objectives and future research directions will be presented briefly in the next section because the energy management strategies (EMSs) and optimization algorithms such as GMPP tracking algorithms will be presented in detail in Chap. 3.

2.9   Objectives and Directions of Research for the Hybrid Power System

The common objectives of the EMSs proposed in the literature are as follows:

i.   Preventing deep battery discharge.
ii. Reducing the dynamics of DC energy exchanged with battery by using a backup energy source (with consequence on battery capacity and battery life).
iii. Improving system efficiency.
iv. Stabilization of voltage on the DC bus.
v.   Reduction of operating and maintenance costs.
vi. Minimization of the system operational costs.

The HPS architecture using the PEMFC as backup energy responds to the aforementioned objectives as follows: The power following control that is implemented for the fueling regulators of the PEMFC system ensures a charge sustained mode for the battery, preventing deep battery discharge (i) and reducing the dynamics of the battery power (ii); the MEP, MPP, and GMPP algorithms, which are implemented for the PEMFC system, WT, and PV panels, respectively, increase the system efficiency (iii); the DC voltage regulation, which is implemented on the ESS side (to ultracapacitors or battery), stabilizes the voltage on the DC bus (iv); the implementation of the fuel economy strategy will reduce the total fuel consumed, lowering the operating costs (v); implementation of the charge-sustained mode for the battery will increase their lifetime and reduce its size, reducing the maintenance (v) and system operational costs as well (vi).

The energy flow split strategies (such as the EMSs based on rules to power the load [109], filtering of the load demand [110], model predictive control [111–113], and artificial intelligence concepts using fuzzy logic [114, 115], neural networks [115], and genetic operators [116]) have partially responded to these objectives, so advanced EMSs have recently been proposed in the literature [117, 118].

It is worth mentioning that an active topology for the EES (Fig. 2.21) may offer flexibility in control of the power flows to offset the power flow balance and mitigate the load pulses as well [71], but the mitigation of the load pulses and power ripples of low frequency will not be completely approached in this book.

Only modeling for load impulse mitigation and proposed control schemes will be presented in this chapter. The reader can read [71, 80] if he is interested in seeing the proposed control for both bidirectional DC-DC power converters used for battery and ultracapacitors stacks, along with an analyze where is better to apply control loops for load-following control, impulse mitigation control, and DC voltage regulation (choosing from available variants: on the side of the PEMFC, battery, or ultracapacitors).

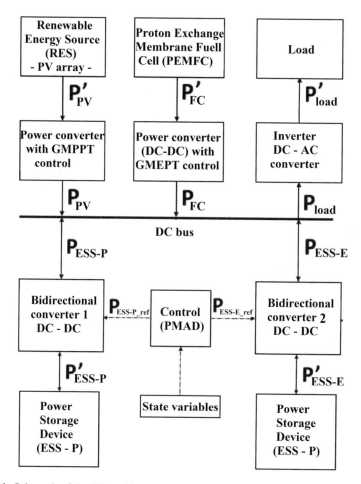

Fig. 2.21 Schematic of the HPS architecture with an active EES topology [71]

2.10   Conclusion

This chapter is focused on modeling of the subsystems that compose a hybrid power system, which produce energy based on a combination of renewable energy generation systems such as wind turbines and photovoltaic panels. When the energy from renewable sources is lower than the load demand, the lack of energy on the common bus will be provided by the energy storage system, which will be operated in the discharge mode. In the opposite case, when the energy from renewable sources is higher than the load demand, the excess energy on the common bus will be stored by the energy storage system, which will be operated in the charge mode. Thus, the variability of the renewable energy and load demand will force many charge–discharge cycles to offset the power flow balance on the common bus, reducing the lifetime of the battery. In addition, a large capacity for the battery stack is necessary.

Consequently, in order to reduce the battery capacity and increase its lifetime, this chapter proposes the use of a non-polluting source of backup energy. The proton exchange membrane fuel cell system can be used as backup energy source, instead of a pollutant solution based, for example, on a diesel generator.

So, the architecture of the hybrid power system based on a common DC bus with semi-active topology for the energy storage system has been chosen to analyze its performance in use. Performance indicators for the efficient use of a proton exchange membrane fuel cell system (such as energy efficiency and fuel economy) and renewable energy sources (such as tracking efficiency and hit count of the searches for global maximum) have been presented, along with the main objectives and research directions in this field. So, the most used technologies for energy and power storage devices used in the energy storage system have been discussed in this chapter. Also, for performance analysis, the models used for the proton exchange membrane fuel cell system, battery, ultracapacitors, wind turbine, photovoltaic panel, power converters, and load demand have been presented.

References

1. Zohuri B (2018) Hybrid energy systems. Driving reliable renewable sources of energy storage. Springer, London, UK
2. Phap VM (2019) Innovative configuration concept for solar-wind hybrid power system. Scholars' Press, Riga, Latvia
3. Bizon N, Tabatabaei NM, Blaabjerg F, Kurt E (2017) Energy harvesting and energy efficiency: technology, methods and applications. Springer, London, UK. http://www.springer.com/us/book/9783319498744
4. Lu Z, Zhou S (2018) Integration of large scale wind energy with electrical power systems in China. Wiley, Hoboken, NJ
5. Bizon N, Tabatabaei NM, Shayeghi H (2013). Analysis, control and optimal operations in hybrid power systems—advanced techniques and applications for linear and nonlinear systems. Springer, London, UK. http://dx.doi.org/10.1007/978-1-4471-5538-6
6. Azzopardi B (2017) Sustainable development in energy systems. Springer, London, UK
7. Teshale A (2012) Modelling and simulation of intelligent hybrid power system: modelling of solar/wind hybrid power Matlab simulation of solar/wind hybrid power fuzzy logic control. LAP LAMBERT Academic Publishing, Riga, Latvia
8. Ferrari ML, Damo UM, Turan A (2017) Hybrid systems based on solid oxide fuel cells: modelling and design. Wiley, Hoboken, NJ
9. Toepfer C (2009) The hybrid electric home: clean * efficient * profitable. Schiffer, Atglen, PA
10. Bizon N, Mazare AG, Ionescu LM, Enescu FM (2018) Optimization of the proton exchange membrane fuel cell hybrid power system for residential buildings. Energy Convers Manage 163:22–37
11. Tabatabaei NM, Bizon N, Aghbolaghi AJ, Blaabjerg F (2017) Fundamentals and contemporary issues of reactive power control in AC power systems. Springer, London, UK. http://www.springer.com/gp/book/9783319511177. Springer, London, UK, 2018; https://www.springer.com/in/book/9783319944418

12. Bizon N, Oproescu M, Raceanu M (2015) Efficient energy control strategies for a standalone renewable/fuel cell hybrid power source. Energ Convers Manage 90:93–110
13. Liu W (2017) Hybrid electric vehicle. System modeling and control. Wiley, Hoboken, NJ
14. Bizon N, Dascalescu L, Tabatabaei NM (2014) Autonomous vehicles: intelligent transport systems and smart technologies, Nova Science Publishers Inc., NY
15. Husain I (2010) Electric and hybrid vehicles: design fundamentals, 2nd edn. CRC Press, Boca Raton, FL
16. Komarnicki P, Lombardi P, Styczynski Z (2017) Electric energy storage systems. Flexibility options for smart grids. Springer, London, UK
17. Tabatabaei NM, Kabalci E, Bizon N (2019) Microgrid architectures, control and protection methods. Springer, London, UK. https://www.springer.com/in/book/9783030237226
18. Tabatabaei NM, Ravadanegh SN, Bizon N (2018) Power systems resiliency: modeling, analysis and practice. Springer, London, UK. https://www.springer.com/in/book/9783319944418
19. Teske S (2019) Achieving the Paris Climate Agreement Goals. Global and regional 100% renewable energy scenarios with non-energy GHG pathways for +1.5 and +2 °C. Springer, London, UK
20. Nikolova S, Causevski A, Al-Salaymeh A (2013) Optimal operation of conventional power plants in power system with integrated renewable energy sources. Energy Convers Manage 65:697–703
21. Tascikaraoglu A, Boynuegri AR, Uzunoglu M (2014) A demand side management strategy based on forecasting of residential renewable sources: a smart home system in Turkey. Energ Build 80:309–320
22. Tascikaraoglu A, Erdinc O, Uzunoglu M, Karakas A (2014) An adaptive load dispatching and forecasting strategy for a virtual power plant including renewable energy conversion units. Appl Energ 119:445–453
23. García-Trivino P et al (2014) Long-term optimization based on PSO of a grid-connected renewable energy/battery/hydrogen hybrid system. Int J Hydrogen Energy 9(21):10805–10816
24. Lior N (2010) Sustainable energy development: the present (2009) situation and possible paths to the future. Energy 35:3976–3994
25. Ahmed NA, Miyatake M, Al-Othman AK (2008) Power fluctuations suppression of standalone hybrid generation combining solar photovoltaic/wind turbine and fuel cell systems. Energy Convers Manage 49:2711–2719
26. NASA (2015) Technology roadmaps—TA 3: space power and energy storage. https://www.nasa.gov/sites/default/files/atoms/files/2015_nasa_technology_roadmaps_ta_3_space_power_energy_storage_final.pdf. Accessed Jan 2019
27. Carapellucci R, Giordano L (2012) Modeling and optimization of an energy generation island based on renewable technologies and hydrogen storage systems. Int J Hydrogen Energy 37(3):2081–2093
28. Eroglu M, Dursun E, Sevencan S, Song J, Yazici S, Kilic O (2011) A mobile renewable house using PV/wind/fuel cell hybrid power system. Int J Hydrogen Energy 36(13):7985–7992
29. Erdinc O, Uzunoglu M (2012) A new perspective in optimum sizing of hybrid renewable energy systems: consideration of component performance degradation issue. Int J Hydrogen Energy 37(14):10479–10488
30. Onur E, Ugur SS (2012) A comparative sizing analysis of a renewable energy supplied standalone house considering both demand side and source side dynamics. Appl Energ 96:400–408
31. Castaneda M, Cano A, Jurado F, Sanchez H, Fernandez LM (2013) Sizing optimization, dynamic modeling and energy management strategies of a stand-alone PV/hydrogen/battery-based hybrid system. Int J Hydrogen Energy 38(10):3830–3845
32. Trifkovic M, Sheikhzadeh M, Nigim K, Daoutidis P (2014) Modeling and control of a renewable hybrid energy system with hydrogen storage. IEEE Trans Control Syst Technol 22(1):169–179
33. Bizon N (2014) Load-following mode control of a standalone renewable/fuel cell hybrid power source. Energ Convers Manage 77:763–772

34. Bizon N, Lopez-Guede JM, Kurt E, Thounthong P, Mazare AG, Ionescu LM, Iana G (2019) Hydrogen economy of the fuel cell hybrid power system optimized by air flow control to mitigate the effect of the uncertainty about available renewable power and load dynamics. Energ Convers Manage 179:152–165
35. Bizon N (2018) Real-time optimization strategy for fuel cell hybrid power sources with load-following control of the fuel or air flow. Energ Convers Manage 157:13–27
36. Bizon N, Iana VG, Kurt E, Thounthong P, Oproescu M, Culcer M, Iliescu M (2018) Air flow real-time optimization strategy for fuel cell hybrid power sources with fuel flow based on load-following. Fuel Cell 18(6):809–823. https://doi.org/10.1002/fuce.201700197
37. Nikiforow K, Koski P, Ihonen J (2017) Discrete ejector control solution design, characterization, and verification in a 5 kW PEMFC system. Int J Hydrogen Energy 42:16760–16772
38. Etxeberria A, Vechiu I, Camblong H, Vinassa J-M (2012) Comparison of three topologies and controls of a hybrid energy storage system for microgrids. Energy Convers Manage 54:113–121
39. Bajpai P, Dash V (2012) Hybrid renewable energy systems for power generation in stand-alone applications: a review. Renew Sustain Energy Rev 16(5):2926–2939
40. Auld AE, Smedley KM, Mueller F, Brouwer J, Samuelsen GS (2010) Load-following active power filter for a solid oxide fuel cell supported load. J Power Sources 195(7):1905–1913
41. Smith AD, Mago PJ (2014) Effects of load-following operational methods on combined heat and power system efficiency. Appl Energ 115:337–351
42. Gou B, Na WK, Diong B (2010) Fuel cells: modeling, control, and applications. CRC Press, Boca Raton, FL
43. Pukrushpan JT, Stefanopoulou AG, Peng H (2004) Control of fuel cell power systems: principles, modeling, analysis, and feedback design. Springer, London, UK
44. Sasaki K, Li H-W, Hayashi A, Yamabe J, Ogura T, Lyth SM (2016) Hydrogen energy engineering. A Japanese perspective. Springer, London, UK
45. Töpler J, Lehmann J (2016) Hydrogen and fuel cell technologies and market perspectives. Springer, London, UK
46. Gandiglio M, Lanzini A, Santarelli M, Leone P (2014) Design and optimization of a proton exchange membrane fuel cell CHP system for residential use. Elsevier
47. Feng L, Mears L, Beaufort C, Schulte J (2016) Energy, economy, and environment analysis and optimization on manufacturing plant energy supply system. Energ Convers Manage 117:454–465
48. Marszal AJ, Heiselberg P, Bourrelle JS, Musall E, Voss K, Sartori I, Napolitano A (2011) Zero energy building—a review of definitions and calculation methodologies. Energy Build 43:971–979
49. Dagdougui H, Minciardi R, Ouammi A, Robba M, Sacile R (2012) Modeling and optimization of a hybrid system for the energy supply of a "green" building. Energ Convers Manage 64:351–363
50. Cao S (2016) Comparison of the energy and environmental impact by integrating a H2 vehicle and an electric vehicle into a zero-energy building. Energ Convers Manage 123:153–173
51. Salata F, Golasi I, Domestico U, Banditelli M, Lo Basso G, Nastasi B, de Lieto Vollaro A (2017) Heading towards the nZEB through CHP + HP systems. A comparison between retrofit solutions able to increase the energy performance for the heating and domestic hot water production in residential buildings. Energ Convers Manage 138:61–76
52. Torreglosa JP, García P, Fern_andez LM, Jurado F (2014) Hierarchical energy management system for stand-alone hybrid system based on generation costs and cascade control. Energy Convers Manage 77:514–26
53. Singh A, Baredar P, Gupta B (2017) Techno-economic feasibility analysis of hydrogen fuel cell and solar photovoltaic hybrid renewable energy system for academic research building. Energ Convers Manage 145:398–414
54. Hemmati R, Saboori H (2016) Emergence of hybrid energy storage systems in renewable energy and transport applications—a review. Renew Sust Energ Rev 65:11–23

55. Wang H, Wang Q, Baozan HuB (2017) A review of developments in energy storage systems for hybrid excavators. Autom Constr 80:1–10
56. Olabi AG (2017) Renewable energy and energy storage systems. Energy 136:1–6
57. Bizon N (2018) Effective mitigation of the load pulses by controlling the battery/SMES hybrid energy storage system. Appl Energ 229:459–473
58. Elsisi M, Soliman M, Aboelela MAS, Mansour W (2017) Optimal design of model predictive control with superconducting magnetic energy storage for load frequency control of nonlinear hydrothermal power system using bat inspired algorithm. J Energy Storage 12:311–318
59. Zhang C, Wei YL, Cao PF, Lin MC (2017) Energy storage system: current studies on batteries and power condition system. Renew Sust Energ Rev. https://doi.org/10.1016/j.rser.2017.10.030
60. Li J, Gee AM, Zhang M, Yuan W (2015) Analysis of battery lifetime extension in a SMES-battery hybrid energy storage system using a novel battery lifetime model. Energy 86:175–185
61. Guneya MS, Tepe Y (2017) Classification and assessment of energy storage systems. Renew Sust Energ Rev 75:1187–1197
62. Hannana MA, Hoque MM, Mohamed A, Ayob A (2017) Review of energy storage systems for electric vehicle applications: issues and challenges. Renew Sust Energ Rev 69:771–789
63. Al-falahi MDA, Jayasinghe SDG, Enshaei H (2017) A review on recent size optimization methodologies for standalone solar and wind hybrid renewable energy system. Energ Convers Manage 143:252–274
64. Naimaster EJ, Sleiti AK (2013) Potential of SOFC CHP systems for energy-efficient commercial buildings. Energy Build 61:153–160
65. Arsalis A, Nielsen MP, Kaer SK (2011) Modeling and off-design performance of a 1 kW HT-PEMFC (high temperature proton exchange membrane fuel cell)-based residential micro-CHP (combined-heat-and-power) system for Danish single family households. Energy 36:993–1002
66. Zafar S, Dincer I (2014) Energy, exergy and exergo economic analyses of a combined renewable energy system for residential applications. Energy Build 71:68–79
67. Corbo P, Migliardini F, Veneri O (2007) Experimental analysis and management issues of a hydrogen fuel cell system for stationary and mobile application. Energy Convers Manage 48(8):2365–2374
68. Zhang C, Yu T, Yi J, Liu Z, Raj KAR, Xia L, Tu Z, Chan SH (2016) Investigation of heating and cooling in a stand-alone high temperature PEM fuel cell system. Energ Convers Manage 129:36–42
69. Iturriaga E, Aldasoro, Campos-Celador A, Sala JM (2017) A general model for the optimization of energy supply systems of buildings. Energy 138:954–966
70. Kwak Y, Huh J-H (2016) Development of a method of real-time building energy simulation for efficient predictive control. Energ Convers Manage 113:220–229
71. Bizon N (2019) Hybrid power sources (HPSs) for space applications: analysis of PEMFC/battery/SMES HPS under unknown load containing pulses. Renew Sustain Energy Rev 105:14–37. https://doi.org/10.1016/j.rser.2019.01.044
72. Julien C, Mauger A, Vijh A, Zaghib K (2016) Lithium batteries—science and technology. Springer, London, UK
73. Reddy T (2010) Linden's handbook of batteries, 4th edn. McGraw-Hill Education, NY, USA
74. Chauhan A, Saini RP (2014) A review on integrated renewable energy system based power generation for stand-alone applications: configurations, storage options, sizing methodologies and control. Renew Sust Energ Rev 38:99–120
75. Warner J (2015) The handbook of lithium-ion battery pack design—chemistry, components, types and terminology. Elsevier Science, Amsterdam, Netherlands
76. Sarasketa-Zabala E, Martinez-Laserna E, Berecibar M, Gandiaga I, Rodriguez-Martinez LM, Villarreal I (2016) Realistic lifetime prediction approach for Li-ion batteries. Appl Energy 162:839–852
77. Tao L, Ma J, Cheng Y, Noktehdan A, Chong J, Lu C (2017) A review of stochastic battery models and health management. Renew Sust Energ Rev 80:716–732

78. Bizon N (2013) Energy efficiency for the multiport power converters architectures of series and parallel hybrid power source type used in plug-in/V2G fuel cell vehicles. Appl Energy 102(12 February):726–734
79. Bizon N (2012) Energy efficiency of multiport power converters used in plug-In/V2G fuel cell vehicles. Appl Energy 96:431–443
80. Penthia T, Panda AK, Sarangi SK (2018) Implementing dynamic evolution control approach for DC-link voltage regulation of superconducting magnetic energy storage system. Int J Electr Power Energy Syst 95:275–286
81. SimPowerSystems TM Reference (2010) Hydro-Québec and the MathWorks, Inc., Natick, MA
82. Andrews J, Doddathimmaiah AK (2008) Regenerative fuel cells. In: Gasik M (ed) Fuel cell materials. Woodhead Publishing, Cambridge
83. Paul P, Andrews J (2017) PEM unitised reversible/regenerative hydrogen fuel cell systems: state of the art and technical challenges. Renew Sust Energ Rev 79:585–599
84. Emadi A, Ehsani M, Miller JM (2003) Vehicular electric power systems land, sea, air, and space vehicles. CRC Press
85. Gongadze E, Petersen S, Beck U, van Rienen U (2009) Classical models of the interface between an electrode and an electrolyte. COMSOL conference, Milano, Italy. https://www.comsol.com/paper/6594
86. Jin JX (2007) HTS energy storage techniques for use in distributed generation systems. Physica C: Supercond Appl 460–462(Part 2):1449–1450
87. Ali MH, Wu B, Dougal RA (2010) An overview of SMES applications in power and energy systems. IEEE Trans Sustain Energy 1(1):38–47
88. Patel MR (2004) Spacecraft power systems. CRC Press, Boca Raton, FL
89. Shimizu T, Underwood C (2013) Super-capacitor energy storage for micro-satellites: feasibility and potential mission applications. Acta Astronaut 85:138–154
90. Shawyer R (2015) Second generation EmDrive propulsion applied to SSTO launcher and interstellar probe. Acta Astronaut 116:166–174
91. Gubser DU (1995) Superconductivity research and development: department of defense perspective. Appl Supercond 3(1–3):157–161
92. Hedlund M, Lundin J, de Santiago J, Abrahamsson J, Bernhoff H (2015) Flywheel energy storage for automotive applications. Energies 8:10636–10663
93. Wang B (2017) Current flywheels moving to Superconducting flywheels using carbon fiber or carbon nanotubes. https://www.nextbigfuture.com/2017/01/uk-building-38-million-combat-laser.html. Accessed Jan 2019
94. Ha SK, Kim MH, Han SC, Sung TH (2006) Design and spin test of a hybrid composite flywheel rotor with a split type hub. J Compos Mater 40:2113–2130
95. Thomas CE. Hydrogen-powered fuel cell electric vehicles compared to the alternatives. http://www.azocleantech.com/article.aspx?ArticleID=214. Accessed Jan 2018
96. Stolten D, Samsun RC, Garland N (2016) Fuel cells: data, facts, and figures. Wiley, Hoboken, NJ
97. Sasaki K, Li HW, Hayashi A (2016) Hydrogen energy engineering: a Japanese perspective. Springer, London, UK
98. Tirnovan R, Giurgea S (2012) Efficiency improvement of a PEMFC power source by optimization of the air management. Int J Hydrogen Energ 37:7745–7756
99. Ramos-Paja CA, Spagnuolo G, Petrone G, Mamarelis E (2014) A perturbation strategy for fuel consumption minimization in polymer electrolyte membrane fuel cells: analysis, design and FPGA implementation. Appl Energ 119:21–32
100. Bizon N (2018) Optimal operation of fuel cell / wind turbine hybrid power system under turbulent wind and variable load. Appl Energ 212:196–209
101. Heier S (1998) Grid integration of wind energy conversion system. Wiley, Hoboken, NJ
102. Lineykin S, Averbukh M, Kuperman A (2014) IEEE Trans Ind Electron 61:6785–6793
103. Kadri R, Gaubert JP, Champenois G (2012) IEEE trans power electron 27:1249–1258

104. Bizon N (2010) Development of a fuel cell stack macro-model for inverter current ripple evaluation. Rev Roum Sci Techn – Électrotechn et Énerg 55:405–415. http://revue.elth.pub. ro/viewpdf.php?id=256
105. Itoh J-I, Hayashi F (2010) Ripple current reduction of a fuel cell for a single-phase isolated converter using a DC active filter with a center tap. IEEE T Power Electr 25(3):550–556
106. Bizon N (2011) A new topology of fuel cell hybrid power source for efficient operation and high reliability. J Power Sour 196(6):3260–3270
107. Dwivedi S, Jain S, Gupta KK, Chaturvedi P (2018) Modeling and control of power electronics converter system for power quality improvements. Academic Press, Cambridge, Massachusetts, USA
108. Bacha S, Munteanu I, Bratcu AI (2014) Power electronic converters modeling and control with case studies. Springer, London, UK
109. Wang Y, Sun Z, Chen Z (2019) Rule-based energy management strategy of a lithium-ion battery, supercapacitor and PEM fuel cell system. Energy Procedia 158:2555–2560
110. Erdinc O, Vural B, Uzunoglu M (2009) A wavelet-fuzzy logic based energy management strategy for a fuel cell/battery/ultra-capacitor hybrid vehicular power system. J Power Sour 194(1):369–380
111. Ziogou C, Papadopoulou S, Georgiadis MC, Voutetakis S (2013) On-line nonlinear model predictive control of a PEM fuel cell system. J Process Contr 23(4):483–492
112. Arce A, Alejandro J, Bordons C, Daniel R (2010) Real-time implementation of a constrained MPC for efficient airflow control in a PEM fuel cell. IEEE Trans Ind Electron 57(6):1892–1905
113. Barzegari MM, Alizadeh E, Pahnabi AH (2017) Grey-box modeling and model predictive control for cascade-type PEMFC. Energy 127:611–622
114. Cano MH, Mousli MIA, Kelouwani S, Agbossou K, Hammoudi M, Dubéc Y (2017) Improving a free air breathing proton exchange membrane fuel cell through the maximum efficiency point tracking method. J Power Sources 345:264–274
115. Baroud Z, Benmiloud M, Benalia A, Ocampo-Martinez C (2017) Novel hybrid fuzzy-PID control scheme for air supply in PEM fuel-cell-based systems. Int J Hydrogen Energy 42(15):10435–10447
116. Hasikos J, Sarimveis H, Zervas PL, Markatos NC (2009) Operational optimization and real-time control of fuel-cell systems. J Power Sources 193(1):258–268
117. Nejad HC, Farshad M, Gholamalizadeh E, Askarian B, Akbarimajd A (2019) A novel intelligent-based method to control the output voltage of proton exchange membrane fuel cell. Energ Convers Manage 185:455–464
118. Ziogou C, Papadopoulou S, Pistikopoulos E, Georgiadis M, Voutetakis S (2017) Model-based predictive control of integrated fuel cell systems—from design to implementation. Advances in energy systems engineering. Springer, pp 387–430

Chapter 3
Optimization Algorithms and Energy Management Strategies

3.1 Introduction

In recent years, different strategies and algorithms for efficient operation of hybrid power systems (HPSs) based on fuel cell (FC) systems [1–4] and renewable energy sources (RESs) [5–7] are designed and analyzed to be implemented. It is known that FC HPS strategies can be classified mainly into two classes as follows [8, 9]: (class 1) rule-based strategies and (class 2) optimization-based strategies.

The optimal solution cannot be found using a class 1 strategy based on deterministic rules, but they can easily find a suboptimal solution, so they are already implemented and marketed in hybrid systems [10]. The optimal solution can be found using a class 2 strategy [7, 11], but they are more difficult to implement in real time [1, 12], so they are intensively studied in order to be simplified and improved.

Real-time optimization (RTO) strategies have recently been implemented for the proton exchange membrane fuel cell (PEMFC) system to improve air control via the maximum efficiency point tracking (MEPT) control [13], maximum power point tracking (MPPT) control [14], robust control [15], or other advanced RTO strategies analyzed in [16]. For example, the MPPT control proposed in [17] can optimize in real time the operation of photovoltaic and PEMFC systems, using the PEMFC system as backup energy source [18, 19].

An effective HPS control can be obtained using artificial intelligence concepts, such as fuzzy logic [20], neural networks [21] or genetic algorithms [22], or advanced techniques based on the data fusion [23] or data combinations [24], or metaheuristic methods [25].

It is worth mentioning that improved variants of the classical control solutions such as model reference adaptive control [26], predictive control [27], equivalent consumption minimization strategy [28], or static Feed-Forward (sFF) strategy [29] can be successfully applied for FC HPSs or FC vehicles [30].

© The Editor(s) (if applicable) and The Author(s), under exclusive license to Springer Nature Switzerland AG 2020
N. Bizon, *Optimization of the Fuel Cell Renewable Hybrid Power Systems*,
Green Energy and Technology, https://doi.org/10.1007/978-3-030-40241-9_3

The performance indicators used to evaluate RTO strategies can be related to maximizing the net power generated by the FC system (P_{FCnet}) or to fuel economy. In the first case, the PEMFC system will operate close to the MEP [31–33]. In the second case, the fuel economy can be maximized using an optimization function defined as a weighting combination of the fuel consumption efficiency and PEMFC system efficiency indicators [34–36]. In both cases, the battery is recommended to work in a sustainable way (such as the charge-sustained mode) to take advantage of the long lifetime and ensure a small size of the energy storage system (ESS). For this, the load-following [37–40] and FC power-following control [41–44] techniques have been proposed for the PEMFC system.

The optimal operation of the PEMFC system close to MEP may face many difficulties [45–47], such as: (1) MEP and other local maximums are located on the plateau of the surface $P_{FCnet} = f(AirFr, FuelFr)$, where the $AirFr$ and $FuelFr$ are the air and fuel flow rates [48]; (2) the positions of the MEP and local maxima depend in time due to variations in load demand, temperature, and so on [7]; (3) the MEP and highest local maximum could be close, so the hit count for MEP search may be less than 100% [34, 49]. Thus, the algorithm for MEP search must be of global type, with a searching time lower than FC time constant, a searching resolution lower than 1%, a searching accuracy higher than 99.9% and requires a simple adjustment of the design parameters or just resizing the normalization gains [11, 50–54].

The chapter is structured as follows. Besides a brief comparison of the current energy management strategies, Sect. 3.2 presents the strategies for a PEMFC system based on the load-following (LFW) control and power-following (PFW). Optimization of the FC HPS operation using well-known optimization algorithms such as the global maximum power point tracking (GMPPT) algorithms, global maximum efficiency point tracking (GMEPT) algorithms, or fuel economy algorithms is presented in Sect. 3.3. The fuel economy under constant and variable load is analyzed in Sect. 3.4 for an advanced fuel economy strategy based on switching fueling regulators. A pulse mitigation strategy using an anti-pulse control of the bidirectional DC-DC power converter, interfacing the ultracapacitors' stack from the hybrid storage system (HSS) with the DC bus, is presented in Sect. 3.5. The last section concludes the chapter.

3.2 Energy Management Strategies

The objective of any energy management strategy is to optimally and safely operate the RES/FC HPSs under a variable power from renewable energy sources and a dynamic load demand [1–7, 55–59]. To better fit this objective or not, the rules, constraints, and optimization functions have been changed, mixed, or improved year after year [8–11, 60–63]. The rules-based strategies have difficulties for search of global optimum, so the research has been focused in the developing of advanced optimization-based strategies [64, 65]. The optimization algorithms can use control based on concepts from artificial intelligence [66], predictive control [67], robust control [68], sliding control [69], adaptive control [70], multi-scheme control [71], and extremum seeking control [72, 73]. The strategies based on equivalent consumption minimization [74, 75] are improved variants of the optimization-based strategies previously proposed [76, 77], using dynamic programming instead of Pontryagin's minimal principle, and it is worth mentioning that research is still focused in improving these strategies as performance and operation in real time due to good results reported in experiments [78].

Due to the proved performance in various applications and operation in real time, classical [79], modified [80–82], and advanced [83–88] extremum seeking algorithms have been developed and applied for RES/FC HPSs. The robustness and dither's persistence of the extremum seeking algorithms can be improved by using a band-pass filter in the searching loop with the cutoff frequency higher than the dither's frequency [80, 81]. The tracking accuracy of the extremum seeking algorithms can be improved by using a sinusoidal dither with the amplitude modulated by a signal proportional to the fundamental of the probing signal (the output of the process to be optimized) [83]. The proposals have been applied for RES/FC HPSs and validated [82, 84]. The performance reported for indicators such as tracking speed, tracking accuracy, and power ripple is better than that obtained with classical extremum seeking algorithms [52, 85]. It is worth mentioning that the performance is better for the extremum seeking algorithm based on two band-pass filters [53] instead of one [54] due to use different cutoff frequencies for the band-pass filters from the search and localization loops. This will speed-up the tracking process of the load variation or both RESs' variation and load dynamic using the load-based strategy or power-based strategy, respectively, as it will be explained in next sections.

If the FC system used in RES/FC HPSs will operate in variable power mode at MEP (or optimum point for the best fuel economy), then this FC power can be estimated based on power flow balance (3.1) in order to operate the battery in charge-sustained mode or close to this regime.

$$C_{DC} u_{dc} \frac{du_{dc}}{dt} = \eta_{boost} p_{FCnet} + p_{ESS} + p_{RES} - p_{Load} \qquad (3.1)$$

where p_{FCnet}, p_{ESS}, p_{RES}, and p_{Load} are the net FC power, ESS power, RES power, and load power, respectively, and η_{boost} is the energy efficiency of the boost DC-DC power converter that interfaces the FC system with the DC bus (having the voltage u_{DC} filtered by the capacitor C_{DC}).

This regime will reduce the battery capacity needed to sustain the lack or excess in energy on the DC bus during the transitory regimes of the FC system for MEP search. So, the ESS size and maintenance costs will decrease, and the battery's lifetime will increase (avoiding frequent charge-discharge cycles).

But, before to model and analyze the proposed energy management strategies, a brief overview of most relevant and recent energy management strategies will be presented.

For example, the extremum seeking algorithm is proposed to optimize the operation of the air regulator using a real-time re-scaling routine based on the load demand

[89]. The load-following strategy sets in real time the airflow rate (AirFr), and the optimization of the FC system is performed using the extremum seeking algorithm to control the boost power converter.

Even if two loops are used (one to ensure the load demand and other to optimize the operation of the FC system) by the proposed load-following strategy, this is simpler and more effective because the optimization loop sets the FC current that is the input of the fuel regulator (which will control the fuel flow rate (FuelFr), so it is indirectly involved in the fuel economy). Furthermore, the search for an optimum of the optimization function is performed using two searching variables (AirFr and FuelFr), extending the searching space for a global optimum, instead of a local one.

The search for a global optimum has been used to improve the air control using the MEPT [13] and MPPT [14] algorithms, robust control [15], or other control techniques [16]. The MPP-based optimization of both renewable and FC energy sources is proposed in [17]. It is needed to power an electrolyzer during the stages when the renewable energy exceeds the load demand in order to harvest the available RES power instead of reducing it to the load demand level [18, 19]. The hydrogen produced will be stored in pressurized tanks for a later use. This architectural solution is considered for the proposed power strategy. The performance of proposed strategies will be compared in Chaps. 5 and 6 with the static Feed-Forward (sFF) control [29] used as a reference strategy by most strategy validated using the hardware-in-loop system method [26, 90–99].

For example, the sliding control [90, 91] and load governor [92] are proposed to optimally control the air regulator, avoiding oxygen starvation. The fueling starvation appears during sharp changes in the load demand when the fueling stoichiometry is difficult to be maintained. Besides the air and fuel slope limiters used by fueling regulators to ensure a safe supply of the FC system, different other control techniques have been proposed and implemented, such as control based on the disturbance rejection [93], differential flatness [94], proportional-integral (PI) error [95], P-I-Derivative (PID) error [96], fuzzy PID error [97], time delay [26], and adaptive techniques [98] (including extremum seeking algorithms [51]), in order to increase the FC lifetime [99] and better control the fueling stoichiometry [98].

Optimal control of the fuel regulator to avoid fuel starvation was also in the attention of the researchers, and many control techniques have been proposed and implemented, such as control based on the intelligent concept (fuzzy logic [100], adaptive fuzzy control [101], and genetic algorithms [22]) or improved of classical ones such as adaptive equivalent consumption minimization strategy (A-ECMS) [102].

It is worth mentioning that the research in the field of optimal control of the FC system will continue by implementing and testing the new advanced strategies proposed recently in the literature, using the performance objectives related to fuel economy and FC electrical efficiency of the FC system such as maximization of the fuel consumption efficiency ($Fuel_{eff}$) and/or FC net power (P_{FCnet}), or minimization of the total fuel consumption.

The optimization function $f(x, AirFr, FuelFr, P_{load}) = P_{FCnet} + k_{fuel}Fuel_{eff}$ is used for searching the MEP point or the best fuel economy if $k_{fuel} = 0$ or $k_{fuel} \neq 0$,

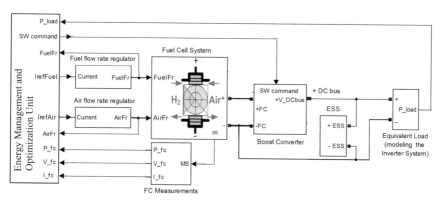

Fig. 3.1 FC HPS based on the load-based strategy [44]

as will be shown in Chaps. 5 and 6. In the last case, a sensitivity analysis based on the k_{fuel} parameter is necessary to be performed in order to define the optimization function for best fuel economy [103].

It is worth mentioning that the FC power generated on the DC bus can be controlled via the fueling regulators and the boost power converter by using the references $I_{ref(Fuel)}$ and $I_{ref(Air)}$, and the switching command $SW_{command}$ (see Figs. 3.1 and 3.4).

The load-based strategy and power-based strategy use the same inputs in the energy management and optimization (EMO) unit (such as the fueling flow rates AirFr and FuelFr, and FC measurements (V_{FC}, I_{FC}, and P_{FC}) related to voltage, current, and power), except the input variable mentioned in the top, which is the load demand (P_{load}) and power requested on the DC bus (P_{DCreq}), both of which need to be supported by the FC system using the load-based strategy and the power-based strategy, respectively.

3.2.1 Load-Based Strategy

Figure 3.1 presents the architecture of the FC HPS based on the load-based strategy [44]. The PEMFC is used as main energy source to sustain the load based on the average (AV) power flow balance (3.1), where the power exchanged by the ESS is zero on average ($P_{ESS(AV)} \cong 0$) and $P_{RES(AV)} = 0$ (because the energy support from RESs is not considered). Thus, the AV power flow balance will be given by (3.2):

$$0 = \eta_{boost(AV)} P_{FCnet(AV)} - P_{load(AV)} \Rightarrow P_{FCnet(AV)} = P_{load(AV)}/\eta_{boost(AV)}. \quad (3.2)$$

Thus, the load-based strategy will use the load-following (LFW) reference ($I_{ref(LFW)}$) to control the power generated by the FC system:

$$I_{ref(LFW)} = I_{FC(AV)} = P_{FCnet(AV)}/V_{FC(AV)} = P_{load(AV)}/(V_{FC(AV)}\eta_{boost(AV)}) \quad (3.3)$$

The FC net power will be approximated by (3.4):

$$P_{FCnet} \cong P_{FC} - P_{cm} \tag{3.4}$$

if the losses (P_{cm}) will be given mainly by the air compressor [78]:

$$P_{cm} = I_{cm} \cdot V_{cm} = \left(a_2 \cdot AirFr^2 + a_1 \cdot AirFr + a_0\right) \cdot \left(b_1 \cdot I_{FC} + b_0\right) \tag{3.5}$$

where $a_0 = 0.6$, $a_1 = 0.04$, $a_2 = -0.00003231$, $b_0 = 0.9987$, and $b_1 = 46.02$.

As mentioned above, the optimization using two searching variables needs two references, $I_{ref(GES1)}$ and $I_{ref(GES2)}$, generated by the global extremum seeking (GES) controllers 1 and 2 based on the aforementioned optimization function (see Fig. 3.2). The reference $I_{ref(LFW)}$ will be generated using (3.3) by the LFW controller shown in Fig. 3.2 and detailed in Fig. 3.3, where the low-pass filter (LPF) will smooth out the high harmonics of load demand.

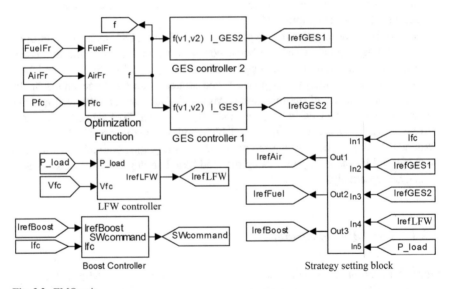

Fig. 3.2 EMO unit

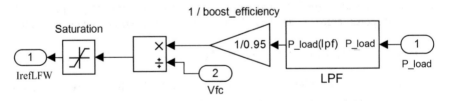

Fig. 3.3 LFW controller

Thus, the inputs of the strategy setting block are as follows: FC current (I_{FC}), load demand (P_{load}), and the aforementioned references ($I_{ref(GES1)}$, $I_{ref(GES2)}$, and $I_{ref(LFW)}$). This block will generate the appropriate selection for the references ($I_{ref(Air)}$, $I_{ref(Fuel)}$, and $I_{ref(Boost)}$) that can control the power generated by the FC system via the fueling regulators and the boost controller (see Fig. 3.2). The boost controller will generate the switching command $SW_{command}$ for the boost power converter using, for example, a hysteresis or PI control.

For example, the strategy called Air-GES1-Fuel-GES2-Boost-LFW-RTO strategy [31] uses the load-following-based control for the DC-DC boost converter and the GES-based control for both air and fuel regulators:

$$I_{ref(Air)} = I_{ref(GES1)} + I_{FC}$$
$$I_{ref(Fuel)} = I_{ref(GES2)} + I_{FC}$$
$$I_{ref(Boost)} = I_{ref(LFW)} \tag{3.6}$$

All variants will be analyzed in Chap. 5.

3.2.2 Power-Based Strategy

Figure 3.1 presents the architecture of the FC/RES HPS based on the power-based strategy [43].

The PEMFC is used as the main energy source to sustain the average (AV) power flow balance (3.1), where the power exchanged by the ESS is zero on average ($P_{ESS(AV)} \cong 0$) and $P_{RES} < P_{load}$ (the energy support from RESs is not enough to sustain the load demand).

Thus, the AV power flow balance will be given by (3.7):

$$0 = \eta_{boost(AV)} P_{FCnet(AV)} + P_{RES(AV)} - P_{Load(AV)} \Rightarrow P_{FCnet(AV)} = P_{DCreq}/\eta_{boost(AV)} \tag{3.7}$$

where the power requested on the DC bus to be generate by the FC system via the boost power converter is given by (3.8):

$$P_{DCreq(AV)} = P_{Load(AV)} - P_{RES(AV)} \tag{3.8}$$

Thus, the power-based strategy will use the power-following (PFW) reference ($I_{ref(PFW)}$) to control the power generated by the FC system:

$$I_{ref(PFW)} = I_{FC(AV)} = P_{FCnet(AV)}/V_{FC(AV)} = P_{DCreq}/(V_{FC(AV)}\eta_{boost(AV)}) \tag{3.9}$$

The reference $I_{ref(PFW)}$ will be generated using (3.9) by the PFW controller, which has the diagram similar to that of the LPF controller shown in Fig. 3.3, except that

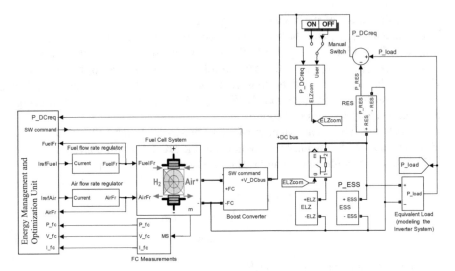

Fig. 3.4 FC/RES HPS using the power-based strategy

the input will be the power requested on the DC (3.8) (Fig. 3.4). The diagram of the FC/RES HPS and EMO unit using the power-based strategy is presented in Fig. 3.5.

For example, the fuel economy obtained with the strategies using settings (3.10), (3.11), and (3.12) has been analyzed in [43] (see Fig. 3.6) using variable RES power and dynamic load.

$$
\begin{aligned}
I_{\text{ref(Air)}} &= I_{\text{ref(PFW)}} \\
I_{\text{ref(Fuel)}} &= I_{\text{ref(GES2)}} + I_{\text{FC}} \\
I_{\text{ref(Boost)}} &= I_{\text{ref(GES1)}}
\end{aligned}
\tag{3.10}
$$

$$
\begin{aligned}
I_{\text{ref(Air)}} &= I_{\text{ref(GES2)}} + I_{\text{FC}} \\
I_{\text{ref(Fuel)}} &= I_{\text{ref(PFW)}} \\
I_{\text{ref(Boost)}} &= I_{\text{ref(GES1)}}
\end{aligned}
\tag{3.11}
$$

$$
\begin{aligned}
I_{\text{ref(Air)}} &= I_{\text{ref(GES1)}} + I_{\text{FC}} \\
I_{\text{ref(Fuel)}} &= I_{\text{ref(GES2)}} + I_{\text{FC}} \\
I_{\text{ref(Boost)}} &= I_{\text{ref(PFW)}}
\end{aligned}
\tag{3.12}
$$

The strategies (3.10), (3.11), and (3.12) use one GES controllers to search for best fuel economy and the PFW control for the air regulator, fuel regulator, and boost controller. So, these strategies will be called below as Air-PFW strategy, Fuel-PFW strategy, and Boost-PFW strategy (see Fig. 3.6). All variants will be analyzed in Chap. 6 in terms of the performance obtained.

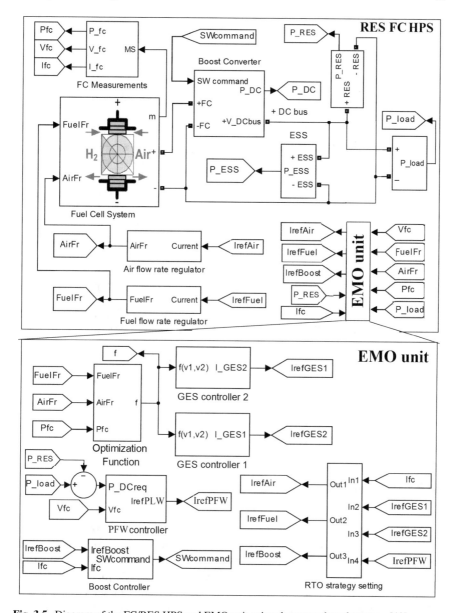

Fig. 3.5 Diagram of the FC/RES HPS and EMO unit using the power-based strategy [43]

3.3 Optimization Algorithms

The FC/RES HPS will be optimized in real time by quickly finding a set of values for the input variables ($I_{ref(Air)}$, $I_{ref(Fuel)}$, and $I_{ref(Boost)}$) of the FC system that will optimally produce the power corresponding to the load demand. This set will be

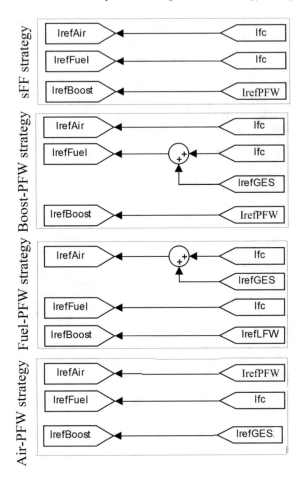

Fig. 3.6 Diagram of the strategy setting block [43]

found by a search algorithm for the global optimum for the optimization function. The search algorithm must have very good performance related to tracking speed and accuracy, but also a searching resolution lower than 1 and 100% hit count for the success in finding the global optimum instead of a local one [5, 6, 35, 100].

During the years, different types of search algorithms based on the global extremum seeking (GES) [52–54] or classical extremum seeking control [72–74] have been tested for the FC/RES HPS. Thus, Chap. 4 will be dedicated to GES-based search algorithms, with a short introduction of classical extremum seeking control schemes.

3.3.1 Performance of the Global Maximum Power Point Tracking Algorithms

The goal of any global optimization algorithm is to always find the optimization function's optimum, but also to be fast and accurate (which means that the tracking time must be comparable with the response time of the process to be optimized and the accuracy (T_{acc}) higher than 99.9% during stationary regimes) [52, 104]. The tracking efficiency (T_{eff}) is closely related to aforementioned performance indicators, being a measure of how close the GMPP is followed [105, 106]. Note that the search resolution (S_R) defines how close GMPP and local maximum power point (LMPP) can be on a multimodal pattern in order to obtain a 100% success rate to find the GMPP during repetitive tests (this indicator being called hit count). The performance indicators for a global optimization algorithm are defined as follows:

$$T_{acc} = \frac{y_{GMPP}}{y^*_{GMPP}} \cdot 100[\%] \tag{3.14}$$

$$T_{eff} = \frac{\int_0^t p_{PV} dt}{\int_0^t p^*_{PV} dt} \cdot 100[\%] \tag{3.15}$$

$$S_R = \frac{\min_i |y_{GMPP} - y_{LMPP i}|}{y_{GMPP}} \cdot 100[\%] \tag{3.16}$$

where y_{GMPP}, y_{LMPP}, and y^*_{GMPP} are the GMPP value, the LMPP, and the founded value, respectively, and p_{PV} and p^*_{PV} are the PV power and the value of power tracked by a GMPP algorithm.

Besides the aforementioned indicators, the performance of a global optimization algorithm may be measured by robustness to perturbations using appropriate indicators for this transitory regime and the power ripple during stationary regime [107, 108].

It is worth mentioning that the inverter ripple on the DC bus is high and can affect the safety of the PEMFC system [11, 50]. Thus, the global maximum efficiency point tracking algorithms must not further produce a low frequency power ripple due to searching for the FC net power's maximum (a point called MEP) [80].

3.3.2 Performance of the Global Maximum Efficiency Point Tracking Algorithms

The FC net power is computed using (3.4), and the MEP algorithms will search the maximum on the optimization surface $P_{FCnet} = f(AirFf, FuelFr)$. The surface has a plateau with many peaks (local maximum efficiency points—LMEPs) on it [3]; thus, the MEP algorithms must be of global type with a searching resolution lower than 1% in order to search for the global MEP (GMEP).

So, besides the tracking time that now must be comparable with the PEMFC time constant (hundreds of milliseconds), the performance indicators for a GMEP algorithm can be defined in the same manner with the performance indicators for the GMPP algorithms, as follows:

$$T_{acc} = \frac{y_{GMEP}}{y^*_{GMEP}} \cdot 100[\%] \tag{3.14'}$$

$$T_{eff} = \frac{\int_0^t p_{FCnet} dt}{\int_0^t p^*_{FCnet} dt} \cdot 100[\%] \tag{3.15'}$$

$$S_R = \frac{\min_i |y_{GMEP} - y_{LMEPi}|}{y_{GMEP}} \cdot 100[\%] \tag{3.16'}$$

where y_{GMEP}, y_{LMEP}, and y^*_{GMEP} are the GMEP value, the LMEP, and the found value, respectively, and p^*_{FCnet} and p^*_{FCnet} are the FC net power and the value of FC net power tracked by a GMEP algorithm.

Besides the maximization of the FC net power or FC electrical efficiency $\left(\eta_{sys} = \frac{P_{FCnet}}{P_{FC}}\right)$, other hydrogen consumption indicators can be used, such as the total hydrogen consumption $\left(Fuel_T = \int FuelFr(t)dt\right)$, fuel consumption efficiency $\left(Fuel_{eff} = \frac{P_{FCnet}}{FuelFr}\right)$, or hydrogen consumption efficiency $\left(eff_{H2} = \frac{100 \cdot P_{FCnet}}{LHV \cdot Fuel_T}\right)$, where LHV is the lower heating value for hydrogen [109, 110]. Also, a mix of these indicators related to FC system and other indicators related to the battery can define an optimization function and the appropriate constraints for optimal operation of the HPS [111]. In Chap. 6, the optimization function $f(x, AirFr, FuelFr, P_{load}) = P_{FCnet} + k_{fuel} Fuel_{eff}$ is used for searching the best fuel economy. Thus, the fuel economy algorithms will be presented briefly in next subsection, and then, a fuel economy strategy using real-time switching of the fueling regulators [44] will be detailed in Sect. 3.4.

3.3.3 Fuel Economy Algorithms

Fuel economy algorithms are GMPP algorithms applied to search for a maximum of an optimization function defined for best fuel economy using as penalty function [112, 113] the fuel consumption efficiency $\left(Fuel_{eff} = \frac{P_{FCnet}}{FuelFr}\right)$ and using as constrain an imposed window for the battery State of charge (SoC) variation [114]:

$$\text{Maximize}: f(x, AirFr, FuelFr, P_{load}) = P_{FCnet} + k_{fuel} Fuel_{eff} \tag{3.17}$$

$$\text{Subject to FC system dynamics}: \dot{x} = g(x, AirFr, FuelFr, P_{load}), x \in X \tag{3.18}$$

and battery SoC constraints:

$$SOC_{min} < SOC < SOC_{max} \qquad (3.19)$$

where the load demand P_{Load} represents the perturbation, x is state vector, and X is the search space.

The value for best fuel economy of the weighting coefficient k_{fuel} will be determined based on sensitivity analysis [103]. If $k_{fuel} = 0$, then $f(x, AirFr, FuelFr, P_{load}) = P_{FCnet}$ and the FC net power will be maximized. Thus, this definition of the optimization function ensures an easy change of the optimization objective based on available data (fuel in tank, time, and distance to destination, etc.) and information acquired from V2V network related to next route (load cycle profile, delays and traffic jams, fuel price for next stations in the route, and so on) [114, 115]. For example, if the FC vehicle climbs a hill, then both net power FC and fuel economy must be maximized, but if the FC vehicle is on the highway or the fuel alarm signals a low fuel level then the best fuel economy must be obtained. Furthermore, besides the limited space and weight of the FC system and batteries, their safety, life, and maintenance should be considered as optimization constraints [116].

The performance of the fuel economy strategies is compared using the fuel economy obtained under a given load cycle. If the final SoC is different to initial SoC, the energy needed to make final SoC equal with initial SoC is converted in fuel and added to the fuel economy obtained. The fuel economy strategy using real-time switching of the input references of the fueling regulators to $I_{ref(GES2)} + I_{FC}$ and $I_{ref(LFW)}$ (or $I_{ref(PFW)}$).

3.4 Fuel Economy Strategy Using Switching of the Fueling Regulators

Besides the strategies (3.10) and (3.11) that use one GES controller (called Air-PFW strategy and Fuel-PFW strategy in Fig. 3.6), the strategies (3.20) and (3.21) use two GES controllers to search for best fuel economy using two searching variable and the PFW control for the air regulator and fuel regulator. So, these strategies will be called below as Air-PFW-2GES strategy and Fuel-PFW-2GES strategy (see Fig. 3.7) [44].

$$I_{ref(Fuel)} = I_{FC} + I_{ref(GES2)}, \; I_{ref(Air)} = I_{ref(LFW)}, \; \text{and} \; I_{ref(boost)} = I_{ref(GES1)} \qquad (3.20)$$

$$I_{ref(Air)} = I_{FC} + I_{ref(GES2)}, \; I_{ref(Fuel)} = I_{ref(LFW)}, \; \text{and} \; I_{ref(boost)} = I_{ref(GES1)} \qquad (3.21)$$

The fuel economy strategy performs a real-time switching of the fueling regulators' reference used in the Air-PFW-2GES strategy and Fuel-PFW-2GES strategy using (3.22) and the threshold P_{ref}. So, this strategy will be called below as Air/Fuel-PFW-2GES strategy.

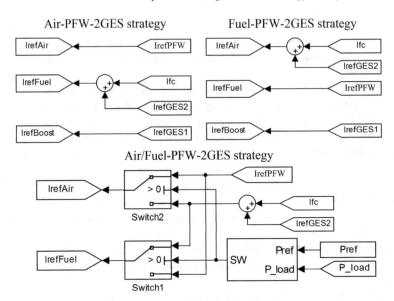

Fig. 3.7 Setting block for the strategies Air-PFW-2GES, Fuel-PFW-2GES, and Air/Fuel-PFW-2GES [44]

$$I_{\text{ref(Fuel)}} = \begin{cases} I_{\text{ref(LFW)}}, & \text{if } P_{\text{load}} \le P_{\text{ref}} \\ I_{\text{FC}} + I_{\text{ref(GES2)}}, & \text{if } P_{\text{load}} > P_{\text{ref}} \end{cases} \tag{3.22a}$$

$$I_{\text{ref(Air)}} = \begin{cases} I_{\text{FC}} + I_{\text{ref(GES2)}}, & \text{if } P_{\text{load}} \le P_{\text{ref}} \\ I_{\text{ref(LFW)}}, & \text{if } P_{\text{load}} > P_{\text{ref}} \end{cases} \tag{3.22b}$$

$$I_{\text{ref(boost)}} = I_{\text{ref(GES1)}} \tag{3.22c}$$

The threshold P_{ref} for best fuel economy may be set in the range of the load power (P_{load}) using a sensitivity analysis for a variable load with a stair profile.

The strategies Air-PFW-2GES, Fuel-PFW-2GES, and Air/Fuel-PFW-2GES proposed in [44] will be compared with the reference sFF strategy [29] that uses the setting (3.23):

$$\begin{aligned} I_{\text{ref(Air)}} &= I_{\text{FC}} \\ I_{\text{ref(Fuel)}} &= I_{\text{FC}} \\ I_{\text{ref(Boost)}} &= I_{\text{ref(LFW)}} \text{ or } I_{\text{ref(Boost)}} = I_{\text{ref(PFW)}} \end{aligned} \tag{3.23}$$

The reference $I_{\text{ref(LFW)}}$ (3.3) [or the reference $I_{\text{ref(PFW)}}$ (3.9)] will set the FC power close to the DC power requested on the DC bus, so the strategy sFF will set the FC current to this reference:

$$I_{\text{FC}} \cong I_{\text{ref(LFW)}} \text{ or } I_{\text{FC}} \cong I_{\text{ref(PFW)}} \tag{3.24}$$

The best fuel economy will be found close to that value of the FC current by searching using (3.22). Thus, the search range of the reference $I_{ref(GES2)}$ can be limited to $\pm 10\%$ from the FC current, resulting, based on (3.22), a soft switching for the input variables ($I_{ref(Air)}$ and $I_{ref(Fuel)}$) of the FC system due to close levels $I_{FC} + I_{ref(GES2)}$ and $I_{ref(LFW)}$:

$$I_{ref(Fuel)} = I_{FC} + I_{ref(GES2)} \cong I_{FC} \cong I_{ref(LFW)} \qquad (3.25a)$$

$$I_{ref(Air)} = I_{FC} + I_{ref(GES2)} \cong I_{FC} \cong I_{ref(LFW)} \qquad (3.25b)$$

In addition, the slope limiters from the fueling regulators ensure a safe switching for the FC system that will optimally produce the power corresponding to the load demand based on (3.22).

The strategies Air-PFW-2GES, Fuel-PFW-2GES, and Air/Fuel-PFW-2GES use $I_{ref(boost)} = I_{ref(GES1)}$ as the second searching variable. Thus, due the 0.1 A hysteresis boost controller:

$$I_{FC} \cong I_{ref(boost)} = I_{ref(GES1)} \qquad (3.26)$$

Considering (3.25) and (3.26), the references $I_{ref(GES1)}$, $I_{ref(GES2)}$, and $I_{ref(LFW)}$ are close to FC current (3.24) set by the sFF strategy, but these differences may give a lot of fuel economy [44]. The setting block shown in Fig. 3.7 will set the references ($I_{ref(Air)}$, $I_{ref(Fuel)}$, and $I_{ref(Boost)}$) to optimally control the generated FC power via the fueling regulators and the boost controller.

3.4.1 Design for Fuel Economy

The strategies Air-PFW-2GES, Fuel-PFW-2GES, and Air/Fuel-PFW-2GES use the optimization function (3.17) that is fuel economy oriented due to the use of the weighted term related to the fuel consumption efficiency, in addition to the FC net power.

The subsystems of the EMO unit shown in Fig. 3.2 are as follows: (1) the optimization subsystem that will give based on (3.20)–(3.23) the references $I_{ref(GES1)}$ and $I_{ref(GES2)}$ using two GES controllers based on (3.27); (2) the LFW subsystem that will give, based on (3.3), the references $I_{ref(LFW)}$ (or, based on (3.9), the references $I_{ref(PFW)}$); (3) the 0.1 A *hysteresis* boost controller; (4) the setting block detailed in Fig. 3.7.

The searching references $I_{ref(GES1)}$ and $I_{ref(GES2)}$ are generated by two GES controllers based on (3.27) [52–54]:

$$y = f(v_1, v_2), \quad y_N = k_{Ny} \cdot y \qquad (3.27a)$$

$$\dot{y}_f = -\omega_h \cdot y_f + \omega_h \cdot y_N, \quad y_{\text{HPF}} = y_N - y_f, \quad \dot{y}_{\text{BPF}} = -\omega_l \cdot y_{\text{BPF}} + \omega_l \cdot y_{\text{HPF}}$$
$$(3.27\text{b})$$

$$\omega_h = b_h \omega, \quad \omega_l = b_l \omega, \quad s_d = \sin(\omega t), \quad \omega = 2\pi f_d, \quad (3.27\text{c})$$

$$y_{\text{DM}} = y_{\text{BPF}} \cdot s_d, \quad \dot{y}_{\text{Gradient}} = y_{\text{DM}}, \quad p_1 = k_1 \cdot y_{\text{Gradient}} \quad (3.27\text{d})$$

$$y_M = \left| \frac{1}{T_d} \cdot \int y_{\text{BPF}} dt \right|, \quad p_2 = k_2 \cdot y_M \cdot s_d \quad (3.27\text{e})$$

$$I_{\text{ref(GES)}} = k_{Np} \cdot (p_1 + p_2) \quad (3.27\text{f})$$

Searching for the best fuel economy will use two search variables, v_1 and v_2, and two sinusoidal dithers $s_d = \sin(\omega t)$ with the frequencies $f_d = 100$ Hz and $2f_d = 200$ Hz. The f normalization (3.27a) uses the parameter k_{Ny}. The band-pass filter (3.27b) uses the cutoff frequencies $\omega_l = b_l \omega$ and $\omega_h = b_h \omega$ that are set by the parameters b_l and β_h (3.27c). The output of the band-pass filter is demodulated resulting the signal y_{DM}, which is then integrated in order to obtain the searching gradient y_{Gradient}. The search signal p_1 is set proportional with y_{Gradient} using the tuning parameter k_1 [see (3.27d)]. To speed-up the localization of the optimum, the dither's amplitude is modulated with the y_M. Thus, the localization signal p_2 is set proportional with the modulated dither ($y_M \cdot s_d$) using the tuning parameter k_2 [see (3.27e)]. The sum of search and localization signals is normalized using the parameter k_{Np} (3.27f). This signal represents the searching output of the GES controller.

The values for aforementioned parameters can be obtained using the design rules given in [50] as follows: $k_1 = 1$ and $k_2 = 2$, $b_h = 0.1$ and $b_l = 1.5$; $k_{Np} = 20$ and $k_{Ny} = 1/1000$.

The fuel economy using sFF strategy as reference will be estimated based on (3.28):

$$\Delta \text{Fuel}_{T(S1)} = \text{Fuel}_{T(S1)} - \text{Fuel}_{T(\text{sFF})} \quad (3.28\text{a})$$

$$\Delta \text{Fuel}_{T(S2)} = \text{Fuel}_{T(S2)} - \text{Fuel}_{T(\text{sFF})} \quad (3.28\text{b})$$

$$\Delta \text{Fuel}_{T(\text{SW})} = \text{Fuel}_{T(\text{SW})} - \text{Fuel}_{T(\text{sFF})} \quad (3.28\text{c})$$

where the subscript indices $S1$, $S2$, and SW will be used for the strategies Air-PFW-2GES, Fuel-PFW-2GES, and Air/Fuel-PFW-2GES, respectively.

3.4.2 *Fuel Economy*

3.4.2.1 Constant Load

The fuel economy is recorded in Tables 3.1, 3.2, and 3.3 for $k_{fuel} = 0$, 25, and 50 lpm/W, using for the Air/Fuel-PFW-2GES strategy the threshold $P_{ref} = 5$ kW, which is the middle of the considered load range (5 kW = (2 kW + 8 kW)/2) [44].

It is worth mentioning that the same fuel economy is obtained for the Air/Fuel-PFW-2GES strategy with that obtained with the strategies Air-PFW-2GES and Fuel-PFW-2GES in the ranges $2 \, \text{kW} \leq P_{load} \leq P_{ref}$ and $8 \, \text{kW} \geq P_{load} > P_{ref}$, respectively, due to switching rules (3.22). The fuel economy (ΔFuel) for strategies Air-PFW-2GES, Fuel-PFW-2GES, and Air/Fuel-PFW-2GES using the data recoded in Tables 3.1, 3.2, 3.3 ($k_{fuel} = 0$, 25 and 50 lpm/W) is represented in Fig. 3.8a–c.

Table 3.1 Fuel economy for $k_{eff} = 0$

P_{load}	Fuel $_{T(sFF)}$	Fuel $_{T(SW)}$	ΔFuel $_{T(SW)}$	ΔFuel $_{T(S1)}$	ΔFuel $_{T(S2)}$
[kW]	[liters]	[liters]	[liters]	[liters]	[liters]
2	34.02	33.6	−0.42	8	−0.42
3	56.3	54.6	−1.7	6.16	−1.7
4	74.88	71.78	−3.1	1.94	−3.1
5	98.6	93.42	−5.18	−5.18	−5.24
6	125.58	114.02	−11.56	−11.56	−8.48
7	158.34	133.86	−24.48	−24.48	−14.04
8	176	132.66	−43.34	−43.34	−27.36

Table 3.2 Fuel economy for $k_{fuel} = 25$ lpm/W

P_{load}	Fuel $_{T(sFF)}$	Fuel $_{T(SW)}$	ΔFuel $_{T(SW)}$	ΔFuel $_{T(S1)}$	ΔFuel $_{T(S2)}$
[kW]	[liters]	[liters]	[liters]	[liters]	[liters]
2	34.02	33.46	−0.56	6.78	−0.56
3	56.3	54.3	−2	1.76	−2
4	74.88	71.12	−3.76	−3.72	−3.76
5	98.6	87.18	−11.42	−11.42	−6.52
6	125.58	107.76	−17.82	−17.82	−11.28
7	158.34	128.1	−30.24	−30.24	−20.76
8	176	128.28	−47.72	−47.72	−37.98

Table 3.3 Fuel economy for $k_{fuel} = 50$ lpm/W

P_{load}	Fuel $_{T(sFF)}$	Fuel $_{T(SW)}$	Δ Fuel $_{T(SW)}$	Δ Fuel $_{T(S1)}$	Δ Fuel $_{T(S2)}$
[kW]	[liters]	[liters]	[liters]	[liters]	[liters]
2	34.02	33.6	−0.42	8.56	−0.42
3	56.3	54.3	−2	4	−2
4	74.88	71.22	−3.66	1.1	−3.66
5	98.6	92.26	−6.34	−6.34	−6.28
6	125.58	112.58	−13	−13	−9.42
7	158.34	134.44	−23.9	−23.9	−14.48
8	176	130.48	−45.52	−45.52	−23.44

Note that the best fuel economy is obtained for $k_{fuel} = 25$ in the entire load demand range, so this value will be considered in the next simulations.

3.4.2.2 Variable Load

Case 1 Load cycle with different $P_{load(AV)}$ values
The variable load cycle uses on each 4 s the levels of $0.75 \cdot P_{load(AV)}$, $1.25 \cdot P_{load(AV)}$ and $1.00 \cdot P_{load(AV)}$ in order to obtain for this variable load cycle the average value (AV) of $P_{load(AV)}$. The fuel economy is recorded in Tables 3.4 and is represented in Fig. 3.9 for the strategies Air-PFW-2GES, Fuel-PFW-2GES, and Air/Fuel-PFW-2GES [44]. The threshold $P_{ref} = 5$ kW is also used for the Air/Fuel-PFW-2GES strategy.

Note that the best fuel economy is obtained for the Air/Fuel-PFW-2GES strategy in the entire load demand range.

Case 2 Load cycle with pulses
The load cycle with pulses uses on each 3 s the levels of 3 and 7 kW (see the first plot of Fig. 3.10). The fuel economy is recorded in Tables 3.5 for the strategies Air-PFW-2GES, Fuel-PFW-2GES, and Air/Fuel-PFW-2GES [44]. The threshold $P_{ref} = 5$ kW (3 kW $< P_{ref} < 7$ kW) is also used for the Air/Fuel-PFW-2GES strategy.

It is worth mentioning the following findings: (1) the LFW control is applied in the strategy Air-PFW-2GES to the air regulator, so *AirFr* will follow the pulses (see the fourth plot in Fig. 3.10a); (2) the LFW control is applied in the strategy Fuel-PFW-2GES to the fuel regulator, so *FuelFr* will follow the pulses (see the fifth plot in Fig. 3.10b); (3) the LFW control is applied in the strategy Air/Fuel-PFW-2GES based on the rules (3.22) to the air or fuel regulator, so minor spikes in *AirFr* and *FuelFr* can be observed; (4) the LFW control ensure for battery's charge-sustained mode of operation, with a dynamic compensation of the power flow balance on the DC bus by the ultracapacitors (see the third plot in Fig. 3.10); (5) Fuel$_{eff}$ varies from

Fig. 3.8 Fuel economy for constant load [44]. **a** $k_{\text{eff}} = 0$, **b** $k_{\text{eff}} = 25$, **c** $k_{\text{eff}} = 50$

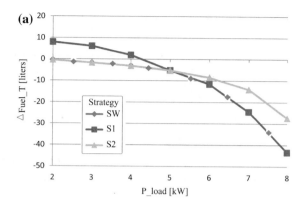

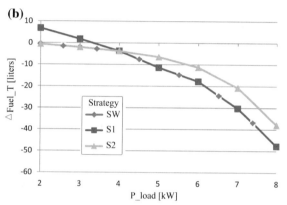

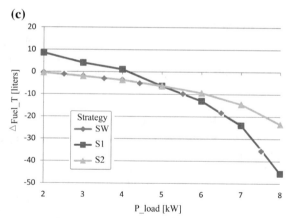

Table 3.4 Fuel economy for load cycles with different $P_{load(AV)}$ values

$P_{load(AV)}$	Fuel $_{T(sFF)}$	Fuel $_{T(SW)}$	ΔFuel $_{T(SW)}$	ΔFuel $_{T(S1)}$	ΔFuel $_{T(S2)}$
[kW]	[liters]	[liters]	[liters]	[liters]	[liters]
2	34.14	34.04	−0.1	10.8	−0.1
3	53.92	52.88	−1.04	8.74	−1.04
4	75.8	72.7	−3.1	−0.26	−3.84
5	100.62	89.22	−11.4	−12.96	−9.3
6	130.2	87.66	−42.54	−42.54	−18.56

Fig. 3.9 Fuel economy for load cycles with different $P_{load(AV)}$ values [44]

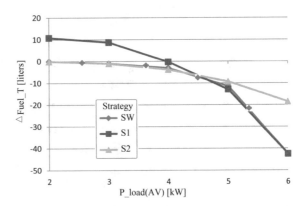

96 to 145 W/lpm (see the seventh plot in Fig. 3.10); (6) η_{sys} varies from 86 to 95% (see the eighth plot in Fig. 3.10). Note that the fuel economy for the Air/Fuel-PFW-2GES is 1.63 times and 3.67 times higher compared to strategies Air-PFW-2GES and Fuel-PFW-2GES. The optimal threshold P_{ref} can be obtained using a sensitivity analysis.

Case 3: Sensitivity analysis for threshold P_{ref} using a load with a stair profile

The load cycle with a stair profile used in the sensitivity analysis for the threshold P_{ref} is presented in the first plot of Fig. 3.11. The fuel economy is recorded in Tables 3.6 for the strategies Air-PFW-2GES, Fuel-PFW-2GES, and Air/Fuel-PFW-2GES [44].

It is worth mentioning that the best fuel economy has been obtained for the strategy Air/Fuel-PFW-2GES using $P_{ref} = 5.5$ kW, equal to 14% of Fuel$_{T(sFF)}$. Also, the fuel economy for the strategy Air/Fuel-PFW-2GES using $P_{ref} = 5.5$ kW is 2.07 times and 2.54 times higher compared to other strategies for the strategy Air-PFW-2GES and for the strategy Fuel-PFW-2GES (see Table 3.6 and Fig. 3.12 as well).

(a)

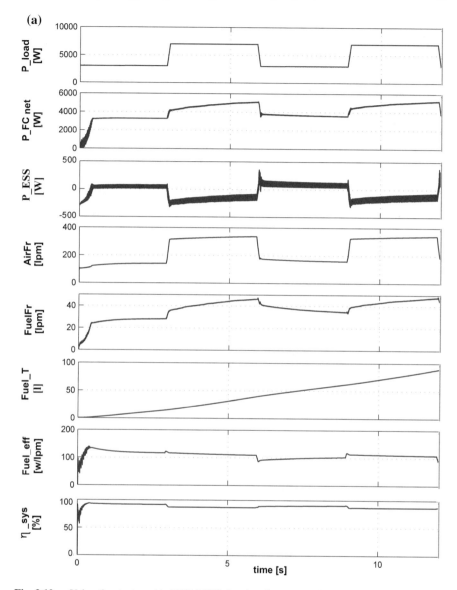

Fig. 3.10 a Using the strategy Air-PFW-2GES, **b** using the strategy fuel-PFW-2GES, **c** using the strategy Air/Fuel-PFW-2GES

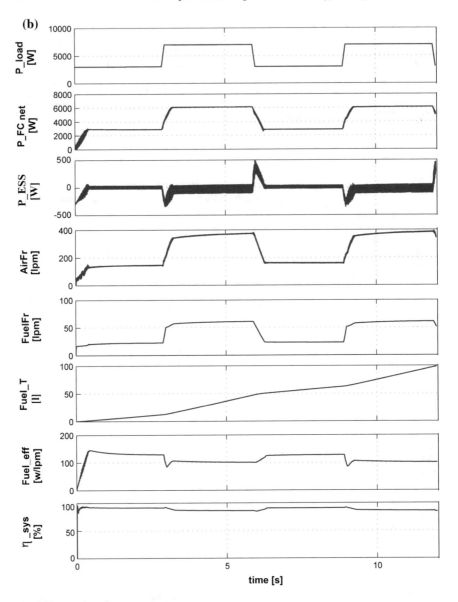

Fig. 3.10 (continued)

3.4.3 Discussion

The fuel economies $\Delta\mathrm{Fuel}_{T(\mathrm{SW})A/B/C}$ and $\mathrm{Fuel}_{T(\mathrm{sFF})}$ for the SW-strategy and the sFF-strategy and the percentage increase are registered in Table 3.7 for cases A ($k_{\mathrm{eff}} = 0$), B ($k_{\mathrm{eff}} = 25$), and C ($k_{\mathrm{eff}} = 50$). Note that the percentage increases

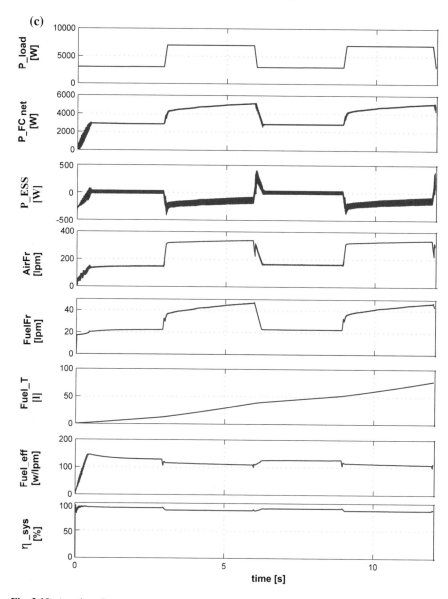

Fig. 3.10 (continued)

Table 3.5 Fuel economy for a load cycle with pulses

$P_{\text{load(pulse)}}$	$\text{Fuel}_{T(\text{sFF})}$	$\text{Fuel}_{T(\text{SW})}$	$\Delta\text{Fuel}_{T(\text{SW})}$	$\Delta\text{Fuel}_{T(S1)}$	$\Delta\text{Fuel}_{T(S2)}$
(kW)	(l)	(l)	(l)	(l)	(l)
3/7 kW	105.9	78.5	−27.4	−16.79	−7.43

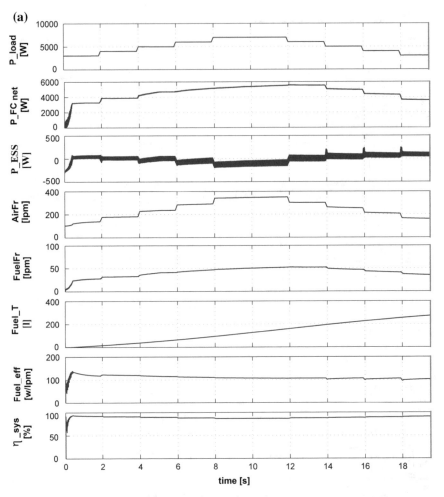

Fig. 3.11 FC HPS behavior for a load cycle with stair profile [44]. **a** Using the strategy Air-PFW-2GES, **b** using the strategy Fuel-PFW-2GES, **c** using the strategy Air/Fuel-PFW-2GES with $P_{\text{ref}} = 4.5\,\text{kW}$

with the load level. This is normal as fuel consumption increases with load demand, so the fuel optimization will be proportional (more or less) to the fuel consumed. The best fuel economy has been obtained for case C ($k_{\text{eff}} = 25$).

The fuel economies $\Delta\text{Fuel}_{T(\text{SW})j}$ and the percentage increase are registered in Table 3.8 for variable load cycle ($j = 1$), load cycle with pulses ($j = 2$), and load cycle with a stair profile ($j = 3$). Note that the fuel economy's nonlinear increase with the load demand is also observed for a variable load (see case of variable load cycle, $j = 1$). So, the sensitivity analysis for a load cycle with stair profile highlights a fuel economy that is half from that obtained for a load cycle with pulses:

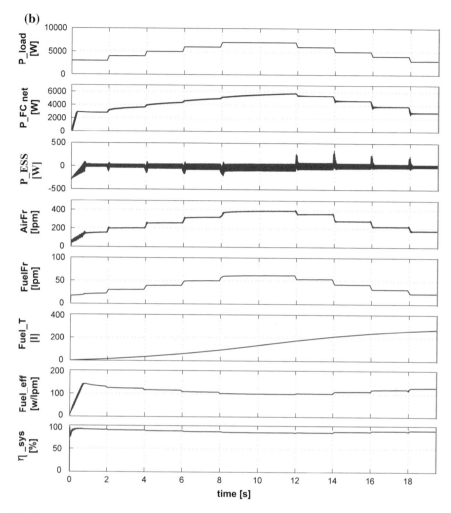

Fig. 3.11 (continued)

$$\left|\frac{\Delta \text{Fuel}_{T(\text{SW})2}}{\text{Fuel}_{T(\text{sFF})}}\right| = 25.87\%, \quad \left|\frac{\Delta \text{Fuel}_{T(\text{SW})3}}{\text{Fuel}_{T(\text{sFF})}}\right| = 13.72\% \qquad (3.29)$$

Anyway, 13.72% represents approximately a quarter from the fuel flow rate of 84.5 lpm at maximum load demand, because 39.3 L/20 s = 19.65 lpm ≅ 84.5/4 lpm.

The main findings of this study regarding the fuel economy strategy using switching of the fueling regulators are as follows: (1) the fuel economy strategy called Air/Fuel-PFW-2GES ensures the best fuel economy by switching the PFW (or LFW) reference to the fueling regulators; (2) the switching of the fueling regulators ensures the FC power requested on the DC bus due to PFW (or LFW) control of the FC system, so the battery will operate in charge-sustained mode; and (3) the best fuel economy

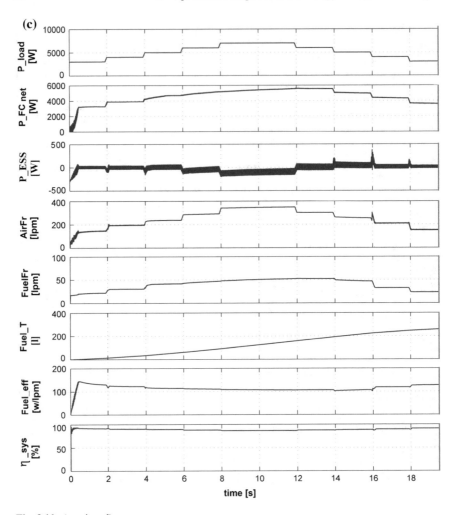

Fig. 3.11 (continued)

Table 3.6 Fuel economy for a load cycle with stair profile

P_{ref}	Fuel$_{T(sFF)}$	Fuel$_{T(SW)}$	ΔFuel$_{T(SW)}$	ΔFuel$_{T(S1)}$	ΔFuel$_{T(S2)}$
(kW)	(l)	(l)	(l)	(l)	(l)
2.5	286.5	267.1	−19.4	−19	−15.5
3.5	286.5	257.8	−28.7	−19	−15.5
4.5	286.5	251.3	−35.2	−19	−15.5
5.5	286.5	247.2	−39.3	−19	−15.5
6.5	286.5	252.8	−33.7	−19	−15.5

Fig. 3.12 Fuel economy for a load cycle with stair profile [44]

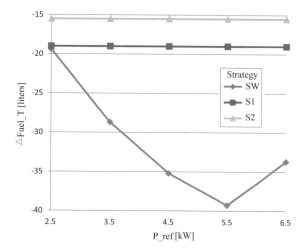

has been obtained for the strategy Air/Fuel-PFW-2GES using $P_{ref} = 5.5\,\text{kW}$ due to the search using two searching variables.

So, besides the pulse mitigation techniques that can damage the FC system (see next section), the implementation of the Air/Fuel-PFW-2GES strategy in FC vehicles may be of great interest in the coming years.

3.5 Strategy to Mitigate Load Impulses and Low Frequency Power Ripples

Load impulses and low frequency power ripples are very frequent on the DC bus, and these perturbations may propagate back to the FC system and damage the PEM membrane. This perturbation can be mitigated using active control by generating an anti-signal with almost same shape as the perturbation but opposite sign [87, 88]. The anti-signal generator may be a bidirectional DC-DC power converter supplied from the hybrid ESS, which will use active or semi-active topology to mix the energy and power storage devices. Besides the batteries [117, 118], different other energy storage technologies have been proposed such as flywheel energy systems (FES) [119], pumped hydroelectric storage (PHS) systems [120], FC systems [121], super-conducting magnetic energy storage (SMES) systems [122], compressed air energy storage (CAES) systems [123], and seasonal storage using hybrid storage systems (HSS) [124]. The hybrid ESS will use power storage devices such as ultracapacitors (UC), small SMES devices, and high-speed FES to ensure the appropriate genera-tion of the anti-signal. From these, the SMES systems have the smallest response time and have been used in power pulses applications which are commonly found in domains such as space missions [125], communication [126], military [127, 128], and microgrids for RES integration [129], energy storage [130], and stabilization

Table 3.7 Fuel economy for constant load demand

P_{load}	$Fuel_{T(sFF)}$	$\Delta Fuel_{T(SW)A}$	$\Delta Fuel_{T(SW)B}$	$\Delta Fuel_{T(SW)C}$	$\left|\frac{\Delta Fuel_{T(SW)A}}{Fuel_{T(sFF)}}\right|$	$\left|\frac{\Delta Fuel_{T(SW)B}}{Fuel_{T(sFF)}}\right|$	$\left|\frac{\Delta Fuel_{T(SW)C}}{Fuel_{T(sFF)}}\right|$
(kW)	(l)	(l)	(l)	(l)	(%)	(%)	(%)
2	34.02	−0.42	−0.56	−0.42	1.23	1.65	1.23
3	56.3	−1.7	−2	−2	3.02	3.55	3.55
4	74.88	−3.1	−3.76	−3.66	4.14	5.02	4.89
5	98.6	−5.18	−11.42	−6.34	5.25	11.58	6.43
6	125.58	−11.56	−17.82	−13	9.21	14.19	10.35
7	158.34	−24.48	−30.24	−23.9	15.46	19.10	15.09
8	176	−43.34	−47.72	−45.52	24.63	27.11	25.86

Table 3.8 Fuel economy for variable load demand

j	Case	$\text{Fuel}_{T(\text{sFF})}$	$\Delta\text{Fuel}_{T(\text{SW})j}$	$\left\| \dfrac{\Delta\text{Fuel}_{T(\text{SW})j}}{\text{Fuel}_{T(\text{sFF})}} \right\|$
–	–	(l)	(l)	(%)
1	$P_{\text{load(AV)}} = 2$	34.1	−0.1	0.3
1	$P_{\text{load(AV)}} = 3$	53.9	−1.0	1.9
1	$P_{\text{load(AV)}} = 4$	75.8	−3.1	4.1
1	$P_{\text{load(AV)}} = 5$	101	−11.4	11.3
1	$P_{\text{load(AV)}} = 6$	130	−42.5	32.7
2	Load pulses	106	−27	25.9
3	Stair profile	287	−39	13.7

[131, 132]. Both pulse mitigation and fuel economy optimization will be shown in next section using the strategies Air-PFW and Fuel-PFW.

3.5.1 Strategies Air-PFW and Fuel-PFW

The strategies called Air-PFW strategy and Fuel-PFW strategy (see Fig. 3.6) will use the following settings [72].

$$I_{\text{ref(Fuel)}} = I_{\text{FC}},\ I_{\text{ref(Air)}} = I_{\text{ref(LFW)}},\ \text{and}\ I_{\text{ref(boost)}} = I_{\text{ref(GES)}} \qquad (3.30)$$

$$I_{\text{ref(Air)}} = I_{\text{FC}},\ I_{\text{ref(Fuel)}} = I_{\text{ref(LFW)}},\ \text{and}\ I_{\text{ref(boost)}} = I_{\text{ref(GES)}} \qquad (3.31)$$

The FC/Battery/SMES HPS will use the HPS architecture based on the load-based strategy represented in Fig. 3.1. The difference is given by the power storage device used in the ESS active topology, which now is an HSS based on Battery/SMES instead of Battery/UC.

The load model (containing power pulses and low-frequency ripples) is presented in Fig. 3.13 [88] and has been detailed in Chap. 2.

The search for the fueling economy function's optimum (3.17) using the strategies Air-PFW and Fuel-PFW under a 6-kW load demand is presented in Figs. 3.14 and 3.15, respectively.

It is worth mentioning the following aspects: (1) the good regulation of the DC voltage to the reference of 200 V obtained in both cases (see the third plot in Figs. 3.14 and 3.15); (2) the operation of the battery in charge-sustained mode in both cases (see the forth plot in Figs. 3.14 and 3.15); (3) the search for the fueling economy function's optimum (3.17) based on the strategies Air-PFW and Fuel-PFW is performed using the inputs FuelFr and AirFr, respectively (see sixth and eighth plots in Figs. 3.14 and 3.15); (4) the values of the performance indicators such as the

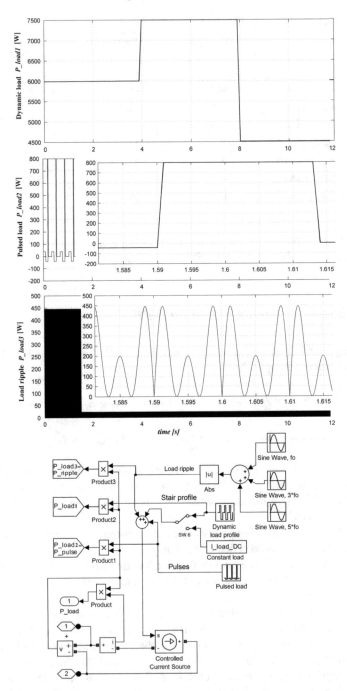

Fig. 3.13 Modeling load demand containing power pulses and low-frequency ripples [88]

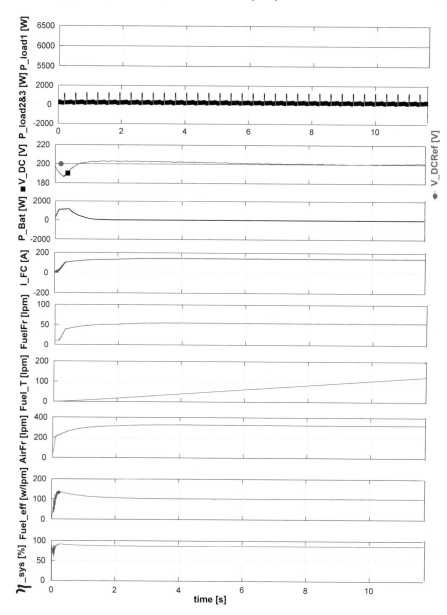

Fig. 3.14 Searching for the best fuel economy using the Air-PFW strategy [88]

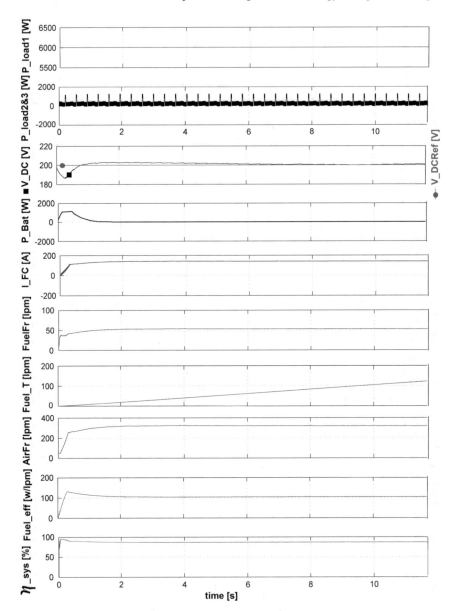

Fig. 3.15 Searching for the best fuel economy using the Fuel-PFW strategy [88]

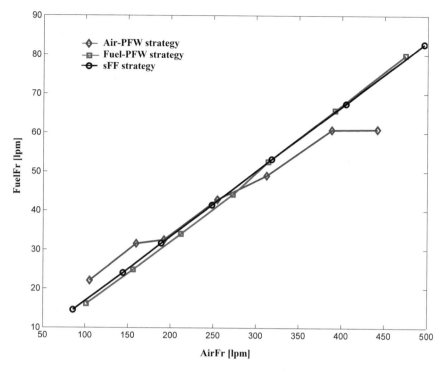

Fig. 3.16 Position of the operating points for the strategies Air-PFW (◆), Fuel-PFW (□), and sFF strategy (○) [88]

total hydrogen consumption $\left(\text{Fuel}_T = \int \text{FuelFr}(t)\mathrm{d}t\right)$, fuel consumption efficiency $\left(\text{Fuel}_{\text{eff}} = \frac{P_{\text{FCnet}}}{\text{FuelFr}}\right)$, and FC electrical efficiency $\left(\eta_{\text{sys}} = \frac{P_{\text{FCnet}}}{P_{\text{FC}}}\right)$ differ very little using the strategies Air-PFW and Fuel-PFW (see the seventh, tenth, and eleventh plots in Figs. 3.14 and 3.15).

The last finding is clearly shown in Fig. 3.16, where the position of the operating points for the strategies Air-PFW (◆) and Fuel-PFW (□) is compared with those obtained using the sFF strategy (○). Note that under nominal operating conditions for all strategies the operating points are very close.

Also, note that the position of the operating points and the values of the performance indicators do not depend on the type of the power storage device used by HSS. The design of the battery/SMES HSS will be presented in next section.

3.5.2 HSS Design

The battery/SMES HSS is presented in Fig. 3.17. An active topology was chosen for the battery/SMES HSS in order to implement the pulse mitigation on the SMES

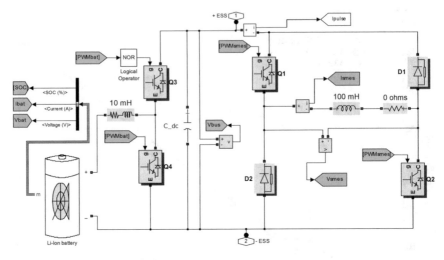

Fig. 3.17 Battery/SMES HSS [87]

side and the DC voltage regulation on the battery side via the asymmetric full-bridge DC-DC converter and bidirectional buck-boost DC-DC converter.

The load 2 represented in Fig. 3.13 contains $n_{\text{cycle}}(i)$ pulses during a load cycle, having amplitude $\Delta P_{\text{load}}(i)$ and duration $\Delta t_{\text{load}}(i)$. So, the energy of these type of pulses, pulse(i), is given by (3.32):

$$\Delta E_{\text{pulse}(i)} = \Delta P_{\text{load}}(i) \cdot \Delta t_{\text{load}}(i) \cdot n_{\text{cycles}}(i) \tag{3.32}$$

If the load contains different types of pulses, then the energy of all pulses is given by (3.33):

$$\Delta E_{\text{load}} = \sum_i \Delta P_{\text{load}}(i) \cdot \Delta t_{\text{load}}(i) \cdot n_{\text{cycles}}(i) \tag{3.33}$$

Considering the voltage ripple of $\pm \Delta V_{\text{DC}}$, the energy of the capacitor C_{DC} that filter the DC voltage will be given by (3.34):

$$\Delta E_{C\text{dc}} = \frac{1}{2} \cdot C_{\text{DC}} \cdot V_{\text{DC}}^2 \cdot \left[\left(1 + \frac{\Delta V_{\text{DC}}}{V_{\text{DC}}} \right)^2 - \left(1 - \frac{\Delta V_{\text{DC}}}{V_{\text{DC}}} \right)^2 \right] \tag{3.34}$$

Considering $\Delta E_{C\text{dc}} \cong \Delta E_{\text{load}}$, then:

$$C_{\text{DC}} \cong \frac{2 \cdot \Delta E_{\text{load}}}{V_{\text{DC}}^2 \cdot \left[\left(1 + \frac{\Delta V_{\text{DC}}}{V_{\text{DC}}} \right)^2 - \left(1 - \frac{\Delta V_{\text{DC}}}{V_{\text{DC}}} \right)^2 \right]} \tag{3.35}$$

For example, $C_{DC} = 1.1667\ F$ if $V_{DC} = 200$ V, $\Delta V_{DC} = 1,5\%\ V_{DC} = 3$ V, and for only one type of pulse $\Delta P_{load} = 4P_{load} = 400$ W, $\Delta t_{load} = 1\%\ T_{cycle} = 3,5$ ms, and $n_{cycle} = 1000$.

In the same manner, considering the current ripple of $\pm \Delta I_{SMES}$, the SMES energy will be given by (3.36):

$$\Delta E_{LSMES} = \frac{1}{2} \cdot L_{SMES} \cdot I_{SMES}^2 \cdot \left[\left(1 + \frac{\Delta I_{SMES}}{I_{SMES}} \right)^2 - \left(1 - \frac{\Delta I_{SMES}}{I_{SMES}} \right)^2 \right] \quad (3.36)$$

and:

$$L_{SMES} = \frac{2 \cdot \Delta E_{load}}{I_{SMES}^2 \cdot \left[\left(1 + \frac{\Delta I_{SMES}}{I_{SMES}} \right)^2 - \left(1 - \frac{\Delta I_{SMES}}{I_{SMES}} \right)^2 \right]} \quad (3.37)$$

For example, $L_{SMES} = 112$ mH if $I_{SMES} = 500$ A and $\Delta I_{SMES} = 2.5\%\ I_{SMES} = 12.5$ A.

If $\Delta E_{Bat} \cong \Delta E_{load}$ and $\Delta V_{bat} = 15\% V_{bat}$, the 100 Ah battery's capacity (C_{Bat}) will be obtained using (3.38):

$$C_{Bat} = \frac{\Delta P_{load} \cdot n_{cycles} \cdot T_{cycle(Bat)}}{\Delta V_{Bat}} \quad (3.38)$$

The 10 mH/0.01 Ω inductance (L_{Bat}) of the bidirectional buck-boost DC-DC converter (operating in continuous current mode with the duty cycle of $D_{Bat} = V_{DC}/V_{Bat} = 0.5$, $\Delta I_{bat} = 20\% I_{bat}$, $I_{bat} < I_{bat(max)} = 100$ A, and $f_{min} = 250$ kHz) will be obtained using (3.39):

$$L_{Bat} = \frac{V_{DC} \cdot (1 - D_{Bat})}{2 \cdot f_{min} \cdot \Delta I_{Bat}} \quad (3.39)$$

Considering the aforementioned values, the HSS control will be explained in next section.

3.5.3 HSS Control

The HSS control usually uses strategies to split the energy flow between the energy and power devices [133–135] using filtering based on low-pass filter (LPF) and high-pass filter (HPF) [136], rules-based filtering to select the weighting parameters [137] and filtering based on artificial intelligence concepts [138, 139]. The LPF/HPF-based filtering method is the most simple and effective, so will be considered in the particular design of the HSS control presented in Fig. 3.18.

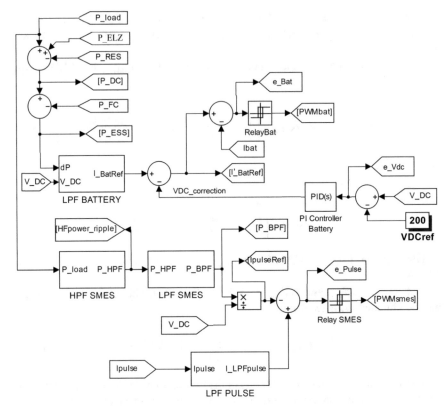

Fig. 3.18 HSS control [87]

The input signal P_{load} and dP given by (3.40) and (3.41) will generate the reference currents $I_{Pulse(ref)}$ and $I_{Bat(ref)}$ by (3.42) and (3.42) based on the filters BPF$_{SMES}$ and LPF$_{Bat}$:

$$P_{load} = P_{load1} + P_{load2} + P_{load3} \tag{3.40}$$

$$dP = p_{ESS} = p_{load} - p_{FC} - p_{PV} \tag{3.41}$$

$$I_{Pulse(ref)} = BPF_{SMES}(P_{load}/V_{DC}) \tag{3.42}$$

$$I_{Bat(ref)} = LPF_{Bat}(dP/V_{DC}) \tag{3.43}$$

A band-pass filter (BPF$_{SMES}$) is used instead of a HPF$_{SMES}$ to filter the switching noise. The filters LPF$_{SMES}$ and LPF$_{Pulse}$ have the cutoff frequency of 1000 Hz in order to generate the reference current $I_{Pulse(ref)}$ (3.42) and to not distort the pulses generating $I_{Pulse(LPF)}$:

$$I_{\text{Pulse(LPF)}} = \text{LPF}_{\text{Pulse}}\left(I_{\text{pulse}}\right) \tag{3.44}$$

where I_{Pulse} is the current sensed at the power line of the SMES power converter (see Fig. 3.17). The error e_{Pulse} given by (3.45) is the input for the 0.1 hysteresis control of the SMES power converter:

$$e_{\text{Pulse}} = I_{\text{Pulse(ref)}} - I_{\text{Pulse(LPF)}} \tag{3.45}$$

So, considering (3.42), (3.44), and $e_{\text{Pulse}} \cong 0$, the pulse current will be given by (3.46):

$$I_{\text{Pulse}} \cong \frac{P_{\text{load2}} + P_{\text{load3}}}{V_{\text{DC}}} \tag{3.46}$$

Thus, the HSS control implemented for the SMES power converter will generate an anti-pulse I_{Pulse} based on the command signal *PWMsmes* that will mitigate the perturbation $\frac{P_{\text{load2}} + P_{\text{load3}}}{V_{\text{DC}}}$.

The DC voltage regulation has been implemented at the battery side using the $V_{\text{DC(correction)}}$ (which is the proportional-integral (PI) correction of the voltage errors $e_{\text{Vdc}} = V_{\text{DC}} - V_{\text{DC(ref)}}$).

The error e_{Bat} given by (3.47) is the input for the 0.1 hysteresis control of the battery power converter:

$$e_{\text{Bat}} = I'_{\text{Bat(ref)}} - I_{\text{Bat}} \tag{3.47}$$

where

$$I'_{\text{Bat(ref)}} = I_{\text{Bat(ref)}} - V_{\text{DC(correction)}} \tag{3.48}$$

Considering the proportional and integral gain of the PI controller of 5 and 1, the behavior of the HSS will be presented in next section.

3.5.4   HSS Behavior

The load profile shown in Fig. 3.13 contains power pulses (P_{load2}) and low-frequency ripples (P_{load3}). The HSS behavior will be highlighted using the Air-PFW strategy, but note that almost same performance in the mitigation of pulses is obtained using the Fuel-PFW strategy because both strategies operate the FC system to sustain the power flow balance (3.1), with minor consequences on the battery control based on (3.41) (see Fig. 3.18).

Figure 3.19 illustrates the HSS behavior under two profiles of pulses, P_{load2a} and P_{load2b}, with low-frequency ripples (P_{load3}) added on them, as it is shown in the last plots of Fig. 3.19.

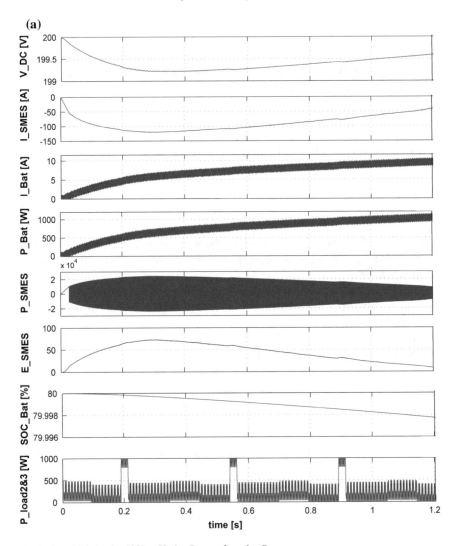

Fig. 3.19 HSS behavior [88]. **a** Under P_{load2a}, **b** under P_{load2b}

P_{load2a} contains an 800 W pulse with 0.0225 s width and also a positive and negative 40 W pulse with 0.1 s width. P_{load2b} contains an 800 W pulse with 0.0225 s width, but the positive and negative amplitudes of 0.1 s width pulses are higher (of 400 W) to evaluate the performance on the proposed mitigation algorithm. The plots of Fig. 3.19 are as follows: The DC voltage is shown in the first plot; the SMES and battery currents are shown in the second plot and the third plot, respectively; the battery and SMES power profiles are shown in the fourth plot and fifth plot, respectively; the SMES energy (3.49) is shown in the sixth plot; the battery SoC is represented in the seventh plot.

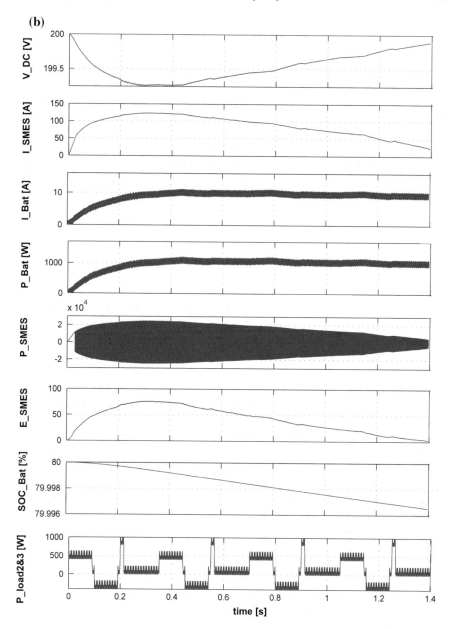

Fig. 3.19 (continued)

$$E_{SMES} = \int p_{SMES} dt \qquad (3.49)$$

Note a voltage drop of 0.8 V during start-up stage. The shapes of the generated anti-pulse (I_{pulse}) under P_{load2a} and P_{load2b} are represented in Figs. 3.20a and 3.21a, respectively, during the start-up stage. A zooming of the generated anti-pulse (I_{pulse}) under P_{load2a} and P_{load2b} is represented in Figs. 3.20b and 3.21b, respectively. Figures 3.20a and 3.21a show the mitigation of the low frequency ripple (P_{load3}). The performance of mitigation is given by the error = $I_{Pulse(ref)} - I_{Pulse}$, represented in the last plots of Figs. 3.20 and 3.21. Note that the error is less than 1 A, so the high frequency power ripples on the DC bus will be lower than 200 W, being filtered by the C_{DC} capacitor.

3.6   Conclusion

This chapter briefly presents and compares the energy management strategies and optimization algorithms currently used for hybrid power systems (HPS). Two new strategies that can operate the battery operating in charge-sustaining mode based on the load-following (LFW) control and the power-following (PFW) control are analyzed for the FC HPS without and with energy support from the renewable energy sources (RES). The performance indicators of the well-known optimization algorithms such as the global maximum power point tracking (GMPPT) algorithms, global maximum efficiency point tracking (GMEPT) algorithms, or fuel economy algorithms are briefly presented.

The fuel economy of Air/Fuel-PFW-2GES strategy, which performs a real-time switching of the fueling regulators' reference used in the Air-PFW-2GES strategy and Fuel-PFW-2GES strategy, is analyzed under constant and variable load. Better fuel economy compared to the strategies Air-PFW-2GES and Fuel-PFW-2GES is obtained in the entire load range due to the fueling regulators' switching using a load threshold that was optimally determined using a sensitivity analysis in that parameter.

The fueling regulator which does not operate under PFW control is optimally controlled in the first loop of the fuel economy optimization strategy. The second optimization loop controls the DC-DC boost converter interfacing the FC system with the DC bus. Thus, a real-time search for best fuel economy using two optimization variables is proposed for the switching strategy called Air/Fuel-PFW-2GES strategy and the reference strategies Air-PFW-2GES and Fuel-PFW-2GES.

The fuel economy of the Air/Fuel-PFW-2GES strategy compared to the reference strategies is 1.63 times and 3.67 times higher for a pulsed load, and 2.07 times and 2.54 times higher for a stair load. This means about 25.9 and 13.7% from the total fuel consumption of the best reference strategy operating FC HPS under load pulses and stair load, respectively.

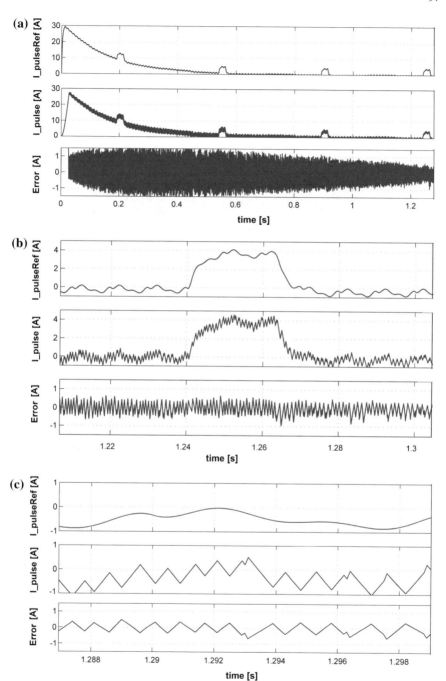

Fig. 3.20 I_{pulse} generated using P_{load2a} [88]. **a** Start-up phase, **b** zooming of I_{pulse}, **c** low-frequency ripple

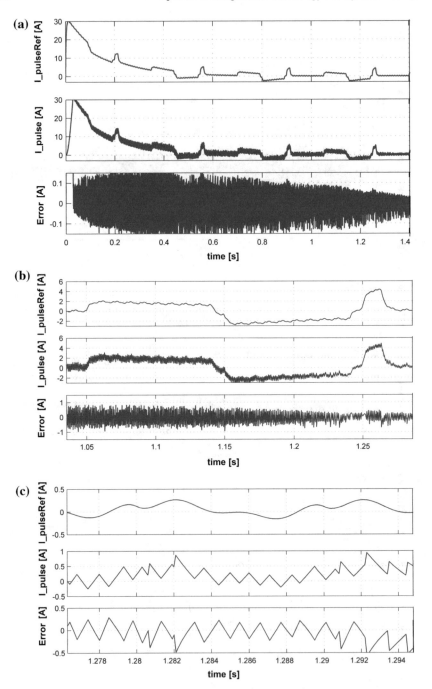

Fig. 3.21 I_{pulse} generated using P_{load2b} [88]. **a** Start-up phase, **b** zooming of I_{pulse}, **c** low-frequency ripple

Therefore, the Air/Fuel-PFW-2GES strategy combined with the mitigation strategy of the perturbed pulses on the DC bus may be applied in FC vehicles in order to improve their performance.

The mitigation performance of the anti-pulse control applied to the bidirectional DC-DC power converter, interfacing the ultracapacitors' stack from the hybrid storage system (HSS) with the DC bus, is highlighted in this chapter using different pulse profiles. The highest pulse has amplitude of 800 W, which represents about 13.33% from the 6-kW nominal FC power.

The remaining perturbation (the difference between the perturbed pulse and generated anti-pulse and the high frequency ripple added on it due to the switching control of the bidirectional DC-DC power converter) has a peak-to-peak amplitude lower than 1 A. Therefore, the low-frequency power ripple on the 200 V bus will be less than 200 W, which represents 3.33% from the 6-kW nominal FC power. Note that this is further mitigated by the FC boost converter, so this low-frequency power ripple is lower than the recommended value of 5%.

Also, it is worth mentioning the DC voltage regulation at battery side that ensures a voltage drop less than 0.6 and 0.1 V during the start-up stage and the perturbed pulse.

References

1. Bizon N, Tabatabaei NM, Blaabjerg F, Kurt E (2017) Energy harvesting and energy efficiency: technology, methods and applications. Springer; available on http://www.springer.com/us/book/9783319498744
2. Madaci B, Chenni R, Kurt E, Hemsas KE (2016) Design and control of a stand-alone hybrid power system. Int J Hydrogen Energ 41(29):12485–12496
3. Han I-S, Park S-K, Chung C-B (2015) Modeling and operation optimization of a proton exchange membrane fuel cell system for maximum efficiency. Int J Hydrogen Energ 113:52–65
4. Sikkabut S, Mungporn P, Ekkaravarodome C, Bizon N, Tricoli P, Nahid-Mobarakeh B, Pierfederici S, Davat B, Thounthong P (2017) Control of high-energy high-power densities storage devices by Li-ion battery and supercapacitor for fuel cell/photovoltaic hybrid power plant for autonomous system applications. IEEE T Ind Appl 52(5):4395–4407
5. Ishaque K, Salam Z (2013) A review of maximum power point tracking techniques of PV system for uniform insolation and partial shading condition. Renew Sustain Energy Rev 19:475–488
6. Liu Z-H, Chen J-H, Huang J-W (2015) A review of maximum power point tracking techniques for use in partially shaded conditions. Renew Sustain Energy Rev 41:436–453
7. Bizon N, Tabatabaei NM, Shayeghi H (2013) Analysis, control and optimal operations in hybrid power systems—advanced techniques and applications for linear and nonlinear systems. Springer; available on http://dx.doi.org/10.1007/978-1-4471-5538-6
8. Olatomiwa L, Mekhilef S, Ismail MS, Moghavvemi M (2016) Energy management strategies in hybrid renewable energy systems: a review. Renew Sustain Energy Rev 62:821–835
9. Zhang P, Yan F, Du C (2015) A comprehensive analysis of energy management strategies for hybrid electric vehicles based on bibliometrics. Renew Sustain Energy Rev 48:88–104
10. Peng J, He H, Xiong R (2017) Rule based energy management strategy for a series–parallel plug-in hybrid electric bus optimized by dynamic programming. Appl Energ 185(2):1633–1643

11. Bizon N, Kurt E (2017) Performance analysis of the tracking of global extreme on multimodal patterns using the asymptotic perturbed extremum seeking control scheme. Int J Hydrogen Energ 42(28):17645–17654

12. Das V, Padmanaban S, Venkitusamy K, Selvamuthukumaran R, Siano P (2017) Recent advances and challenges of fuel cell based power system architectures and control—a review. Renew Sustain Energy Rev 73:10–18

13. Cano MH, Mousli MIA, Kelouwani K, Agbossou K, Dubé Y (2017) Improving a free air breathing proton exchange membrane fuel cell through the maximum efficiency point tracking method. J Power Sources 345:264–274

14. Ahmadi S, Abdi Sh, Kakavand M (2017) Maximum power point tracking of a proton exchange membrane fuel cell system using PSO-PID controller. Int J Hydrog Energy 42(32):20430–20443

15. Hernández-Torres D, Riu D, Sename O (2017) Reduced-order robust control of a fuel cell air supply system. IFAC-PapersOnLine 50(1):96–101

16. Daud WRW, Rosli RE, Majlan EH, Hamid SAA, Mohamed R, Husaini T (2017) PEM fuel cell system control: a review. Renew Energ 113:620–638

17. Fathabadi H (2017) Novel fast and high accuracy maximum power point tracking method for hybrid photovoltaic/fuel cell energy conversion systems Renew. Energ 106:232–242

18. Eriksson ELV, Mac E, Gray A (2017) Optimization and integration of hybrid renewable energy hydrogen fuel cell energy systems—a critical review. Appl Energ 202:348–364

19. Vivas FJ, De las Heras A, Segura F, Andújar JM (2017) A review of energy management strategies for renewable hybrid energy systems with hydrogen backup. Renew Sust Energ Rev 82(1):126–155

20. Tiar M, Betka A, Drid S, Abdeddaim S, Tabandjat A (2017) Optimal energy control of a PV-fuel cell hybrid system. Int J Hydrogen Energy 42(2):1456–1465

21. Muñoz PM, Correa G, Gaudiano ME, Fernández D (2017) Energy management control design for fuel cell hybrid electric vehicles using neural networks. Int J Hydrogen Energy. https://doi.org/10.1016/j.ijhydene.2017.09.169

22. Fernández RÁ, Caraballo SC, Cilleruelo FB, Lozano JA (2108) Fuel optimization strategy for hydrogen fuel cell range extender vehicles applying genetic algorithms. Renew Sust Energ Rev 81(1):655–668

23. Zhou D, Al-Durra A, Gao F, Ravey A, Matraji I, Simões MG (2017) Online energy management strategy of fuel cell hybrid electric vehicles based on data fusion approach. J Power Sources 366:278–291

24. Caux S, Gaoua Y, Lopez P (2017) A combinatorial optimisation approach to energy management strategy for a hybrid fuel cell vehicle. Energy 133:219–230

25. Koubaa R, krichen L (2017) Double layer metaheuristic based energy management strategy for a fuel cell/ultra-capacitor hybrid electric vehicle. Energy 133:1079–1093

26. Han J, Yu S, Yi S (2017) Adaptive control for robust air flow management in an automotive fuel cell system. Appl Energ 190:73–83

27. Carignano M, Costa-Castelló R, Nigro N, Junco S (2017) A Novel energy management strategy for fuel-cell/supercapacitor hybrid vehicles. IFAC-PapersOnLine 50(1):10052–10057

28. Zhang W, Li L, Xu L, Ouyang M (2017) Optimization for a fuel cell/battery/capacity tram with equivalent consumption minimization strategy. Energy Convers Manage 134:59–69

29. Pukrushpan JT, Stefanopoulou AG, Peng H (2004) Control of fuel cell power systems. Springer, New York

30. Haneda T, Ono Y, Ikegami T, Akisawa A (2017) Technological assessment of residential fuel cells using hydrogen supply systems for fuel cell vehicles. Int J Hydrogen Energy 42(42):26377–26388

31. Bizon N, Thounthong P (2018) Real-time strategies to optimize the fueling of the fuel cell hybrid power source: A review of issues, challenges and a new approach. Renew Sustain Energy Rev 91:1089–1102

32. Bizon N, Radut M, Oproescu M (2015) Energy control strategies for the fuel cell hybrid power source under unknown load profile. Energy 86:31–41

33. Bizon N, Mazare AG, Ionescu LM, Enescu FM (2018) Optimization of the proton exchange membrane fuel cell hybrid power system for residential buildings. Energy Convers Manage 163:22–37
34. Bizon N, Thounthong P (2018) Fuel economy using the global optimization of the fuel cell hybrid power systems. Energ Convers Manage 173:665–678
35. Bizon N (2019) Real-time optimization strategies of FC hybrid power systems based on load-following control: a new strategy, and a comparative study of topologies and fuel economy obtained. Appl Energ 241C:444–460
36. Bizon N, Hoarcă CI (2019) Hydrogen saving through optimized control of both fueling flows of the fuel cell hybrid power system under a variable load demand and an unknown renewable power profile. Energ Convers Manage 184:1–14
37. Bizon N (2018) Real-time optimization strategy for fuel cell hybrid power sources with load-following control of the fuel or air flow. Energ Convers Manage 157:13–27
38. Bizon N, Oproescu M (2018) Experimental comparison of three real-time optimization strategies applied to renewable/FC-based hybrid power systems based on load-following control. Energies 11(12):3537–3569. https://doi.org/10.3390/en11123537
39. Bizon N, Iana VG, Kurt E, Thounthong P, Oproescu M, Culcer M, Iliescu M (2018) Air flow real-time optimization strategy for fuel cell hybrid power sources with fuel flow based on load-following. Fuel Cell 18(6):809–823. https://doi.org/10.1002/fuce.201700197
40. Bizon N (2014) Load-following mode control of a standalone renewable/fuel cell hybrid power source. Energ Convers Manage 77:763–772
41. Bizon N, Lopez-Guede JM, Kurt E, Thounthong P, Mazare AG, Ionescu LM, Iana G (2019) Hydrogen economy of the fuel cell hybrid power system optimized by air flow control to mitigate the effect of the uncertainty about available renewable power and load dynamics. Energ Convers Manage 179:152–165
42. Bizon N, Stan VA, Cormos AC (2019) Optimization of the fuel cell renewable hybrid power system using the control mode of the required load power on the DC bus. Energies 12(10):1889–1904. https://doi.org/10.3390/en12101889
43. Bizon N (2019) Efficient fuel economy strategies for the fuel cell hybrid power systems under variable renewable/load power profile. Appl Energ 251:113400–113518. https://doi.org/10.1016/j.apenergy.2019.113400
44. Bizon N (2019) Fuel saving strategy using real-time switching of the fueling regulators in the proton exchange membrane fuel cell system. Appl Energy 252:113449–113453. https://doi.org/10.1016/j.apenergy.2019.113449
45. Harrag A, Bahri H (2017) Novel neural network IC-based variable step size fuel cell MPPT controller: performance, efficiency and lifetime improvement. Int J Hydrogen Energ. https://doi.org/10.1016/j.ijhydene.2016.12.079
46. Thounthong P, Luksanasakul A, Koseeyaporn P, Davat B (2013) Intelligent model-based control of a standalone photovoltaic/fuel cell power plant with supercapacitor energy storage. IEEE T Sustain Energ 4(1):240–249
47. Huang Y, Wang H, Khajepour A, He H, Ji J (2017) Model predictive control power management strategies for HEVs: a review. J Power Sources 341:91–106
48. Bizon N (2017) Energy optimization of fuel cell system by using global extremum seeking algorithm. Appl Energ 206:458–474
49. Bizon N (2017) Searching of the extreme points on photovoltaic patterns using a new asymptotic perturbed extremum seeking control scheme. Energ Convers Manage 144:286–302
50. Bizon N, Thounthong P, Raducu M, Constantinescu LM (2017) Designing and modelling of the asymptotic perturbed extremum seeking control scheme for tracking the global extreme. Int J Hydrogen Energy 42(28):17632–17644
51. Bizon N (2014) Tracking the maximum efficiency point for the FC system based on extremum seeking scheme to control the air flow. Appl Energ 129:147–157
52. Bizon N (2016) Global maximum power point tracking (GMPPT) of photovoltaic array using the extremum seeking control (ESC): a review and a new GMPPT ESC scheme. Renew Sustain Energy Rev 57:524–539

53. Bizon N (2016) Global extremum seeking control of the power generated by a photovoltaic array under partially shaded conditions. Energy Convers Manage 109:71–85
54. Bizon N (2016) Global maximum power point tracking based on new extremum seeking control scheme. Prog Photovoltaics Res Appl 24(5):600–622
55. Sulaiman N, Hannan MA, Mohamed A, Ker PJ, Majlan EH, Wan Daud WR (2018) Optimization of energy management system for fuel-cell hybrid electric vehicles: Issues and recommendations. Appl Energ 228:2061–2079
56. Sorrentino M, Cirillo V, Nappi L (2019) Development of flexible procedures for co-optimizing design and control of fuel cell hybrid vehicles. Energ Convers Manage 185:537–551
57. Wang F-C, Yi-Shao Hsiao Y-S, Yi-Zhe Yang Y-Z (2018) The optimization of hybrid power systems with renewable energy and hydrogen generation. Energies 11(8):1948. https://doi.org/10.3390/en11081948
58. Lawan Bukar AL, Wei Tan CW (2019) A review on stand-alone photovoltaic-wind energy system with fuel cell: system optimization and energy management strategy. J Clean Prod (in press). https://doi.org/10.1016/j.jclepro.2019.02.228
59. Tabatabaei NM, Kabalci E, Bizon N (2019) Microgrid architectures, control and protection methods. Springer, London, UK; available on https://www.springer.com/in/book/9783030237226
60. Wang Y, Sun Z, Chen Z (2019) Rule-based energy management strategy of a lithium-ion battery, supercapacitor and PEM fuel cell system. Energy Procedia 158:2555–2560
61. Li G-P, Zhan J-L, He H-W (2017) Battery SOC constraint comparison for predictive energy management of plug-in hybrid electric bus. Appl Energ 194:578–587
62. Pan ZF, An L, Wen CY (2019) Recent advances in fuel cells based propulsion systems for unmanned aerial vehicles. Appl Energ 240:473–485
63. Yue M, Jemei S, Gouriveau R, Zerhouni N (2019) Review on health-conscious energy management strategies for fuel cell hybrid electric vehicles: degradation models and strategies. Int J Hydrogen Energy 44(13):6844–6861
64. Zhang T, Wang P, Chen H, Pei P (2018) A review of automotive proton exchange membrane fuel cell degradation under start-stop operating condition. Appl Energy 223:249–262
65. Dijoux E, Steiner NY, Benne M, Péra M-C, Pérez BG (2017) A review of fault tolerant control strategies applied to proton exchange membrane fuel cell systems. J Power Sources 359:119–133
66. Wieczorek M, Lewandow M (2017) A mathematical representation of an energy management strategy for hybrid energy storage system in electric vehicle and real time optimization using a genetic algorithm. Appl Energ 192:222–233
67. Wang H, Huang Y, Khajepour A, Song Q (2016) Model predictive control-based energy management strategy for a series hybrid electric tracked vehicle. Appl Energ 182:105–114
68. Matraji I, Ahmed FS, Laghrouche S, Wack M (2015) Comparison of robust and adaptive second order sliding mode control in PEMFC air-feed systems. Int J Hydrogen Energ 40(30):9491–9504
69. Mungporn P, Poonnoi N, Sikkabut S, Ekkaravarodome E, Thounthong P, Bizon N, Hinaje M, Pierfederici S, Davat B (2015). Model based control of modified four-phase interleaved boost converter for fuel cell power source for mobile based station. In: International telecommunications energy conference INTELEC® 2015, pp1–6; available on https://doi.org/10.1109/intlec.2015.7572414
70. Han J, Park Y, Dongsuk K (2014) Optimal adaptation of equivalent factor of equivalent consumption minimization strategy for fuel cell hybrid electric vehicles under active state inequality constraints. J Power Sources 267:491–502
71. Bassam AM, Phillips AB, Turnock SR, Wilson PA (2017) Development of a multi-scheme energy management strategy for a hybrid fuel cell driven passenger ship. Int J Hydrogen Energ 42(1):623–635
72. Ettihir K, Boulon L, Agbossou K (2016) Optimization-based energy management strategy for a fuel cell/battery hybrid power system. Appl Energ 163:142–153

73. Ettihir K, Cano MH, Boulon L, Agbossou K (2017) Design of an adaptive EMS for fuel cell vehicles. Int J Hydrogen Energ 42(2):1481–1489

74. Zhang W, Li J, Xu L, Ouyang M (2017) Optimization for a fuel cell/battery/capacity tram with equivalent consumption minimization strategy. Energ Convers Manage 134:59–69

75. Cai Y, Ouyang MG, Yang F (2017) Impact of power split configurations on fuel consumption and battery degradation in plug-in hybrid electric city buses. Appl Energ 188:257–269

76. Xu L, Ouyang M, Li J, Yang F, Jianfeng H (2013) Application of Pontryagin's minimal principle to the energy management strategy of plugin fuel cell electric vehicles. Int J Hydrogen Energ 38(24):10104–10115

77. Hou C, Ouyang MG, Xu LF, Wang HW (2014) Approximate Pontryagin's minimum principle applied to the energy management of plug-in hybrid electric vehicles. Appl Energy 115:174–189

78. Ramos-Paja CA, Spagnuolo G, Petrone G, Emilio Mamarelis M (2014) A perturbation strategy for fuel consumption minimization in polymer electrolyte membrane fuel cells: analysis, design and FPGA implementation. Appl Energ 119:21–32

79. Ariyur KB, Krstic M (2003) Real-time optimization by extremum-seeking control. Wiley-Interscience, Hoboken

80. Bizon N (2010) On tracking robustness in adaptive extremum seeking control of the fuel cell power plants. Appl Energ 87(10):3115–3130

81. Bizon N (2013) Energy harvesting from the FC stack that operates using the MPP tracking based on modified extremum seeking control. Appl Energ 104:326–336

82. Bizon N (2014) Improving the PEMFC energy efficiency by optimizing the fueling rates based on extremum seeking algorithm. Int J Hydrogen Energ 39(20):10641–10654

83. Bizon N (2013) FC energy harvesting using the MPP tracking based on advanced extremum seeking control. Int J Hydrogen Energ 38(4):1952–1966

84. Bizon N, Oproescu M, Raceanu M (2015) efficient energy control strategies for a standalone renewable/fuel cell hybrid power source. Energ Convers Manage 90:93–110

85. Zhou D, Ravey A, Al-Durra A, Gao F (2017) A comparative study of extremum seeking methods applied to online energy management strategy of fuel cell hybrid electric vehicles. Energ Convers Manage 151:778–790

86. Bizon N (2018) Optimal operation of fuel cell/ wind turbine hybrid power system under turbulent wind and variable load. Appl Energ 212:196–209

87. Bizon N (2018) Effective mitigation of the load pulses by controlling the battery/SMES hybrid energy storage system. Appl Energ 229:459–473

88. Bizon N (2019) Hybrid power sources (HPSs) for space applications: analysis of PEMFC/Battery/SMES HPS under unknown load containing pulses. Renew Sustain Energy Rev 105:14–37. https://doi.org/10.1016/j.rser.2019.01.044

89. O'Rourke J, Arcak M, Ramani M (2009) Real-time optimization of net power in a fuel cell system. J Power Sources 187(2):422–430

90. Matraji I, Laghrouche S, Jemei S, Wack M (2013) Robust control of the PEM fuel cell air-feed system via sub-optimal second order sliding mode. Appl Energy 104:945–957

91. Kunusch C, Puleston PF, Mayosky MA, Fridman L (2013) Experimental results applying second order sliding mode control to a PEM fuel cell based system. Control Eng Pract 21(5):719–726

92. Laghrouche S, Matraji I, Ahmed FS, Jemei S, Wack M (2013) Load governor based on constrained extremum seeking for PEM fuel cell oxygen starvation and compressor surge protection. Int J Hydrogen Energy 38(33):14314–14322

93. Zhao D, Zheng Q, Gao F, Bouquain D, Dou M, Miraoui A (2014) Disturbance decoupling control of an ultra-high speed centrifugal compressor for the air management of fuel cell systems. Int J Hydrogen Energy 39(4):1788–1798

94. da Fonseca R, Bideaux E, Gerard M, Jeanneret B, Desbois-Renaudin M, Sari A (2014) Control of PEMFC system air group using differential flatness approach: validation by a dynamic fuel cell system model. Appl Energy 113:219–229

95. Zhou N, Yang C, Tucker D, Pezzini P, Traverso A (2015) Transfer function development for control of cathode airflow transients in fuel cell gas turbine hybrid systems. Int J Hydrogen Energy 40(4):1967–1979
96. Beirami H, Shabestari AZ, Zerafat MM (2015) Optimal PID plus fuzzy controller design for a PEM fuel cell air feed system using the self-adaptive differential evolution algorithm. Int J Hydrogen Energy 40:9422–9434
97. Wang YZ, Xuan DJ, Kim YB (2013) Design and experimental implementation of time delay control for air supply in a polymer electrolyte membrane fuel cell system. Int J Hydrogen Energy 38:13381–13392
98. Nikezhadi A, Fantova MA, Kunusch C, Martinez CO (2011) Design and implementation of LQR/LQG strategies for oxygen stoichiometry control in PEM fuel cells based systems. J Power Sources 196(9):4277–4282
99. Ou K, Wang YX, Li ZZ, Shen YD (2015) Feedforward fuzzy-PID control for air flow regulation of PEM fuel cell system. Int J Hydrogen Energy 40:11686–11695
100. Won JS, Langari R (2005) Intelligent energy management agent for a parallel hybrid vehicle-part II: torque distribution, charge sustenance strategies, and performance results. IEEE Trans Veh Technol 54(3):935–953
101. Pisu P, Rizzoni G (2007) A comparative study of supervisory control strategies for hybrid electric vehicles. IEEE Trans Control Syst Technol 15(3):506–518
102. Musardo C, Rizzoni G, Guezennec Y, Staccia B (2005) A-ECMS: an adaptive algorithm for hybrid electric vehicle energy management. Eur J Control 11(4–5):509–524
103. Bizon N, Mazare AG, Ionescu LM, Thounthong P, Kurt E, Oproescu O, Serban G, Lita I (2019) Better fuel economy by optimizing airflow of the fuel cell hybrid power systems using the load-following control based on the fuel flow. Energies 12(14):2792–2810. https://doi.org/10.3390/en12142792
104. Bidram A, Davoudi A, Balog RS (2012) Control and circuit techniques to mitigate partial shading effects in photovoltaic arrays. IEEE J Photovolt 2:532–546
105. Kotti R, Shireen W (2015) Efficient MPPT control for PV systems adaptive to fast changing irradiation and partial shading conditions. Sol Energy 114:397–407
106. Wang L, Chen S, Ma K (2016) On stability and application of extremum seeking control without steady-state oscillation. Automatica 68:18–26
107. Jiang LL, Srivatsan R, Maskell DL (2018) Computational intelligence techniques for maximum power point tracking in PV systems: a review. Renew Sustain Energy Rev 85:14–45
108. Belhachat F, Larbes C (2019) Comprehensive review on global maximum power point tracking techniques for PV systems subjected to partial shading conditions. Sol Energy 183:476–500
109. Ahmadi S, Bathaee SMT, Hosseinpour AH (2018) Improving fuel economy and performance of a fuel-cell hybrid electric vehicle (fuel-cell, battery, and ultra-capacitor) using optimized energy management strategy. Energy Convers Manage 160:74–84
110. Nojavan S, Majidi M, Zare K (2017) Performance improvement of a battery/PV/fuel cell/grid hybrid energy system considering load uncertainty modeling using IGDT. Energ Convers Manage 147:29–39
111. Hu Z, Li J, Xu L, Song Z, Fang C, Ouyang M, Dou G, Kou G (2016) Multi-objective energy management optimization and parameter sizing for proton exchange membrane hybrid fuel cell vehicles. Energ Convers Manage 129:108–121
112. Secanell M, Wishart J, Dobson P (2011) Computational design and optimization of fuel cells and fuel cell systems: a review. J Power Sources 196(8):3690–3704
113. Yeniay O (2005) Penalty function methods for constrained optimization with genetic algorithms. Math Comput Appl 10(1):45–56
114. Kim M-J, Peng H (2007) Power management and design optimization of fuel cell/battery hybrid vehicles. J Power Sources 165(2):819–832
115. Turkmen AC, Solmaz S, Celik C (2017) Analysis of fuel cell vehicles with advisor software. Renew Sustain Energy Rev 70:1066–1071
116. Castaings A, Lhomme W, Trigui R, Bouscayrol A (2016) Comparison of energy management strategies of a battery/supercapacitors system for electric vehicle under real-time constraints. Appl Energ 163:190–200

117. Gallo AB, Simões-Moreira JR, Costa HKM, M.M.Santos MM, Moutinho dos Santos E (2016) Energy storage in the energy transition context: a technology review. Renew Sustain Energy Rev 65:800–822

118. Jing W, Lai CH, Wong WSH, Wong DML (2018) A comprehensive study of battery-supercapacitor hybrid energy storage system for standalone PV power system in rural electrification. Appl Energ 224:340–356

119. Díaz-González F, Sumper A, Gomis-Bellmunt O, Bianchi FD (2013) Energy management of flywheel-based energy storage device for wind power smoothing. Appl Energ 110:207–219

120. de Oliveira e Silva G, Hendrick P (2016) Pumped hydro energy storage in buildings. Appl Energ 179:1242–1250

121. Aneke M, Wang M (2016) Energy storage technologies and real life applications—a state of the art review. Appl Energ 179:350–377

122. Weijia Y (2011) Second-generation high-temperature superconducting coils and their applications for energy storage. PhD Thesis, Springer

123. Budt M, Wolf D, Span R, Yan J (2016) A review on compressed air energy storage: basic principles, past milestones and recent developments. Appl Energ 170:250–268

124. Gabrielli P, Gazzani M, Martelli E, Mazzotti M (2018) Optimal design of multi-energy systems with seasonal storage. Appl Energ 219:408–424

125. Patel MR (2004) Spacecraft power systems. CRC Press

126. Shimizu T, Underwood C (2013) Super-capacitor energy storage for micro-satellites: feasibility and potential mission applications. Acta Astronaut 85:138–154

127. Shawyer R (2015) Second generation EmDrive propulsion applied to SSTO launcher and interstellar probe. Acta Astronaut 116:166–174

128. Gubser DU (1995) Superconductivity research and development: department of defense perspective. Appl Supercond 3(1–3):157–161

129. Dong L, Xu Q, Lu F, Nie X, He Y, Wang Y, Yan Z (2017) Simulation and experimental investigation of a high-temperature superconducting inductive pulsed power supply with time delay effect of the secondary side. Physica C: Supercond Appl 541:16–21

130. Jin JX (2007) HTS energy storage techniques for use in distributed generation systems. Physica C: Supercond Appl 460–462(Part 2):1449–1450

131. Yang J, Zhang L, Wang X, Chen L, Chen Y (2016) The impact of SFCL and SMES integration on the distance relay. Physica C: Supercond Appl 530:151–159

132. Ali MH, Wu B, Dougal RA (2010) An overview of SMES applications in power and energy systems. IEEE Trans Sustain Energy 1(1):38–47

133. Li Q, Yang H, Han Y, Li M, Chen W (2016) A state machine strategy based on droop control for an energy management system of PEMFC—battery—supercapacitor hybrid tramway. Int J Hydrogen Energy 41:16148–16159

134. Bendjedia B, Rizoug N, Boukhnifer M, Bouchafaa F (2017) Hybrid fuel cell/battery source sizing and energy management for automotive applications. IFAC PapersOnLine 50(1):4745–4750

135. Santucci A, Sorniotti A, Lekakou C (2014) Power split strategies for hybrid energy storage systems for vehicular applications. J Power Sour 258:395–407

136. Turpin C, Morin B, Bru E, Rallieres O, Roboam X, Sareni B, Arregui GA, Roux N (2017) Power for aircraft emergencies—a hybrid proton-exchange membrane H2/O2 fuel cell and ultracapacitor system. IEEE Electrif Mag 5(4):72–85. https://doi.org/10.1109/MELE.2017.2758879

137. Gokce K, Ozdemir A (2016) A rule based power split strategy for battery/ultracapacitor energy storage systems in hybrid electric vehicles. Int J Electrochem Sci 11:1228–1246

138. Wu J, Wang X, Li L, Qin C, Du Y (2018) Hierarchical control strategy with battery aging consideration for hybrid electric vehicle regenerative braking control. Energy 145:301–312

139. Chong LW, Wong YW, Rajkumar RK, Isa D (2016) An optimal control strategy for standalone PV system with battery—supercapacitor hybrid energy storage system. J Power Sour 331:553–565

Chapter 4
Global Extremum Seeking Algorithms

4.1 Introduction

The Extremum Seeking Control (ESC) algorithm is an adaptive search technique for extremes of a nonlinear function [1]. Multimodal function is associated with an optimization problem that is generally resolved in an analogous or discrete manner [2].

Two general classes can be identified for the ESC schemes [1, 2]: (1) ESC-based analog optimization (using a dither as perturbation [2], sliding techniques [3], feedback based on the gradient [4] or the system's model [5], and others ESC variants [1, 2]); (2) ESC-based numerical optimization [1, 6]. Perturbation-based ESC (PESC) is the most known and applied ESC technique to optimize the processes and systems [1, 7].

The area of application of the ESC techniques is wide [8]. Some application domains can be mentioned as follows: localization and tracking of a signal source [9, 10]), optimization of the complex processes in vehicles (setting of the throttle angle [11], the breaking force [12], the optimal fueling [13], etc.), in industry [14], in research applications [15, 16], in energy sector based on renewable energy sources and Fuel Cell (FC) systems (minimizing the pollutant emission [17], optimizing the operation of the air compressor [18], tracking the maximum efficiency point (MEA) of the FC systems [19] and the maximum power point (MPP) of the renewable energy sources such as photovoltaic (PV) systems [20–22] and wind turbine (WT) farms [23], optimal operation of the FC/renewable hybrid power systems (HPS) [24], etc.), and other specific applications [1, 2].

The optimization techniques have usually been borrowed to optimally solve the problem of global MPP tracking (GMPPT) in PV hybrid systems due to changes in PV power depending on the irradiance and temperature under different weather and environmental conditions [25]. Thus, the PV power characteristics of a PV array (obtained by serial and/or parallel connection of PV panels) are dynamic multimodal functions depending on these parameters, having a unique maximum (the GMPP)

N. Bizon, *Optimization of the Fuel Cell Renewable Hybrid Power Systems*, Green Energy and Technology, https://doi.org/10.1007/978-3-030-40241-9_4

and more local MPPs (LMPPs) [26]. The changes in PV power due to temperature are much slower than those due to irradiance variation [27]. Consequently, the PV patterns obtained using variable irradiance profiles will be considered in simulations presented in this chapter [28, 29].

The GMPPT algorithm must always track the global maximum of the PV pattern under partial shading conditions (PSCs) [30]. This optimization objective will allow the target achievement in 2030 for the PV power production in the total power generated from renewable sources [31] because the PV power generated increases with up to 45% if the GMPP will be tracked instead of a LMPP [26]. Furthermore, the MPPT algorithms lose the LMPP if the shading coefficient is over to 30% [32].

Anyway, the MPPT algorithms are based on Perturb and Observe (P&O) [33, 34], Incremental Conductance (IC) [35, 36], and Hill climbing (HC) [37, 38] are still implemented in commercial solutions due to their simplicity. Improved MPPT algorithms were proposed using the ripple correlation control (RCC) [38–40], sweeping control [41] or minimization techniques [42, 43] of the PV current or the PV voltage, fractional method based on the short-circuit current [44] or open-circuit voltage [45], feedback control based on the derived PV power [46], model predictive control [47], and other techniques [48–52].

For the first time, the performance of the ESC algorithms is compared with other MPPT algorithms in [23]. Meanwhile, GMPPT algorithms have been proposed based on soft computing techniques and advanced ESC schemes [26–30], which will be presented in this chapter. The firmware-based GMPPT algorithms track the GMPP in two steps: In the first step, the GMPP is located and then it is accurately found. The PESC-based GMPPT algorithms can track the GMPP in less than 10 dither periods, which means a small tracking time for a high frequency of the sinusoidal dither (better than the tracking time of the best firmware-based GMPPT algorithm). In addition, the firmware-based GMPPT algorithms use complex soft computing techniques, such as the chaotic [53] and intelligent search based on fuzzy logic controller (FLC) [54, 55], Artificial neural network (ANN) [56], and Evolutionary Algorithms (EAs) [57–60], which are difficult to be implement in cheap microcontrollers [61].

Consequently, the objectives of this chapter are as follows: (1) to analyze the asymptotical Perturbed-based Extremum Seeking Control (aPESC) schemes; (2) to compare aPESC schemes from topological point of view; (3) to design the aPESC schemes for stability; (4) to evaluate the specific performance indicators of the hybrid power systems using aPESC schemes to optimize its operation.

So, the chapter is structured as follows. Section 4.2 analyzes the asymptotical Perturbed-based Extremum Seeking Control schemes (such as derivative-based PESC scheme and global PESC scheme based on two band-pass filters (GaPESCbpf)) for global search. Stability analysis of the GaPESCbpf scheme is performed in Sect. 4.3. Comparative analysis of the schemes GaPESC and GaPESCd (including their models) is performed in Sect. 4.4. The search gradient estimation in case of the 1-dimension (1D) or 2-dimension (2D) pattern is presented in Sect. 4.5. Designing the GaPESC scheme in case of a nonlinear system with a fast or slow dynamic part is presented in Sect. 4.6. Performance indicators are mentioned in Sect. 4.7 and estimated in different case studies of Sect. 4.8. Modeling for local search based on same multimodal patterns is presented in Sect. 4.9 and the last section concludes the chapter.

4.2 Perturbed-Based Extremum Seeking Control

This section will present the Perturbed-based Extremum Seeking Control (PESC) applied to renewable (PV/WT) Fuel Cell hybrid power systems. The PESC algorithms have started to be used in many other industrial applications since 1922, becoming very known in the last decades. The modeling, stability analysis, and proposal of improved variants for multi-input multi-output (MIMO) nonlinear systems have offered new control solutions for processes optimization [14, 62]:

$$\frac{dx}{dt} = f(x(t), u(t)), \quad y = h(x(t)) \tag{4.1}$$

where $x \in R^n$, $u \in R^m$ and $y \in R$ are the states' variables, inputs and output of MIMO system. The functions $f(x, u)$ and $h(x)$ represent the dynamic and static parts of the MIMO system. In general, the ESC algorithms will find an extreme of a function, but not always the GMPP if the function is of multimodal type (being necessary some assumptions to be verified) [62]. The control law $u(t) = g(x(t), p)$ must be carefully designed to ensure stability of the system to be optimized. The search variable p will converge to the global equilibrium $x_e(p)$, where

$$f(x, g(x, p)) = 0 \Leftrightarrow x = x_e(p), \quad x_e: R^l \rightarrow R^n \tag{4.2}$$

and

$$y = h(x) = h(x_e(p)) = h(p) \tag{4.3}$$

The derivative of (4.3) gives the gradient in time:

$$\frac{dy}{dt} = \frac{\partial h}{\partial p} \cdot \frac{dp}{dt} \tag{4.4}$$

$$\frac{\partial h}{\partial p} = \left[\frac{\partial h}{\partial p_1}, \ldots, \frac{\partial h}{\partial p_l}\right]^T, \frac{dp}{dt} = \left[\frac{dp_1}{dt}, \ldots, \frac{dp_l}{dt}\right]^T \tag{4.5}$$

Thus, the output y will converge to the GMPP defined by (4.3) using the search input p. So, the ESC algorithm is useful in optimization of multimodal function h.

If the system is single-input single-output (SISO), then an ESC of SISO type is necessary. Also, if the system is double-input single-output (DISO), then an ESC of single-input double-output (SIDO) type must be used. These types of ESC algorithms

Fig. 4.1 Scalar PESC
scheme [28]

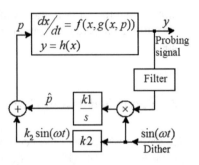

will be analyzed in this chapter for easily understanding the new ESC algorithms proposed in the literature. But it is obvious that MIMO ESC algorithms will be used for MIMO systems.

The scalar PESC (sPESC) scheme (see Fig. 4.1) will be the reference for other PESC algorithms.

The performance of the conventional high-pass filter (HPF) /Low-Pass Filter (LPF) PESC algorithms may be improved if a compensator or fractional-order filter is added in the search loop [52, 63]. The gradient (4.4) results by integration of the signal after demodulation (the product between the filtered signal and the dither; see Fig. 4.1). The $\hat{p}$ signal is the search signal that contains the gradient and dither's harmonics. The harmonics' magnitude is related to the frequency band used for the filters. The dither persistency (high magnitude of harmonics) can be improved by appropriate design of the filters and proper selection of the dithers frequencies in the SIDO ESC schemes.

The ripple from the system to be optimized can be used in PESC algorithms based on ripple or using ripple correlation method. The ripple and its derivative decrease asymptotically, so the derivative of the output y can be used instead of the output y. In this last case, the derivative operator is used instead of band-pass filter used in Fig. 4.1, resulting the derivative-based PESC (dPESC) algorithm. The dPESC algorithm is simple to be modeled and analyzed in order to propose improved variants for GMPP search. The dPESC scheme is sensitive to high-frequency noise and large variations of the input signal. So, the dPESC scheme will not be used for practical implementations. The PESC scheme with a supplementary loop to locate the GMPP based on the first derivative of the system's output will be analyzed in this chapter.

A different topological idea to implement the localization loop by amplitude modulation of the dither is presented in Fig. 4.2 using a signal that decreases to zero in time using an exponential law. These asymptotic PESC (aPESC) scheme must be specifically designed for each multimodal pattern considering the position of the startup point related to the GMPP in order to set appropriately the initial amplitude of the exponential signal that modulate the amplitude of the dither [28].

Other methods use some information to adjust the initial amplitude or decay rate based on irradiation's variation extracted from the measured PV power or using

Fig. 4.2 aPESC scheme
based on ICDA block [28]

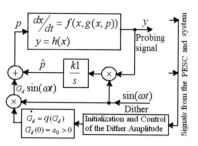

appropriate sensors. For example, the Initialization and Control of the Dither Amplitude (ICDA) sensor-less techniques presented in Fig. 4.2 belongs to this class of aPESC methods. Another proposed aPESC class uses a variable-weather-parameter (VWP) based on signals from the search loop to adjust the dither amplitude during the locating phase. A switching threshold is used to pass from the VWP-based control to the basic ESC. The value of the switching threshold is a difficult task because it depends by many parameters (such as the size and topology of the PV array, irradiance profile, and PV panel temperature).

Finally, the initial amplitude of the dither can be modified using the first harmonic (H_1) of the probing signal from the system under test [64]. The H_1 magnitude is a VWP-based signal that can be obtained using the fast Fourier transform (FFT), or easier with a BPF.

aPESC schemes use only PV power to search for GMPP, instead of more signals from the PV system. Furthermore, the search using the aPESC schemes convergences asymptotically to MPP. Also, these methods are self-tuning with respect to VWPs, having implemented the GMPPT feature as well.

The variants of the scalar PESC (sPESC) algorithm have been proposed in [65, 66] with or without the LPF or HPF, or varying the cutoff frequencies values of the LPF and HPF, $\omega_l = \beta_1\omega$ and $\omega_h = \beta_h\omega$ (where is ω the frequency of the sinusoidal dither), using $0 < \beta_1 < \beta_{1(max)}$ and $0 < \beta_h < 1$ (where $0 < \beta_{1(max)} < 1$ and $3 < \beta_{1(max)} < 6$ for sPESC and modified PESC (mPESC) proposed in [2] and [67], respectively). The mPESC algorithm has better performance compared to the sPESC algorithm due to the increase of the dither persistence based on more harmonics in the search loop. Note that the LPF filters the demodulated signal, but this can be moved before the demodulation circuit without significant loss in dither persistence if the same cutoff frequency is used [27]. In the case of the HPF and the LPF in series configuration, a band-pass filter (BPF) will result with cutoff frequencies $\omega_l = \beta_1\omega$ and $\omega_h = \beta_h\omega$, where $3 < \beta_1 < \beta_{1(max)} < 6$ and $0 < \beta_h < 1$, and the algorithm was called aPESC bpf [68]. The search speed may increase about three times for the sPESCbpf scheme compared to the sPESC scheme. To further improve the performance of the aforementioned PESC algorithms, a compensator is recommended to be used in the search loop [2, 69]. So, the PESC schemes have better performance (about 99.98% tracking accuracy and more the 3 kW/s search speed) compared to other classical MPPT algorithms such as P&O algorithms [70]. It is worth mentioning that the digital implementation is more difficult than the analog implementation, but they are now comparable due to cheap but powerful processors that are commercially available.

As it was mentioned before, the aPESC with GMPP feature implemented needs a supplementary loop (called localization loop) to find the position of the GMPP on the multimodal pattern. The tuning of the new gain of the localization loop will be discussed in this chapter as well. Note that this loop must to generate a signal that approximates the first harmonic (H_1) of the probing signal (the output of the system to be optimized, for example, the power generated by the PV, WT, or FC system). This signal processing can be performed using the same BPF used in the searching loop (called GaPESC algorithm [29]), a separate BPF centered on first harmonic frequency, which is the frequency of the sinusoidal dither (called GaPESCpbf scheme [30]), a separate LPF to filter the first harmonic frequency (which is a variant of the GaPESCpbf scheme) [28], the Fast Fourier Transform (FFT) to obtain the first harmonic H_1 (called GaPESCH1 scheme [70]), or the derivative operator in the derivative-based PESC scheme designed for global searching (called GaPESCd scheme [29]).

The GaPESCd scheme is used in the next section as a theoretical solution to demonstrate the global feature of the aforementioned algorithms, because the derivative operator in signal processing must be avoided due to its susceptibility to noise interference.

4.2.1 PESC Algorithms Designed for Global Search

The design of the GaPESC schemes for global search must use some observations as follows:

O1: The PESC gradient is (4.6):

$$\bar{g} = \frac{d\bar{p}_1}{dt} = \frac{a^2 \cdot k_1}{2} \cdot \frac{dy}{dx} \qquad (4.6)$$

where a is the initial amplitude of the dither amplitude (e.g., $a = 1$ in Fig. 4.1).

O2: The PV power derivative is zero at extremes.

O3: The gain k_1 must be positive or negative for search LMPPs and local minimum power points (LmPPs).

So, this gain must be modified to gain k_{1m} designed to scan all LMPPs and LmPPs (see Fig. 4.3). Firstly, the gain k_{1m} must be positive ($k_{1m} > 0$) to search for GMPP using, for example, the startup point $p(0) = 0$. Thus, the LMPP1 will first be located using a positive gradient $\left(\frac{dy}{dx} > 0\right)$. The search will remain blocked on this LMPP1 if the gain k_{1m} will not become negative ($k_{1m} < 0$) during the locating phase of the LMPP1. If $k_{1m} < 0$, then the LmPP1 will be located and the negative gradient $\left(\frac{dy}{dx} < 0\right)$ will result during this phase and so on. So, the gain k_{1m} must be designed using (4.7) in order to scan all extremes:

Fig. 4.3 PV pattern [28]

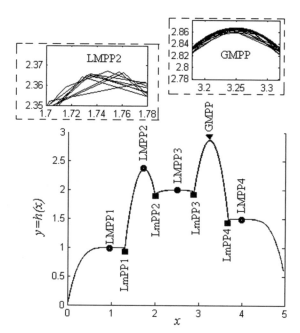

$$\text{sign}(k_{1m}) = \text{sign}\left(\frac{dy}{dx}\right) \Rightarrow k_{1m} = \text{sign}\left(\frac{dy}{dx}\right) \cdot k_1, \, k_1 > 0 \qquad (4.7)$$

At one moment, the GMPP will be located as well, but based on (4.7) the scan of the next extremes will be performed until the search must be back based on the second loop of GMPP localization (see Fig. 4.4), as it will be explained below.

A similar switching rule with (4.7) is given by (4.8):

$$k_{1m} = \frac{dy}{dx} \cdot k_1 \qquad (4.8)$$

Both (4.7) or (4.8) can be used to modulate the dither. The last is implemented for the derivative-based PESC (GaPESCd) scheme presented in Fig. 4.5.

4.2.2 Derivative-Based PESC Scheme for Global Search

Thus, the localization signal $\hat{p}_2$ will be designed using (4.9):

$$\hat{p}_2 = k_2 \left| \frac{dy}{dx} \right| \cdot \sin(\omega t), \, k_2 > 0. \qquad (4.9)$$

Fig. 4.4 Behavior of
searching [28]

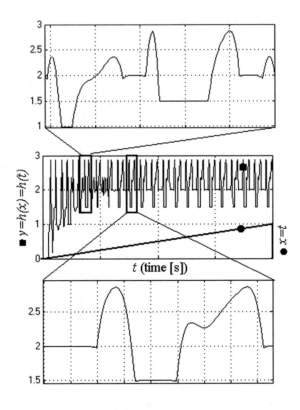

Fig. 4.5 Derivative-based
PESC scheme [28]

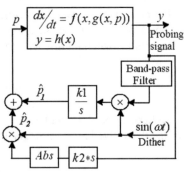

Consequently, the scanning signal for global searching (using the GaPESCd
scheme) is:

$$\hat{p} = A_1 \sin(\omega t) + k_2 \left| \frac{dy}{dx} \right| \cdot \sin(\omega t) \tag{4.10}$$

where A_1 is the magnitude of the first harmonic.

Considering the BPF ideal, the gradient g_S is obtained by averaging the signal $\hat{p}_1$:

$$\dot{\bar{p}}_1 = g_S = \frac{A_1 \cdot k_1}{2} \cdot \frac{dy}{dx} + \frac{k_2 \cdot k_1}{2} \cdot \frac{dy}{dx}\left|\frac{dy}{dx}\right| \qquad (4.11)$$

Thus:

$$\hat{p}_1 = \bar{p}_1 + p_{1(AC)} \Rightarrow p = g_S \cdot t + p_{1(AC)} + p_2 = g_S \cdot t + p_{(AC)} \qquad (4.12)$$

where $p_{(AC)}$ represents the harmonic.

The g_S gradient (4.11) is presented in Fig. 4.6 using the multimodal pattern (4.13):

$$y = \left|1 - (x-1)^4\right| + \left|2 - (x-2.5)^4\right| + \left|1.5 - (x-4)^4\right| \qquad (4.13)$$

The $\hat{p}_2$ amplitude decreases to zero after the GMPP has been located (see Fig. 4.7). Consequently, the schemes GaPESC, GaPESCbpf, and GaPESCH1 can track the GMPP with very good accuracy. In this chapter, the schemes GaPESC, GaPESCbpf, and GaPESCH1 are derived from the GaPESCd scheme (see Fig. 4.5) by using other signal processing techniques to obtain the derivative of the signal probing. For example, see the GaPESCbpf scheme in Fig. 4.8, where the probing signal

Fig. 4.6 g_S gradient for A_1 $= 4$, $k_1 = 1$ and $k_2 = 3$ [28]

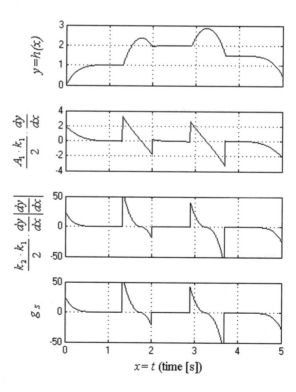

Fig. 4.7 Global extremum searching [28]

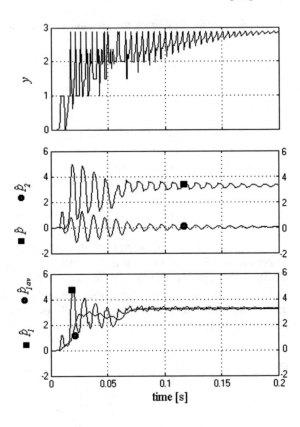

Fig. 4.8 Global PESC scheme based on two BPFs [28]

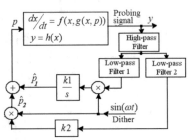

filtered by the BPF2 is a smooth approximation of the absolute value (*Abs*) for the gained derivative of the probing signal. The design and stability analysis of the aforementioned schemes will be performed in this chapter. Also, the validation of the results will be made using an experimental setup. The performance indicators used to compare the schemes GaPESC, GaPESCbpf, and GaPESCH1 are as follows: tracking accuracy, tracking speed, searching resolution, and hit count percent (the percentage to find GMMP in total experiments).

4.2.3 Global PESC Scheme Based on Two Band-Pass Filters

The GaPESCbpf scheme (with two BPFs sharing the same HPF) is presented in Fig. 4.8 and its simulation diagram in Fig. 4.9. The signal p_1 is obtained as in conventional ESC schemes and will be used to accurately track the GMPP in the search loop. The mean value (MV) block is used to smoothen the signal that will modulate the dither used in the localization loop (the signal p_2). The signal p_2 will locate the GMPP. After GMPP localization, its amplitude decays asymptotically to zero. If this is the case (e.g., in case of static multimodal patterns), a dither with small amplitude (A_m) may be used to start the search (the signal p_3). The searching signal p is the sum of aforementioned signals.

The simulation parameters used in the next section to analyze the performance of the GaPESCbpf scheme using 10 Hz sinusoidal dither with unitary amplitude are as follows: $\beta_h = 0.5$ for the HPF, $\beta_{11} = 3.5$ for the LPF1, $\beta_{12} = 1.5$ for the LPF2, and $A_m = 0.001$.

The normalization gain of the input y can be approximately chosen using $k_N \cong 2/y_{max}$, where y_{max} is a maximum value of the multimodal pattern in $h(p)$. For example, k_N may be in range 1–2 for the pattern (4.18). The dither gain in the localization loop can be chosen approximately using $k_2 \cong \Delta p/2$, where $\Delta p = |p_{GMPP} - p_0|$, and $p_0 = p(0)$ is the startup value in searching of the p_{GMPP}. For example, k_2 may be in range 1.5–2 for the pattern (4.18) if $p_0 = 0$. The gain in the search loop, k_1, will be designed using the methodology presented in the next sections. The k_1 gain has been chosen proportional with the dither frequency ($k_1 = \gamma_{sd} \cdot \omega$) in order to improve the dither persistency in the search loop. It is worth mentioning that choosing of the aforementioned parameters is non-critical, but the design of the k_1 gain must ensure the stability of the search loop (as will be shown in the next section).

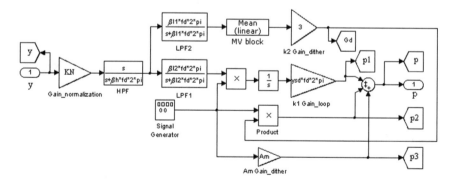

Fig. 4.9 Diagram of the GaPESCbpf scheme [28]

4.3 Stability Analysis of the GaPESCbpf Scheme

The stability analysis of the GaPESCbpf scheme has been performed using the averaging stability analysis method that is recommended for periodic systems [2, 71]. The simulation diagram and the diagram of the averaged signal for the GaPESCbpf schema are presented in Figs. 4.10 and 4.11, respectively. The search signal $\hat{p}_1$ will converge to the GMPP value in less than 10 dithers' periods (see Fig. 4.7). The localization signal $\hat{p}_2$ will allow passing through local extremes to the GMPP.

Any nonlinear map $y = h(p)$ can be approximated close to the GMPP as follows:

$$y \cong h(p_{\text{GMPP}}) + (p - p_{\text{GMPP}}) \cdot \left.\frac{\mathrm{d}h}{\mathrm{d}p}\right|_{p_{\text{GMPP}}} + (p - p_{\text{GMPP}})^2 \cdot \frac{1}{2} \cdot \left.\frac{\mathrm{d}^2h}{\mathrm{d}p^2}\right|_{p_{\text{GMPP}}}$$

$$\cong h(p_{\text{GMPP}}) + (p - p_{\text{GMPP}})^2 \cdot \frac{1}{2} \cdot \left.\frac{\mathrm{d}^2h}{\mathrm{d}p^2}\right|_{p_{\text{GMPP}}} \tag{4.14}$$

If $\left.\frac{\mathrm{d}^2h}{\mathrm{d}p^2}\right|_{p_{\text{GMPP}}} = 2k_3 < 0$ and $\tilde{p} = p - p_{\text{GMPP}}$, then (4.14) become (4.15):

$$\tilde{y} \cong k_3 \cdot \tilde{p}^2 \tag{4.15}$$

where the parameter $|k_3|$ may has values in range of 0.5–5 for PV real patterns under PSC [71].

Fig. 4.10 Simulation diagram of the GaPESCbpf scheme [28]

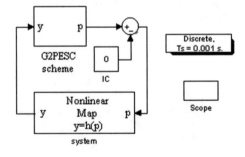

Fig. 4.11 Diagram of the averaged signal for the GaPESCbpf schema [28]

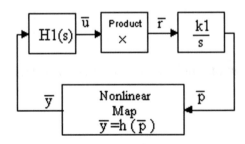

PESC scheme	$H_1(s)$	$H_2(s)=G_d(s)/k_2$	$G_L(s)$
Scalar PESC	$H_{HPF}(s)\cdot H_{LPF}(s)$	1	$1+ k_L\cdot H_{HPF}(s)\cdot H_{LPF}(s)/s$
dPESC	$H_{HPF}(s)\cdot H_{LPF}(s)$	s	$1+ k_L\cdot H_{HPF}(s)\cdot H_{LPF}(s)$
GaPESCbpf	$H_{HPF}(s)\cdot H_{LPF1}(s)$	$H_{HPF}(s)\cdot H_{LPF2}(s)$	$1+ k_L\cdot H^2_{HPF}(s)\cdot H_{LPF1}(s)\cdot H_{LPF2}(s)/s$

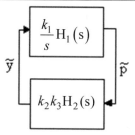

Fig. 4.12 Transfer functions involved in the search loop of the GaPESCbpf scheme [28]

Consequently, the diagram shown in Fig. 4.12 involves the following transfer functions:

$$\frac{k_1}{s}H_1(s) \text{ and } k_3 \cdot G_d(s) = k_2 \cdot k_3 \cdot H_2(s) \tag{4.16}$$

So, the transfer function of the closed loop is given by (4.17):

$$G_L(s) = 1 - \frac{k_1}{s} \cdot H_1(s) \cdot k_2 \cdot k_3 \cdot H_2(s) = 1 + \frac{k_L}{s} \cdot H_1(s) \cdot H_2(s) \tag{4.17}$$

where $k_L = -k_1 \cdot k_2 \cdot k_3 > 0$.

The stability analysis and the results obtained for the schemes sPESC, GaPESCd, and GaPESCbpf are shown in Figs. 4.13, 4.14 and 4.15 as follows: root locus in plot (a), Bode plots in plot (b), and step responses in plot (c) of the aforementioned figures.

The sPESC scheme has a stable search loop (see the plots a and b in Fig. 4.13), but this was already known [1, 2]. The step responses for different values of the k_L gain are represented in plot c in Fig. 4.13. It is worth mentioning that the search speed increases if the k_L gain will increase, but the over-tuning must be avoided ($k_L < k_{L(max)} = 40$).

The closed-loop transfer function of the schemes GaPESCd and GaPESCbpf has a zero in origin (see the plot a in Figs. 4.14 and 4.15). Consequently, both schemes are more sensitive to the high harmonics of the probing signal. This has already been mentioned for the GaPESCd scheme. It results that for the GPESCbpf scheme, the LPFs are mandatory and their design must be made considering $\beta_1 < 6$. Also, an appropriately designed compensator may ensure the stability margin for the phase requested by the GPESCbpf scheme (see the plot b in Figs. 4.15).

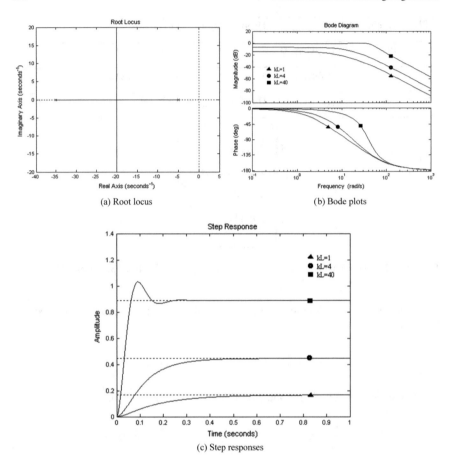

Fig. 4.13 Performance of the sPESC scheme

If the compensator is not included in the search loop, then $k_L < k_{L(max)} = 111$ (see the plot c in Fig. 4.15, where the oscillations appeared for $k_L = 111$.

So, the design methodology is the following: (1) the gain $k_L = k_l \cdot k_2 \cdot k_3$ is selected to ensure stability of the search loop; (2) the gain k_2 is not a critical parameter due to the adaptive feature of the ESs-based search loop, being selected using $k_2 \cong \Delta p/2$, where $\Delta p = |p_{GMPP} - p_0|$ and $p_0 = p(0)$ is the startup value; (3) the k_3 gain is also not a critical parameter, being selected in the range of 0.5–5 for PV real patterns; (4) finally, the gain k_1 of the ESs-based search loop is computed based on (4.18):

$$k_1 = \frac{k_L}{k_2 \cdot k_3} \tag{4.18}$$

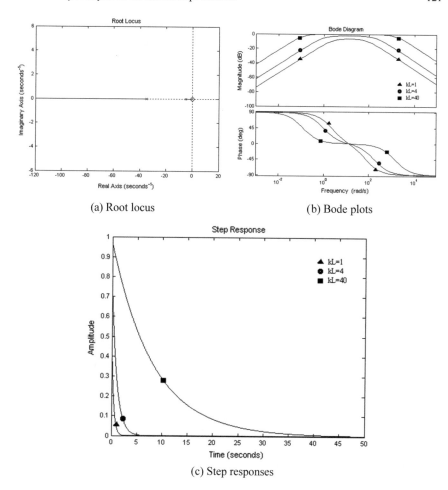

(a) Root locus

(b) Bode plots

(c) Step responses

Fig. 4.14 Performance of the GaPESCd scheme

It is worth mentioning that the performance of the schemes GaPESCbpf with two BPFs sharing the same HPF (see Fig. 4.9) and two different BPFs (see Fig. 4.16) are very similar [30], so they will not be presented in this chapter.

A comparative analysis of the schemes GaPESC and GaPESCd will be briefly presented in the next section.

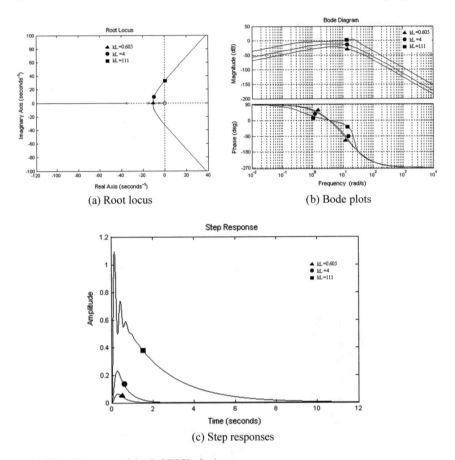

(a) Root locus

(b) Bode plots

(c) Step responses

Fig. 4.15 Performance of the GaPESCbpf scheme

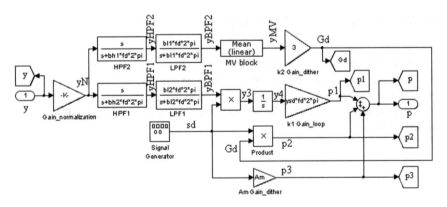

Fig. 4.16 Simulation diagram of the scheme GaPESCbpf with two different BPFs [30]

4.4 Comparative Analysis of the Schemes GaPESC and GaPESCd

The asymptotic Perturbed Extremum Seeking Schemes (aPESC) designed for global searching based on only band-pass filter (called GaPESC scheme) is simpler than the schemes GaPESCbpf (which uses two BPFs) and obviously much simpler than the GaPESCH1 (which is based on the FFT). So, this variant of the aPESC scheme (see Fig. 4.17) will be considered in this chapter for a deep performance analysis, based on different multimodal patterns such as those represented in Figs. 4.17 and 4.18. The operation of locating and searching loops of the GaPESC scheme is principally presented in Fig. 4.17.

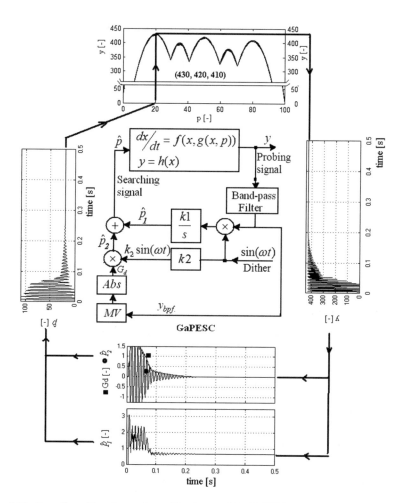

Fig. 4.17 Operation of locating and searching loops of the GaPESC scheme

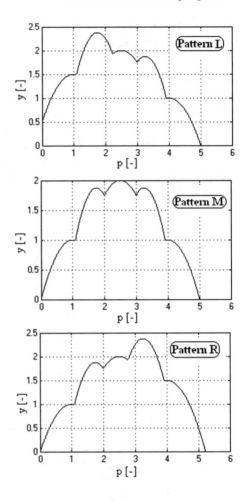

Fig. 4.18 Multimodal patterns with the GMPP positioned in left (pattern L), in middle (pattern M), and in right side (pattern R) [64]

It worth mentioning that the first aPESC scheme with GMPPT feature has used an exponentially decreasing law for the dither's amplitude such as given by (4.19) [62]:

$$G_d = q(G_d), \quad q(t) = -\rho t, \quad G_d(0) = a_0 \Rightarrow G_d = a_0 \exp(-\rho t) \qquad (4.19)$$

where the parameter $\rho > 0$ sets the decay rate from the initial amplitude a_0. The parameters ρ and a_0 can be appropriately set for the most multimodal patterns, but the counter-examples from [72] pointed out that this is not possible for all types of patterns. These patterns will be used in [72] to test the proposed aPESC scheme and to test the performance of the GaPESC scheme.

Although the aPESC technique analyzed in [27] presents high performance on different patterns, this cannot be easily implemented in practice because it needs two identical process units to measure the difference between the responses.

The GaPESC scheme operates using the process's output and generates the search signal that must be normalized to be independent of the process. Some topological changes can be highlighted for the GaPESC scheme (see Fig. 4.19) compared to the GaPESCbpf scheme (see Fig. 4.16): (1) the k_2 gain has been moved after the amplitude modulation of the dither; (2) The normalization gain of the output and input of the process makes the search to be independent of the process by using $k_{Ny} \cong 2/y_{max}$ and $k_{Np} \cong p_{max}/2$, where y_{max} and p_{max} are the maximum values for the output and the input of a specific process or SISO system. It seems that the k_2 gain or the k_{Np} gain is no longer necessary using the same setting values for these gains, and a gain can be set to 1. But note that the k_2 gain amplifies the modulated dither (the localization signal) and the k_{Np} gain amplifies both signals (for search and localization). Also, the k_2 gain may be used for processes with variable structure by changing the value of the k_2 gain based on designed meta-rules. Thus, both are needed and must be appropriately designed.

The simulation diagram of the GaPESC scheme is presented in Fig. 4.20. The GaPESC scheme operates using the localization loop and the search loop, which generates the signals p_2 and p_1.

The p_2 signal is the modulated dither using an asymptotic signal resulted from the process's output by smoothing the output of the BPF via a Mean Value (MV) block. The asymptotic signal is noted G_d in Figs. 4.19 and 4.20. The asymptotic signal in case of GaPESCd scheme (see Fig. 4.8) is obtained by derivation of the output of the process:

Fig. 4.19 Diagram of the GaPESC scheme with normalization gains [64]

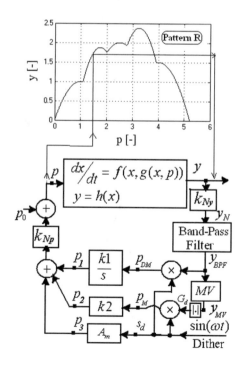

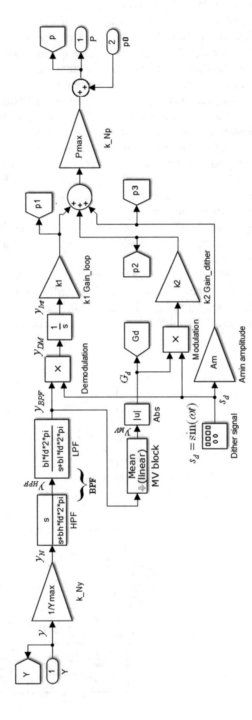

Fig. 4.20 Simulation diagram of the GaPESC scheme [29]

$$G_d = k_2 \cdot \left| \frac{dy}{dp} \right| \qquad (4.20)$$

Both asymptotic signals are represented in Fig. 4.21 and seem to be very similar. The amplitude of the asymptotic signals decreases to about zero in less than 10 periods of the dither, which defines the search time.

The search time is very less dependent on the dither's parameters (such as frequency and shape) in comparison with other parameters (such as tuning values for the gains k_1 and k_2). This aspect will be explored in this chapter as well. To do this, modeling the GaPESC scheme will be presented in the next section.

Fig. 4.21 Asymptotic signals (G_d) for the schemes GaPESC and GaPESCd [64]

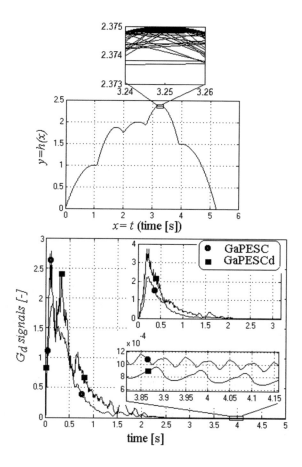

4.4.1 Modeling the GaPESC Scheme

The simulation diagram of the GaPESC scheme presented in Fig. 4.20 has been implemented using (4.21a) [29]:

$$y = h(p), \ y_N = k_{Ny} \cdot y \tag{4.21a}$$

$$\dot{y}_l = -\omega_h \cdot y_f + \omega_h \cdot y_N, \ y_{HPF} = y_N - y_f, \ \dot{y}_{BPF} = -\omega_l \cdot y_{BPF} + \omega_l \cdot y_{HPF} \tag{4.21b}$$

$$p_{DM} = y_{BPF} \cdot s_d, \quad s_d = \sin(\omega t), \tag{4.21c}$$

$$G_d = |y_{MV}|, \ y_{MV} = \frac{1}{T_d} \cdot \int y_{BPF} \, dt \tag{4.21d}$$

$$\dot{p}_{Int} = p_{DM} \tag{4.21e}$$

$$p_1 = k_1 \cdot y_{Int}, \ k_1 = \gamma_d \cdot \omega \tag{4.21f}$$

$$p_2 = k_2 \cdot G_d \cdot s_d \tag{4.21g}$$

$$p_3 = A_m \cdot s_d, \tag{4.21h}$$

$$p = k_{Np} \cdot (p_1 + p_2 + p_3) + p_0. \tag{4.21i}$$

The normalized signal (y) of the process's output (4.21a) is filtered using a BPF (4.21b). The BPF output signal (y_{bpf}) is demodulated through multiplication (4.21c) with the dither (s_d), resulting the demodulated signal (y_{DM}). This signal is integrated (4.21e) and multiplied (4.21f) with the tuning gain (k_1) to obtain the search signal (p_1). The y_{bpf} signal is filtered using the MV techniques (4.21d) and used to modulate the dither's amplitude (4.21g). The tuning gain (k_1) sets the localization signal (p_2). A small dither (p_3) (4.21h) can be added to signals p_1 and p_2 to startup the search from p_0. The normalized signal (p) is applied to the process's input (4.21i). Because the objective is to obtain a good approximation of the first harmonic (H1) without reducing the level of dither persistence in the search loop, the BPF cutoff frequencies ($f_h = \beta_h \cdot f_d$ and $f_l = \beta_l \cdot f_d$) will be set using $\beta_h = 0.5$ and β_l in range of 1.5–5.5. A sensitivity analysis will be performed in this chapter for different values of the dither's frequency and parameter β_l in range of 10–100 Hz and 1.5–5.5, respectively.

The normalization gains (k_{Ny} and k_{Np}) and the tuning gains (k_1 and k_2) are set specifically to a given process. The tuning gain k_1 is set proportional with the dither's frequency (ω) to make the harmonics' amplitude less dependent to the dither's frequency. The design parameter γ_{sd} is set using $\gamma_d = k_1/\omega$ (4.21f). The design, stability analysis, and case studies using the GaPESC scheme will be presented in the next sections of this chapter.

4.4.2 Stability Model of the GaPESC Scheme

The diagrams of the GaPESC scheme for the closed loop, medium signal, and local loop are presented in Fig. 4.22. The stability model of the GaPESC scheme will be obtained in the same manner for GaPESCbpf scheme using a quadratic approximation of the pattern $y = h(p)$:

$$\tilde{y} \cong k_3 \cdot \tilde{p}^2 \tag{4.22}$$

where $\tilde{p} = p - p_{\text{GMPP}}$ and $|k_3|$ estimate the curvature of the pattern $y = h(p)$ close to the GMPP.

Consequently, the diagram shown in Fig. 4.22 involves the following transfer functions:

$$\frac{k_1}{s} H_{\text{BPF}}(s) \text{ and } k_3 \cdot G_d(s) = k_2 \cdot k_3 \cdot H_{\text{BPF}}(s) \tag{4.23}$$

Fig. 4.22 Diagrams of the GaPESC scheme for the closed loop, medium signal, and local loop [29]

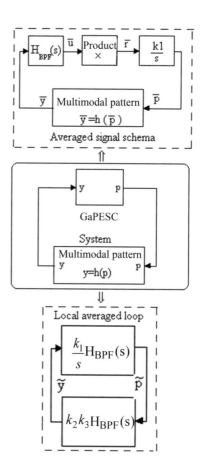

So, the transfer function of the closed loop is given by (4.24):

$$G_L(s) = 1 - k_{\text{Ny}} \cdot k_{\text{Np}} \cdot H_p(s) \cdot (H_{\text{BPF}}(s) \cdot k_1/s) \cdot (k_2 \cdot k_3 \cdot H_{\text{BPF}}(s))$$

$$= 1 + k_L \cdot \frac{H_p(s) \cdot H_{\text{BPF}}(s) \cdot H_{\text{BPF}}(s)}{s} = 1 + k_L \cdot H_d(s) \qquad (4.24)$$

where $H_d(s) := H_p(s) \cdot H_{\text{BPF}}(s) \cdot H_{\text{BPF}}(s)/s$ and $k_L = -k_{\text{Ny}} \cdot k_{\text{Np}} \cdot k_1 \cdot k_2 \cdot k_3 > 0$.

This model will be used below to analyze the performance of the GaPESC scheme applied to PV, WT, and FC systems.

4.5 The Search Gradient Estimation

4.5.1 Case of the 1-Dimension (1D) Pattern

The search gradient estimation in case of 1D pattern will consider the assumption of an ideal BPF with $\beta_h = 0.5$ and $1.5 < \beta_1 < 5.5$, only the first harmonic ($G_d \cong a_{1m}$) for the localization signal (4.25):

$$p_2 = k_2 a_{1m} \cdot \sin(\omega t) \qquad (4.25)$$

and only the first three harmonics of the original p signal (4.26):

$$p_{\text{init}} = \sum_{i=1}^{5} a_i \sin(i\omega t) + k_2 a_{1m} \cdot \sin(\omega t) \qquad (4.26)$$

The 1D pattern $\tilde{y} = y - h(p_{GMPP})$ is linearized using (4.27):

$$\tilde{y} \cong \frac{dy}{dx} \cdot \tilde{x} = \dot{h} \cdot \tilde{x} \qquad (4.27)$$

The processing of the output signal (4.28):

$$y = \dot{h} \cdot p_{\text{init}} \qquad (4.28)$$

will use (4.21a) as follows [29].

The demodulated (4.29):

$$p_{\text{DM}}(t) = \dot{h} \left[\sum_{i=1}^{5} a_i \sin(i\omega t) + k_2 a_{1m} \cdot \sin(\omega t) \right] \cdot \sin(\omega t) \qquad (4.29)$$

The search signal (4.30):

$$p(t) = g_S \cdot t + \sum_{i=1}^{5} b_i \cdot \sin(i\omega t + \phi_i) \qquad (4.30)$$

where the search gradient is given by (4.31):

$$g_S = \dot{h} \cdot \frac{k_1 a_1}{2} + \dot{h} \cdot \frac{k_1 k_2 a_{1m}}{2} \qquad (4.31)$$

It is worth mentioning that the second component $\dot{h} k_1 k_2 a_{1m}/2$ is added to the first component $\dot{h} k_1 a_1/2$ that is obtained with a sPESC scheme [2].

4.5.2 Case of the 2-Dimensions (2D) Pattern

The single-input dual-output (SIDO) ESC scheme will be analyzed in this section using a 2D pattern $y = f(x_1, x_2)$. The SIDO ESC scheme (see Fig. 4.23) will generate two search signals, which will be applied to the inputs (x_1 and x_2) in order to search the GMPP on the pattern's surface. The sinusoidal dithers can be orthogonal signals of same frequency [20] or two sinusoidal signals with different frequencies [73].

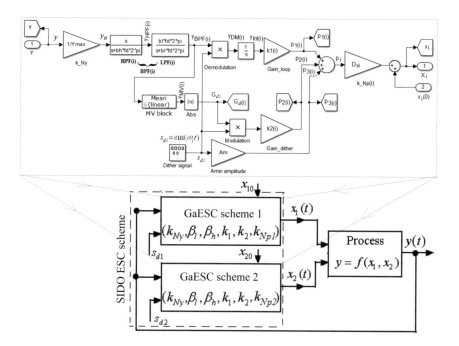

Fig. 4.23 SIDO ESC scheme

The 2D pattern $y = f(x_1, x_2)$ will be normalized using (4.32) and then approximated in point (x_{10}, x_{20}) by the Taylor series (4.33):

$$y_N = f_N(x_1, x_2) = k_{Ny} \cdot y = k_{Ny} \cdot f(x_1, x_2) \tag{4.32}$$

$$f_N(x_1, x_2) = \sum_{i_2=0}^{\infty} \sum_{i_1=0}^{\infty} \frac{(x_1 - x_{10})^{i_1}(x_2 - x_{20})^{i_2}}{i_1! i_2!} \cdot \frac{\partial^{i_1+i_2} f_N}{\partial x_1^{i_1} \partial x_2^{i_2}}(x_{10}, x_{20}) \tag{4.33}$$

The gradients (G_1 and G_2, $i = 1, 2$) are given by (4.34):

$$G_i = \frac{\partial f_N}{\partial x_i}(x_{10}, x_{20}) = \frac{d f_N}{dt} \Big/ \frac{dx_i}{dt}, \quad i = 1, 2 \tag{4.34}$$

4.5.2.1 Orthogonal Sinusoidal Dithers

The orthogonal signals $s_{d1} = \sin(\omega t)$ and $s_{d2} = \cos(\omega t)$ are used in Fig. 4.23 [20].
   The assumptions to evaluate the gradients are as follows:

– Only three components ($h = 1, 2, 3$) of the Taylor series will be considered in signal processing based on (4.21a);
– the BPF is ideal;
– $k_{Ny} = k_{Nx1} = k_{Nx2} = 1$.

The signals x_{iBPF} are given by (4.34) for $i = 1, 2$ (where the superscript "i" will be used for the relationships of the GaESC scheme using the signals $s_{d1} = \sin(\omega t)$ and $s_{d2} = \cos(\omega t)$, respectively):

$$x_{iBPF}(t) \cong D_1 \cdot x_{iLF}(t) + \frac{1}{2} D_2 \cdot x_{iLF}^2(t) + \frac{1}{6} D_3 \cdot x_{iLF}^3(t) \tag{4.34}$$

where

$$D_1 = \frac{\partial f_N}{\partial x_1}(x_{10}, x_{20}) + \frac{\partial f_N}{\partial x_2}(x_{10}, x_{20})$$

$$D_2 = \frac{\partial f_N^2}{\partial x_1^2}(x_{10}, x_{20}) + 2\frac{\partial f_N^2}{\partial x_1 \partial x_2}(x_{10}, x_{20}) + \frac{\partial f_N^2}{\partial x_2^2}(x_{10}, x_{20})$$

$$D_3 = \frac{\partial f_N^3}{\partial x_1^3}(x_{10}, x_{20}) + 3\frac{\partial f_N^3}{\partial x_1^2 \partial x_2}(x_{10}, x_{20}) + 3\frac{\partial f_N^3}{\partial x_1 \partial x_2^2}(x_{10}, x_{20}) + \frac{\partial f_N^3}{\partial x_2^3}(x_{10}, x_{20})$$

$$\tag{4.35}$$

$$x_{1LF}(t) = \sum_{h=1}^{3} a_{1h} \sin(j\omega t + \varphi_{1h}), \quad x_{2LF}(t) = \sum_{h=1}^{3} b_{1h} \sin(j\omega t + \psi_{1h}) \tag{4.36}$$

Only three low frequency (LF) harmonics have been considered in (4.36). The signals x_{iDM} are given by (4.37):

$$x_{iDM}(t) = x_{iBPF}(t) \cdot \sin[\omega t + (i-1)\pi/2], \ i = 1, 2 \qquad (4.37)$$

The form (4.38) is obtained by signal processing of (4.37):

$$x_{iDM}(t) \cong k_{sgi} + x_{iDM(LF)}(t), \ p = 1, 2 \qquad (4.38)$$

where:

$$k_{sg1} = \frac{1}{2} D_1 a_1 R_{SS1} \cos \varphi_1$$

$$k_{sg2} = \frac{1}{2} D_1 b_1 R_{SS2} \sin \psi_1$$

$$R_{SS1} \cong 1 + \frac{1}{8} \frac{D_3}{D_1} \left(a_1^2 + 2a_2^2 + 2a_3^2 \right)$$

$$R_{SS2} \cong 1 + \frac{1}{8} \frac{D_3}{D_1} \left(b_1^2 + 2b_2^2 + 2b_3^2 \right) \qquad (4.39)$$

The signals x_{iDM}, $i = 1, 2$, is integrated and multiplied with gain k_1 in order to obtain the search signals (4.40):

$$x_i(t) \cong k_1 k_{sgi} \cdot t + k_2 H_{1i} \sin[\omega t + (i-1)\pi/2] + x_{pLF}(t), \ i = 1, 2 \qquad (4.40)$$

where only the first harmonic is shown (H_{1i} is the first harmonic of the x_{iBPF} signal, $i = 1, 2$), the rest of harmonics being included in the signal x_{iLF}, $i = 1, 2$.

So, the gradients can be estimated by

$$G_i = k_{sgi} \cdot k_1, \ i = 1, 2 \qquad (4.41)$$

The values of the gradients given by (4.41) have been validated through simulation [20]. The R_{ssi} values, $i = 1, 2$, are the correction added by the GaPESC scheme in comparison with the sPESC scheme [2].

4.5.2.2 Sinusoidal Dithers with Different Frequencies

The signals $s_{d1} = \sin(\omega_1 t)$ and $s_{d2} = \sin(\omega_2 t)$ are used in Fig. 4.23 [73].

The dither persistency will be improved if $\omega_2 = 2\omega_1 = 2\omega$, $k_{1(1)} = \gamma_1 \omega$, and $k_{1(2)} = \gamma_2 2\omega$.

Based on same aforementioned assumptions and signal processing (4.21a), the search signal $p_{1(1)}$ will be given by (4.41):

$$p_{1(1)} \cong \frac{t}{2}\gamma_1 \omega a_{11} \frac{\partial f_N}{\partial x_1}(x_{10}, x_{20}) + \gamma_1 \sum_{j=1}^{6} b_{1j} \sin(j\omega t) \tag{4.41}$$

So, the gradient is given by (4.42):

$$G_1 = \frac{1}{2}\gamma_1 \omega a_{11} \frac{\partial f_N}{\partial x_1}(x_{10}, x_{20}) \tag{4.42}$$

where the coefficients b_{1j} are as follows:

$$2b_{11} = -a_{12}\frac{\partial f_N}{\partial x_1}(x_{10}, x_{20}) - a_{21}\frac{\partial f_N}{\partial x_2}(x_{10}, x_{20})$$

$$4b_{12} = a_{11}\frac{\partial f_N}{\partial x_1}(x_{10}, x_{20})$$

$$6b_{13} = a_{11}\frac{\partial f_N}{\partial x_1}(x_{10}, x_{20}) + a_{21}\frac{\partial f_N}{\partial x_2}(x_{10}, x_{20}) - a_{22}\frac{\partial f_N}{\partial x_2}(x_{10}, x_{20})$$

$$8b_{14} = 0$$

$$10b_{15} = a_{22}\frac{\partial f_N}{\partial x_2}(x_{10}, x_{20})$$

$$12b_{16} = 0$$

If $\omega_2 = 2\omega_1 = 2\omega$, $a_{12} \ll a_{11}$ and $a_{22} \ll a_{21}$, then the signal $p_{2(1)}$ will be approximated by (4.43):

$$\frac{2}{k_{2(1)}}p_{2(1)} \cong a_{11}\frac{\partial f_N}{\partial x_1}(x_{10}, x_{20}) + a_{21}\cos(\omega t)\frac{\partial f_N}{\partial x_2}(x_{10}, x_{20})$$

$$- a_{11}\cos(2\omega t)\frac{\partial f_N}{\partial x_1}(x_{10}, x_{20}) - a_{21}\cos(3\omega t)\frac{\partial f_N}{\partial x_2}(x_{10}, x_{20}) \tag{4.43}$$

where:

$$k_{2(1)} = 2r_1, \quad r_1 \in (0.5, 2) \tag{4.44}$$

If

$$k_{Np(1)} = \left|x_{1(max)} - x_{1(min)}\right| = D_{x1} \text{ and } \frac{2r_1 p_{2(1)}}{k_{2(1)}} \approx 1 \tag{4.45}$$

then

$$k_{Np(1)} \cdot p_{2(1)} = D_{x1}r_1 p_{N2(1)} \approx \left|x_{1(max)} - x_{1(min)}\right| \tag{4.46}$$

The selection of the parameter r_1 is not critical, because any value in range from 0.5 to 2 ensures the convergence of the searching loop. The design of the GaPESC scheme will be detailed in the next section for static and dynamic systems.

4.6   Designing the GaPESC Scheme

4.6.1   Case of a Nonlinear System Modeled by a Multimodal Function

It is well known that a PV array under PSCs has a multimodal power characteristic with a GMPP and a number of LMPPs that depends on the number of PV panels in series and parallel and the specific PSCs for these panels. For example, the simulation diagram of the PV power characteristics is presented in Fig. 4.24 for 5 PV panels SX60 in series. The PV cell and panel parameters are also mentioned in Fig. 4.24. The specific PSCs for these five panels are modeled by a variable profile of the irradiance for each panel (see Fig. 4.24). Thus, the panels' irradiance will be different at the moment mentioned in Fig. 4.25 (1, 2, 3, 4, and 5 s), when the PV power characteristics of this array (with 5 PV panels SX60 in series) are presented.

The PV array has a unique maximum at 0 s (when the irradiance is the same (500 W/m^2) for all PV panels) and multiple peaks in the rest (when specific PSCs are obtained).

The multimodal patterns represented in Fig. 4.18 are obtained using (4.46):

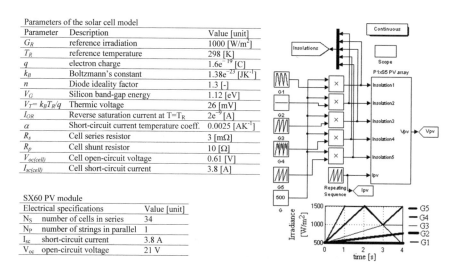

Parameters of the solar cell model

Parameter	Description	Value [unit]
G_R	reference irradiation	1000 [W/m^2]
T_R	reference temperature	298 [K]
q	electron charge	$1.6e^{-19}$ [C]
k_B	Boltzmann's constant	$1.38e^{-23}$ [JK^{-1}]
n	Diode ideality factor	1.3 [-]
V_G	Silicon band-gap energy	1.12 [eV]
$V_T = k_B T_R/q$	Thermic voltage	26 [mV]
I_{OR}	Reverse saturation current at T=T$_R$	$2e^{-9}$ [A]
α	Short-circuit current temperature coeff.	0.0025 [AK^{-1}]
R_s	Cell series resistor	3 [mΩ]
R_p	Cell shunt resistor	10 [Ω]
$V_{oc(cell)}$	Cell open-circuit voltage	0.61 [V]
$I_{sc(cell)}$	Cell short-circuit current	3.8 [A]

SX60 PV module

Electrical specifications	Value [unit]	
N_S	number of cells in series	34
N_P	number of strings in parallel	1
I_{sc}	short-circuit current	3.8 A
V_{oc}	open-circuit voltage	21 V

Fig. 4.24 Simulation diagram of the PV characteristics, and PV cell and panel parameters

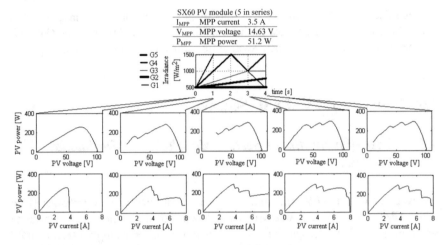

Fig. 4.25 PV power characteristics of 5 PV panels SX60 in series

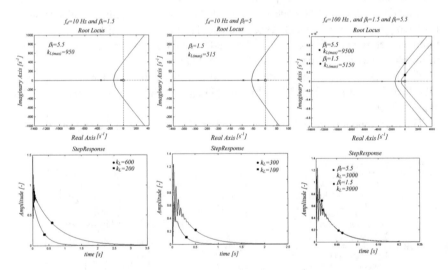

Fig. 4.26 Root locus (top) and the step responses (bottom) for $f_d = 10$ Hz and $\beta_1 = 1.5$ (left), $f_d = 10$ Hz and $\beta_1 = 5$ (middle), and $f_d = 100$ Hz, and $\beta_1 = 1.5$ and $\beta_1 = 5.5$ (right) [64]

$$y = \text{sat}\left(L - (p-1)^2\right) + \text{sat}\left(M - (p-2.5)^2\right) + \text{sat}\left(R - (p-4)^2\right) \qquad (4.46)$$

where sat is the saturation function.

The GMPP will be positioned in the left, middle, and right side of the pattern for the values of the (L, M, R) triplet as follows: (1.5, 2, 1), (1, 2, 1), or (1, 2, 1.5).

For these patterns the parameter $|k_3| = 2$. The gains k_{Ny} and k_{Np} may be set to 1 and 3.

The root locus and the step responses are represented in top and bottom of Fig. 4.26 in the following cases: $f_d = 10$ Hz and $\beta_1 = 1.5$ (left), $f_d = 10$ Hz and $\beta_1 = 5$ (middle), and $f_d = 100$ Hz, and $\beta_1 = 1.5$ and $\beta_1 = 5.5$ (right). The LPF cutoff frequency $(f_1 = \beta_1 \cdot f_d)$ used is set using the aforementioned values of the β_1. The HPF cutoff frequency $(f_h = \beta_h \cdot f_d)$ is set the same using $\beta_h = 0.5$. The step responses are represented in bottom of Fig. 4.26 for aforementioned cases using (in the first and the second case) the values $k_L = 600$ (●) and $k_L = 200$ (■), and $k_L = 300$ (●) and $k_L = 100$ (■) for $\beta_1 = 1.5$ and $\beta_1 = 5.5$, respectively. In the third case, $k_L = 3000$ for $\beta_1 = 1.5$ (●) and $\beta_1 = 5.5$ (■). The selected values $k_L < k_{L(max)}$ ensure a stable operation of the search loop. Some oscillations appear if k_L is closer to $k_{L(max)}$. The oscillations in the search loop increase too much for k_L higher than $k_{L(max)}/2$, affecting the performance of the GaPESC scheme. Consequently, the optimum value for k_L in the design of the GaPESC will be considered $k_{L(optim)} \cong k_{L(max)}/2$.

It is worth mentioning that the ratio $k_{L(max)}/f_d$ has the almost same value in all three cases, so if a higher dither frequency will be chosen to reduce appropriately the search time, then $k_{L(max)}$ will increase and tuning gain k_1 as well.

So, the design condition will be given by (4.47):

$$k_{Ny}k_{Np}k_1k_2|k_3| = k_{L(optim)} \cong k_{L(max)}/2 \tag{4.47}$$

If the tuning gain k_1 is given by (4.48):

$$k_1 = \gamma_d\omega = 2\pi f_d\gamma_d \tag{4.48}$$

then (4.47) will become (4.49):

$$k_{Ny}k_{Np}\gamma_dk_2|k_3| = k_{L(optim)}/2\pi f_d \tag{4.49}$$

For example, if $k_{L(optim)} \cong 120\pi$, then (4.49) will become (4.50):

$$\gamma_dk_2 \cong 1 \tag{4.50}$$

If the startup point is $p_0 = 0$, then $\Delta p_{max} = k_{Np}k_{2(max)} < 5 \Rightarrow k_{2(max)} < 5/3 \cong 1.66$. Also, the conditions to sweep all the GMPPs of the patterns presented in Fig. 4.18 is $\Delta p_{min} = k_{Np}k_{2(min)} > \max(1.75, 2, 3.25) \Rightarrow k_{2(min)} > 1.08$. So, the tuning gain k_2 must be in the range of $k_{2(min)} = 1.1$ to $k_{2(max)} = 1.6$. For example, choosing $k_2 = 1.5 \Rightarrow \gamma_d \leq 2/3$ from (4.50). Thus, $\gamma_d = 0.5 \Rightarrow k_1 = 100\pi$.

It is worth mentioning that GMPP will always be located if $k_{2(min)} < k_2 < k_{2(max)}$ (see Fig. 4.27, where the stages of GMP localization and tracking are highlighted).

To avoid interferences between the dither and the dynamic part of the system (with the natural frequency of $f_n = \omega_n/2\pi$), the dither frequency must be chosen using (4.51):

$$f_d < \omega_n/2\pi = f_n \tag{4.51}$$

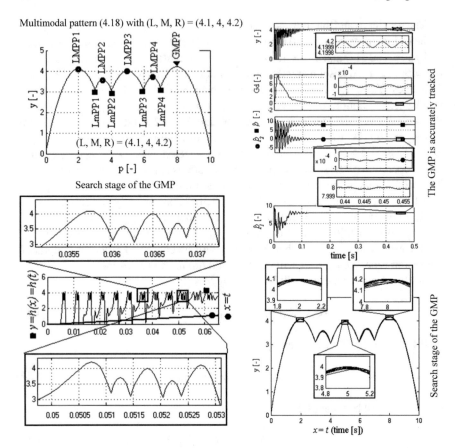

Fig. 4.27 Stages of GMP localization and tracking

The searching signal of the GaPESCd scheme is given by (4.52):

$$p(t) = g_S \cdot t + k_2|\dot{h}| \cdot \sin(\omega t) + \sum_{i=1}^{3} b_i \cdot \sin(i\omega t + \phi_i) \tag{4.52}$$

where the gradient is:

$$g_S = k_1\dot{h} \cdot \left(\frac{a_1}{2} + \frac{k_2}{2} \cdot |\dot{h}|\right) \tag{4.53}$$

The gradient $g_S = k_1\dot{h} \cdot \left(\frac{a_1}{2} + \frac{k_2}{2} \cdot |\dot{h}|\right)$ and the system slope $\frac{dy}{dx} = \frac{dy}{dt}/\frac{dx}{dt} = \dot{h}$ are presented in Fig. 4.28. Both signals go through zero at maximum points and have a similar waveform (Fig. 4.29).

Cases of nonlinear systems with a fast and slow dynamic part will be analyzed in the following two sections.

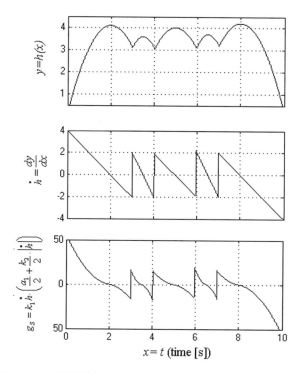

Fig. 4.28 Gradient of the GaPESCd scheme and the derivate of the system output

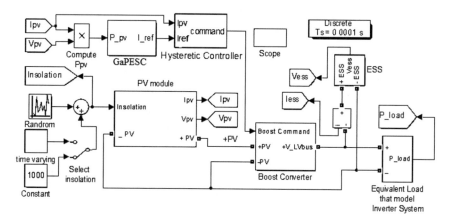

Fig. 4.29 PV hybrid power system [66]

4.6.2 Case of a Nonlinear System with a Fast Dynamic Part

The PV hybrid power system will be considered in this section. The PV system is a nonlinear subsystem and the DC–DC boost converter is the dynamic subsystem. These subsystems are connected in series as in Fig. 4.28 to supply the DC load or charge the 20 Ah/200 V battery from the energy storage system (ESS). When the load demand is higher than the power generated by the PV system, the 20 Ah/200 V battery (operating in the discharge mode) will compensate the difference in power flow balance.

The model of the boost DC–DC converter operating in continuous-current mode is given by (4.54) [70]:

$$H_p(s) = \frac{V_{DC}}{D} = \frac{\omega_n^2}{s^2 + 2\xi\omega_n s + \omega_n^2} \tag{4.54}$$

where the duty cycle D is in range of 0.5–0.9, where the control sensitivity $\partial I_{PV}/\partial D$ is highest) and $V_{DC} = 200$ V. For C $= 100$ μF and $L = 800$ μH $\Rightarrow \omega_n = \frac{1}{\sqrt{LC}} \cong$ 3535 [rad/s] and $\xi \cong 0.7$.

The power flow balance on DC bus is assured by 20 Ah/200 V battery from the ESS.

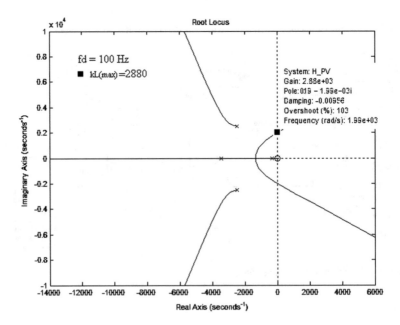

Fig. 4.30 The roots locus for the PV system [64]

A 100 Hz dither frequency is used in order to have $f_d < f_n$ (4.51) and $k_{L(optim)} \cong$ $k_{L(max)}/2 = 1440$ based on Fig. 4.30, where the roots locus of the PV hybrid power system is represented.

So, (4.47) will become (4.47'):

$$k_{Ny}k_{Np}\gamma_d k_2 |k_3| \cong k_{L(optim)}/2\pi f_d \cong 2.3 \tag{4.47'}$$

Choosing $k_{Ny} = 1/P_{MPP} = 1/2400$, $k_{Np} = I_{sc} = 78$, and $|k_3| = 1$ (in range of 0.5–5), (4.49) will become (4.49'):

$$\gamma_d k_2 \cong 14.15 \tag{4.49'}$$

If $p_0 = 0 \Rightarrow \Delta p_{max} = k_{Np}k_{2(max)} \cong |I_{MPP} - p_0| = 60 \Rightarrow k_{2(max)} < 60/78 \cong 0.77$. Also, $\Delta p_{min} = k_{Np}k_{2(min)} > 8 \Rightarrow k_{2(min)} > 0.1$. For example, $k_2 = 0.7 \Rightarrow \gamma_d \cong 20$.

4.6.3 Case of a Nonlinear System with a Slow Dynamic Part

The fuel cell (FC) system is a highly nonlinear system [74]. The FC net power characteristics are represented in Fig. 4.31 for two air compressors (with consumed power of about 10 and 20% from the FC power). The nonlinear surfaces and their projections (using airflow rate (AirFr) as control variable) are shown in the left side and in the right side (top) of Fig. 4.31. The maximum of the FC net power is called

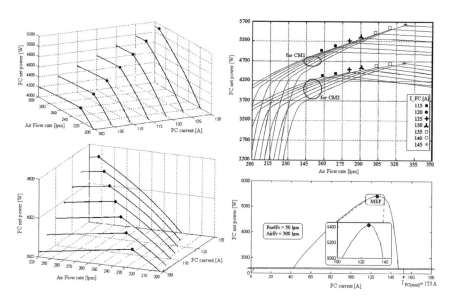

Fig. 4.31 Net power characteristics of the 6 kW FC stack [64]

the maximum efficiency point (MEP). For example, if AirFr = 300 L per minute [lpm] and fuel flow rate (FuelFr) is 50 lpm, then MEP1 $\cong$ 5400 W and MEP2 $\cong$ 4500 W for the compressors 1 (CM1) and 2 (CM2).

The slow dynamics of the FC system has been modeled using a first-order system (4.55):

$$H_{FC}(s) = \frac{1}{T_{FC}s + 1} \tag{4.55}$$

where the FC time constant (T_{FC}) is 100 ms (in general, being in range of 100–200 ms).

The roots locus of the FC system without and with a DC–DC boost converter are represented in top and bottom of Fig. 4.32 for a dither frequency of 1 Hz (see

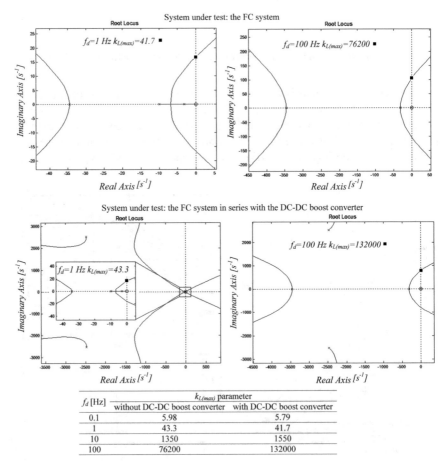

f_d [Hz]	$k_{L(max)}$ parameter	
	without DC-DC boost converter	with DC-DC boost converter
0.1	5.98	5.79
1	43.3	41.7
10	1350	1550
100	76200	132000

Fig. 4.32 Roots locus for the system under test: the FC system without (top) and with (down) the DC–DC boost converter [64]

the left side) and 100 Hz (see the right side) in order to analyze the effect of a fast dynamic part on a nonlinear system with a slow dynamic part. The results for a dither frequency in range of 0.1–100 Hz (also presented in Fig. 4.32) have highlighted the smaller dependence of the design parameter $k_{L(max)}$ on dither frequency in range of 0.1–10 Hz. Note that $f_d = 10$ Hz $\Rightarrow 1/f_d = 100$ ms, which is comparable to the value chosen for the FC time constant. Thus, for $f_d = 10$ Hz the dither will interfere with the FC system via the DC–DC boost converter.

Choosing $f_d = 10$ Hz $\Rightarrow 1/f_d \Rightarrow k_{L(max)} = 1350$, $k_{Ny} = 1/P_{MEP} = 1/5000$, $k_{Np} = I_{FC(max)} \cong 150$, and $|k_3| = 5$ (in range of 3–7), (4.49) will become (4.49″):

$$\gamma_d k_2 \cong 51 \tag{4.49″}$$

If $p_0 = 0 \Rightarrow \Delta p_{max} = k_{Np} k_{2(max)} \cong |I_{MPP} - p_0| = 130 \Rightarrow k_{2(max)} < 130/150 \cong 0.87$. Also, $\Delta p_{min} = k_{Np} k_{2(min)} > 300 \Rightarrow k_{2(min)} > 0.5$, for example, $k_2 = 0.8 \Rightarrow \gamma_d \cong 60$.

The stability and performance of the hybrid power systems using the PV and FC system as main power source will be analyzed in this chapter. The FC hybrid power system will be obtained by replacing the PV system with the FC system in simulation diagram shown in Fig. 4.29. The indicators used to evaluate the performance of the GaPESC algorithm applied to control the PV and FC systems close to the GMPP and MEP, respectively, will be detailed in the next section.

4.7 Performance Indicators

The tracking speed represents the time to accurately find the maximum (GMPP) and is measured by the number of dither's period in case of the PESC algorithms.

The tracking accuracy (T_{acc}) represents how close the GMPP can be found during a stationary regime:

$$T_{acc} = \frac{y_{GMPP}}{y^*_{GMPP}} \cdot 100 \, [\%] \tag{4.56}$$

where y_{GMPP} and y^*_{GMPP} are the GMPP and the founded value.

The tracking efficiency (T_{eff}) measures the tracking performance during the transitory regimes:

$$T_{eff} = \frac{\int_0^t y \, dt}{\int_0^t y^* dt} \cdot 100 \, [\%] \tag{4.57}$$

where y is the PV or FC power, and y^* is the value of this power tracked by the GaPESC algorithm.

It is obvious that a better tracking efficiency will be obtained if the searching time will decrease and/or the transitory tracking accuracy will increase.

The search resolution (S_R) defines how close the GMPP and LMPP can be on a multimodal pattern:

$$S_R = \frac{\min_i |y_{GMPP} - y_{LMPPi}|}{y_{GMPP}} \cdot 100 \, [\%] \tag{4.58}$$

where y_{LMPPi} is the value of the LMPP (number i).

It is obvious that the resolution of search is limited by tracking accuracy:

$$S_R > 100 - T_{acc} \tag{4.59}$$

The hit count represents the success rate to find the GMPP during repetitive tests: the ratio of the positive results to the total number of tests. It is obvious that the hit count is lower than 100% for a low search resolution. So, $S_{R(100\% \, hit)}$ is the search resolution to always find the GMPP (the hit count is 100%).

It is important to evaluate the $S_{R(100\% \, hit)}$ value, which is the resolution for 100% hit count [9].

These indicators will be used to compare the performance of GaPESC algorithm to other GMPPP algorithms on the case studies of a PV, WT and FC hybrid power systems.

4.8 Case Studies for GaPESC Performance Evaluation

4.8.1 PV Hybrid Power System

The following parameters are used in GMPP searching on the patterns (4.46) shown in Fig. 4.18: $f_d = 100$ Hz, $\beta_h = 0.5$ and $\beta_l = 5.5$, $A_m = 0.001$, $k_1 = 200\pi$ ($\gamma_d = 1$) and $k_2 = 2$, $k_{Ny} = 1$ and $k_{Np} = 2$, and $p_0 = 0$.

GMPP searching using the GaPESC scheme is presented in Fig. 4.33 using tuning gains k_1 and k_2 designed to find the GMPP with 100% hit count.

4.8.1.1 Performance Indicators Evaluation

The search resolution (R_S) for the patterns R, M, and L can be evaluated using (4.58) and the values mentioned in Fig. 4.34 as follows:

$$R_{S(R)} = R_{S(L)} = \frac{2.375 - 2}{2.375} \cdot 100 \cong 15.8 \, [\%],$$

$$R_{S(M)} = \frac{2 - 1.875}{2} \cdot 100 = 6.25 \, [\%] \tag{4.60}$$

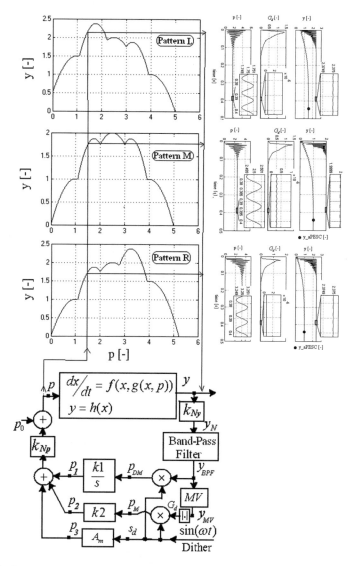

Fig. 4.33 GMPP searching using the GaPESC scheme [27]

The tracking accuracy (T_{acc}) is higher than 99.99% and the output ripple is obviously given by the dither with minimum amplitude $(A_m = 0.001)$. The tracking speed (or the search time) is less than 10 periods of the 100 Hz sinusoidal dither (which means a search time less than 100 ms).

The $S_{R(100 \%hit)}$ is about 0.25%.

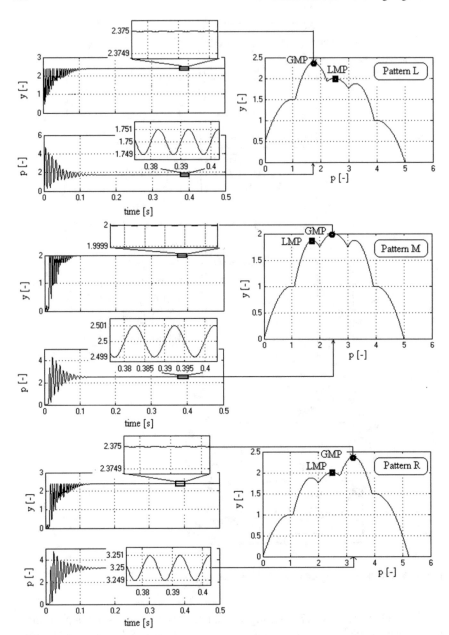

Fig. 4.34 GMPP search on patterns *R*, *M*, and *L* [28]

It is worth mentioning that the maximum tracking accuracy is of 99.96% for the particle swarm optimization (PSO)-based GMPPT algorithms with hit count less than 100% [75].

4.8.1.2   Robustness to a Variable Irradiance

The robustness to a variable irradiance simulated by changing every 0.2 s the pattern, resulting a dynamic sequence of the patterns L, R, and M as in Fig. 4.35.

The GMPP is accurately found in less than 10 dither's periods (see the top of Fig. 4.35). The search signal is represented in the bottom of Fig. 4.35. This converges to the values of 1.75, 3.25, and 2.5, where the GMPP is positioned on the patterns L, R, and M. The dither's gain (G_d) has a high value during the transitory search regimes (see the second plot in Fig. 4.35).

The robustness to variable irradiance for the PV hybrid power system presented in Fig. 4.29 has been tested using the same simulation parameters, except the normalization gains which now have the values $k_{Ny} = 1/2400$ and $k_{Np} = 78$.

The irradiance levels are changed at every 0.5 s (see the first plot in Fig. 4.36). It worth mentioning that the search time is shorter for small variations in the irradiance (see the second plot in Fig. 4.36). The harvested PV power is very close to the maximum power available at the GMP (see the third plot in Fig. 4.36). The bottom plot in Fig. 4.36 represents the GMP search in phase plane (PV power vs. PV current).

Fig. 4.35 Robustness of the GMPP searching for the PV patterns' sequence: (1.5, 2, 1), (1, 2, 1.5) and (1, 2, 1) [27]

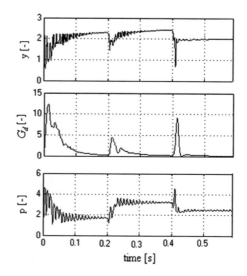

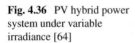

Fig. 4.36 PV hybrid power
system under variable
irradiance [64]

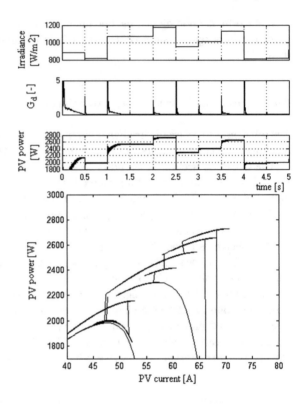

4.8.1.3 Robustness to a Noisy PV Pattern

The robustness to a noisy PV pattern is tested for different levels of noise added to
the pattern R. Because the GMPP has the value of 2.375 and the closest LMP has
the value of 2, the noise amplitude must be lower than the difference between these
values $(2.375 - 2 = 0.373)$ for 100% hit count of the GMPP (which means that
it will always find the GMPP; see the left plot in Fig. 4.37). The search using the
aPESC algorithm is also presented in this plot of Fig. 4.37 (signal y_{aPESC}). Note that
the search remained blocked on left LMPP on the pattern R (positioned at $(1, 1)$. If
the noise is 0.3 or 0.5 (about or higher than $0.373 = y_{GMPP} - y_{LMPP}$) the GaPESC
algorithm still operates, but the aPESC algorithm no longer works (see the middle
and right plots in Fig. 4.37).

4.8.1.4 Robustness to Various Shapes of the Dither

The robustness is shown to various shapes of the dither: 100 Hz sinusoidal dither,
100 Hz rectangular dither with 50% duty cycle, sum of the aforementioned dithers,
100 Hz saw-tooth dither, and random noise with 100 Hz sampling. The search behavior is presented in Fig. 4.38 using the signals given by (4.21a), such as the dither's

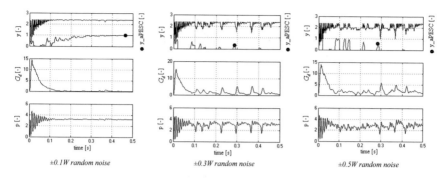

Fig. 4.37 Robustness to a noisy PV pattern [27]

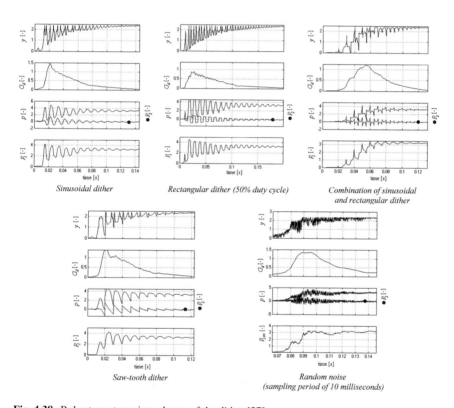

Fig. 4.38 Robustness to various shapes of the dither [27]

gain (G_d), the search signal (p_1), the localization signal (p_2), and the sum of the aforementioned components ($p = p_1 + p_2$).

The results obtained validate the conclusion of previous studies [2, 76] related to the shape of dither and the possibility to use a noisy signal from the switching to be used as dither in the GaPESC scheme. The interested reader can find more results in Chap. 7 dedicated to analysis of the PV hybrid power system under partial shaded conditions. Furthermore, the results highlight that the GaPESC algorithm can be used in WT hybrid power system to find MPP in days with strong variable wind [24], as will be shown in the next section.

4.8.2 WT Hybrid Power System

The simulation diagram of the WT hybrid power system is presented in Fig. 4.39. The results are comparatively presented for both GaPESC and PESC schemes, but other MPPT algorithms may be used as reference as well. The signal y_M that modulate the dither's amplitude will be G_d given by (4.21a) for the GaPESC scheme or $y_M = 1$ for the PESC scheme:

$$y_M = \begin{cases} G_d, & \text{GaPESC} \ (4.21\text{d}) \\ 1, & \text{classic PESC} \end{cases} \tag{4.61}$$

The WT model from the MATLAB–Simulink® [77] has been adapted in order to select the WT mechanical power (P_m) or the WT mechanical torque (T_m) as input of the GaPESC scheme. Except the normalization gains that are now $k_{Ny} = 1/1000$ and $k_{Np} = 1$, the aforementioned parameters of the GaPESC scheme have been used in simulation. Also, these parameters have been used for the PESC scheme.

4.8.2.1   Searching of the Maximum Power Point

The MPP search is presented in Fig. 4.40 for a wind speed with a stair profile (see the first plot in Fig. 4.40). The pu generator speed reference generated by the PESC

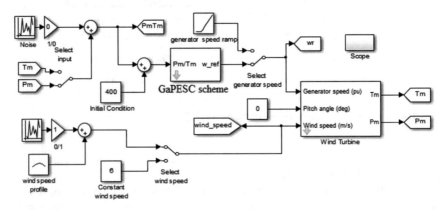

Fig. 4.39 Simulation diagram of the WT hybrid power system [24]

Fig. 4.40 MPP search using the PESC scheme [24]

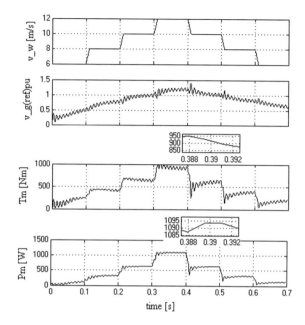

scheme is presented in the second plot of Fig. 4.40. The last two plots present the output signals of the WT (P_m and T_m). It is worth mentioning that the ripple (of about 70 Nm and 7 W for the P_m and T_m, respectively) appears in PESC scheme due to the dither's amplitude that will not decrease asymptotically to zero as in GaPESC scheme.

The behavior of the outputs P_m and T_m is represented in Fig. 4.41 for the stair profile (shown in the first plot of Fig. 4.40).

The tracking accuracy using the PESC scheme is less than 99.6% For example, the tracking accuracy is about 99.59% ($\cong 100 \cdot 1090.5/1095$) during the stationary regimes of 12 m/s wind speed (see the zooms in Fig. 4.40).

MPP search using the GaPESC scheme is presented in Fig. 4.42 for the same profile of the wind speed (see the first plot in Fig. 4.42).

The tracking accuracy using the GaPESC scheme is higher than 99.9%. For example, the tracking accuracy is about 99.93% ($\cong 100 \cdot 1094.25/1095$) during the stationary regimes of 12 m/s wind speed (see the zooms in the last two plots of Fig. 4.42). Also, it is worth mentioning that the ripple of the P_m and T_m decreases to about 0.5 W and 3 Nm, respectively, because the dither's amplitude will decrease asymptotically to zero using the GaPESC scheme. The dither's amplitude is large only during the transitory search regimes (see the second plot in Fig. 4.42) in order to reduce search time. It is known that the search time of the PESC scheme depends on product $k_1 \cdot k_2$, but the increase of the dither's gain (k_2) will increase the ripple to unacceptable values (reducing the tracking efficiency as well). Note that the gain k_1 was designed for both schemes to ensure the stability of the search loop.

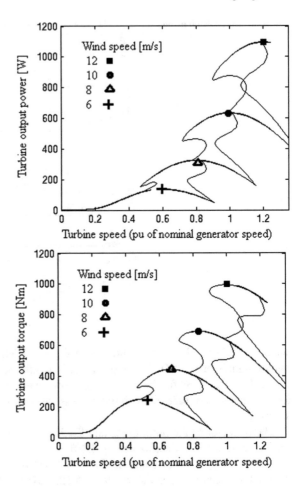

Fig. 4.41 MPP search in the phase plane (using the PESC scheme) [24]

MPP search in the phase plane is presented in Fig. 4.43 using the GaPESC scheme and the stair profile for the wind speed. The MPPs pointed in Fig. 4.43 for wind speed of 6, 8, 10, and 12 m/s are very close to ideal power that is produced by the WT for each wind speed.

4.8.2.2 Searching of the Maximum Torque Point

The maximum torque point (MTP) search can be of interest in some cases [78, 79], especially when wind speed has large variations and the WT must integrate the advantages of horizontal and vertical wind turbines [80].

MTP search using the GaPESC scheme is presented in Fig. 4.44 using the same stair profile for the wind speed. The pu generator speed reference generated by the

Fig. 4.42 MPP search using the GaPESC scheme [24]

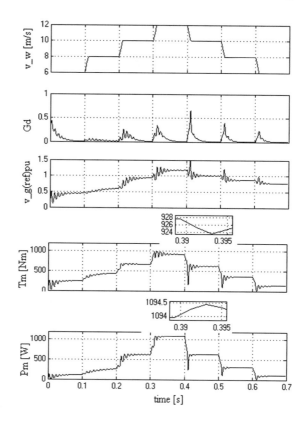

PESC scheme is presented in the second plot of Fig. 4.44. The last two plots present the output signals of the WT (P_m and T_m).

The tracking accuracy using the GaPESC scheme is higher than 99.9%. For example, the tracking accuracy is about 99.62% ($\cong 100 \cdot 994.2/998$) during the stationary regimes of 12 m/s wind speed (see the zooms in the last two plots of Fig. 4.44). Also, it is worth mentioning that the ripple of the T_m and P_m decreases to about 0.7 Nm and 5 W, respectively, but the low value is now obtained for the T_m (the variable to be optimized). This was valid in the previous section for P_m, which was the variable to be optimized there. Almost the same behavior is observed in the search for MPP and MTP. The dither's amplitude is large only during the transitory search regimes (see the second plot in Fig. 4.44) in order to reduce search time.

If the torque is optimized, then the power has values different to those that are obtained when the power has been optimized, and vice versa (see the last plots in Figs. 4.42 and 4.44).

Fig. 4.43 MPP search in the phase plane (using the GaPESC scheme) [24]

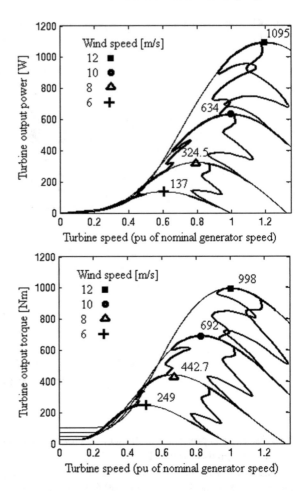

4.8.2.3   Robustness to Noise of Wind or Added by Measuring Power

The noise can appear on the input (the wind speed) or output (WT power) of the WT hybrid power system due to wind turbulence or measurement noise. The behavior of the system is presented in Figs. 4.45 and 4.46 for low and high noise of the wind.

The random noise is sampled at 100 Hz rate. Low and high noise are in range of 0–1 m/s (see the first plot in Fig. 4.45) and, respectively, range of 1–3 m/s (see the first plot in Fig. 4.46). The second plot in Figs. 4.45 and 4.46 highlights that the dither's gain increases for a high perturbed wind. The measurement noise added on the WT power (P_m) is partially filtered by the BPF used in GaPESC scheme (see the fourth plot in Figs. 4.44 and 4.47). The measurement noise is random noise in range of 100–300 W, which is sampled at 10 Hz rate (see the last two plots in Fig. 4.47). This measurement noise does not affect the search and localization loop of the GaPESC scheme due to band-pass filtering of both loops. Also, it was

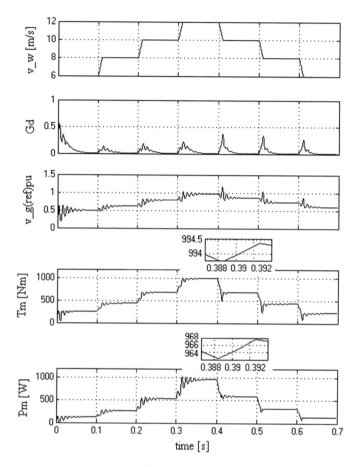

Fig. 4.44 MTP search using the GaPESC scheme [24]

highlighted that the robustness of the PESC scheme is limited to high noise levels. The interested reader can find more results in Chap. 8, where the renewable hybrid power system will be analyzed under variable energy from renewable sources. This variability of renewable power and load demand requests an additional energy source to compensate the power flow balance on the DC bus (if the battery will operate in charge-sustaining mode based on the requested power-following mode control implemented for this additional energy source). The additional energy source may be a green energy source (such as the FC system) or a polluting solution (e.g., based on a diesel engine).

The case of a FC hybrid power system will be analyzed in the next section under variable load demand using the load-following mode control [73].

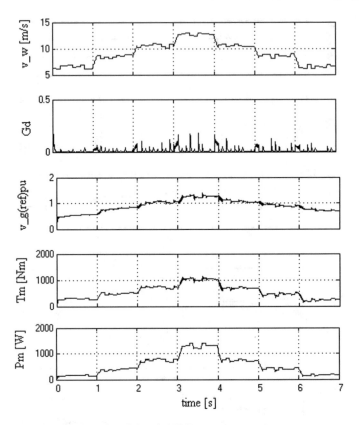

Fig. 4.45 Robustness to low noise of the wind [24]

4.8.3   FC Hybrid Power System

Except the normalization gains and dithers' frequencies that are now $k_{Ny} = 1/2000$ and $k_{Np} = 10$, and $f_1 = f_d = 1$ Hz and $f_2 = 2f_d = 2$ Hz (in order to comply with the rule (4.51), $f_d < f_n$, where the natural frequency f_n of the FC system is about tens of Hz), the aforementioned parameters of the GaPESC scheme have been used in simulation of a 6 kW/45 V Proton Exchange Membrane Fuel Cell (PEMFC) system (see Fig. 4.48). It can be observed that load-following (LF) is applied to the boost controller. Thus, the FC power requested by load demand will be generated by the FC system setting the FC current at reference I_{refLF} of the LF control. An optimization surface is defined using two control variables for any value of the FC current. The control variables are the air and fuel flow rates (AirFR and FuelFr) that are set by the fueling regulators based on the references $I_{refGES1}$ and $I_{refGES2}$. These references are generated by two GE controllers based on the GaPESC algorithm. So, the SIDO ESC scheme shown in Fig. 4.23 will be used to generate the references $I_{refGES1}$ and $I_{refGES2}$, which will be applied to the inputs of the fueling regulators in order to search

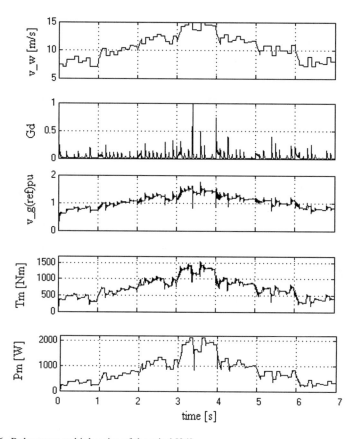

Fig. 4.46 Robustness to high noise of the wind [24]

the GMPP on the optimization surface f. The optimization function f is specified in the block called "Function," being the common input of the SIDO ESC scheme. The SIDO ESC scheme with dithers of different frequencies (see Fig. 4.23) will be used to find the MEP. Thus, the optimization function f used in this section is the FC net power, but other fuel economy functions can also be defined (see Chap. 6). Thus, the FC system will be optimally controlled to operate at MEP using a real-time optimization (RTO) strategy based on the GES control (called GES-RTO strategy) [73]. The performance obtained using FC hybrid power system will be compared with the RTO strategy based on the static Feed-Forward (sFF) control (called sFF-RTO strategy) [74].

    In brief, this is description of the FC hybrid power system to be optimized using the FC hybrid power system. The readers interested by optimization of the FC hybrid power system can read Chaps. 5 and 6, where a deep analysis of the MEP and fuel economy optimization strategies is presented.

    Firstly, the performance of the FC hybrid power system under constant load will be presented using performance indicators such as FC net power, electrical energy

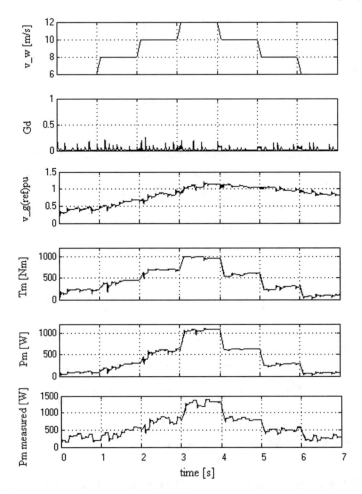

Fig. 4.47 Robustness to measurement noise [24]

efficiency of the FC system ($\eta_{sys} = P_{FCnet}/P_{FC}$), the fuel consumption efficiency (Fuel$_{eff} = P_{FCnet}/$FuelFr) and the total fuel consumption (Fuel$_T = \int$ FuelFr(t)dt) measured in W/lpm (where lpm means liters per minute) and liters. Also, the performance indicators presented in Sect. 4.7 of this chapter (such as the tracking speed and tracking accuracy) will be pointed by presenting dynamic behavior of the FC system in some cases. Of course, the tracking accuracy (T_{acc}) will be estimated using (4.56$'$) instead of (4.56) specifically defined for a PV system:

$$T_{acc} = \frac{P_{FCnet}}{P_{FCnet(max)}} \cdot 100[\%] \qquad (5.46')$$

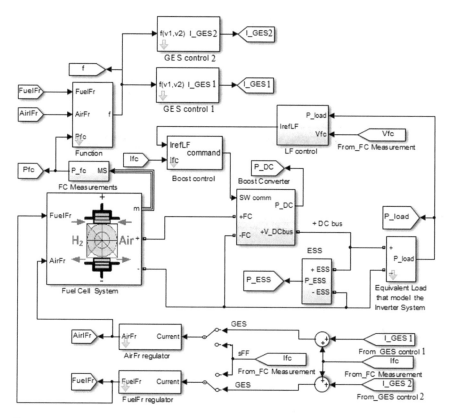

Fig. 4.48 FC hybrid power system [73]

where P_{FCnet} is the FC net power extracted using the GaPESC scheme from the maximum available FC net power of the FC system under given operation conditions.

The behavior of the FC system under an 8 kW load is presented in Fig. 4.49.

The air and fuel flow rates (AirFR and FuelFr presented in the plots fourth and fifth of Fig. 4.49) will find the maximum of the FC net power (presented in the second plot of Fig. 4.49) for a given FC current based on the load-following control of the FC system (which operate the battery in charge-sustained mode; see P_{ESS} presented in the third plot of Fig. 4.49). The aforementioned performance indicators (P_{FCnet}, η_{sys}, Fuel$_{eff}$, and Fuel$_T$) are presented in the last plots of Fig. 4.50 and registered using different levels of load demand for both strategies (the GES-RTO strategy and the sFF-RTO strategy). Better performance across all indicators can be observed for a load power higher than 3 kW. For example, the differences in performance indicators (ΔP_{FCnet}, $\Delta\eta_{sys}$, ΔFuel$_{eff}$, and ΔFuel$_T$) are presented in Fig. 4.50 using the sFF-RTO strategy as reference.

The searching time is about 2 s for the dithers frequencies of $f_1 = 0.5$ Hz and $f_2 = 1$ Hz. By increasing 10-times the dithers frequencies ($f_1 = 5$ Hz and $f_2 = 10$ Hz), the search time will also decrease 10-times (to about 0. 2 s).

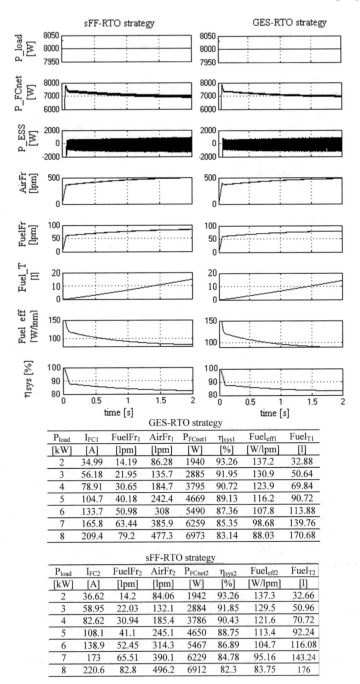

P_{load}	I_{FC1}	$FuelFr_1$	$AirFr_1$	P_{FCnet1}	η_{sys1}	$Fuel_{eff1}$	$Fuel_{T1}$
[kW]	[A]	[lpm]	[lpm]	[W]	[%]	[W/lpm]	[l]
2	34.99	14.19	86.28	1940	93.26	137.2	32.88
3	56.18	21.95	135.7	2885	91.95	130.9	50.64
4	78.91	30.65	184.7	3795	90.72	123.9	69.84
5	104.7	40.18	242.4	4669	89.13	116.2	90.72
6	133.7	50.98	308	5490	87.36	107.8	113.88
7	165.8	63.44	385.9	6259	85.35	98.68	139.76
8	209.4	79.2	477.3	6973	83.14	88.03	170.68

sFF-RTO strategy

P_{load}	I_{FC2}	$FuelFr_2$	$AirFr_2$	P_{FCnet2}	η_{sys2}	$Fuel_{eff2}$	$Fuel_{T2}$
[kW]	[A]	[lpm]	[lpm]	[W]	[%]	[W/lpm]	[l]
2	36.62	14.2	84.06	1942	93.26	137.3	32.66
3	58.95	22.03	132.1	2884	91.85	129.5	50.96
4	82.62	30.94	185.4	3786	90.43	121.6	70.72
5	108.1	41.1	245.1	4650	88.75	113.4	92.24
6	138.9	52.45	314.3	5467	86.89	104.7	116.08
7	173	65.51	390.1	6229	84.78	95.16	143.24
8	220.6	82.8	496.2	6912	82.3	83.75	176

Fig. 4.49 FC system under 8 kW load [73]

Fig. 4.50 Differences in performance indicators [73]

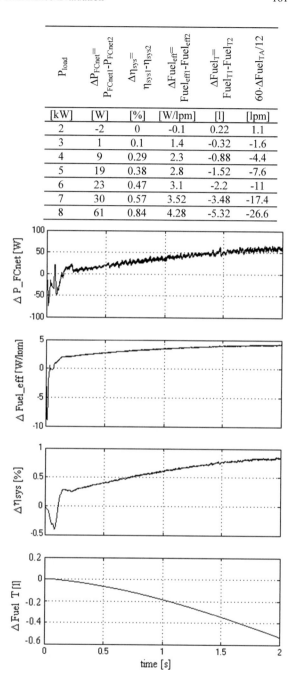

P_{load}	$\Delta P_{FCnet} = P_{FCnet1}-P_{FCnet2}$	$\Delta\eta_{sys} = \eta_{sys1}-\eta_{sys2}$	$\Delta Fuel_{eff} = Fuel_{eff1}-Fuel_{eff2}$	$\Delta Fuel_T = Fuel_{T1}-Fuel_{T2}$	$60\cdot\Delta Fuel_{TA}/12$
[kW]	[W]	[%]	[W/lpm]	[l]	[lpm]
2	-2	0	-0.1	0.22	1.1
3	1	0.1	1.4	-0.32	-1.6
4	9	0.29	2.3	-0.88	-4.4
5	19	0.38	2.8	-1.52	-7.6
6	23	0.47	3.1	-2.2	-11
7	30	0.57	3.52	-3.48	-17.4
8	61	0.84	4.28	-5.32	-26.6

Note that $f_2 = 10$ Hz $\Rightarrow 1/f_2 = 100$ ms, which is comparable to the value chosen for the FC time constant of 200 ms, so the dither will interfere with the FC system via the DC–DC boost converter (see Figs. 4.51 and 4.52, where the search for the FC net power and the fuel consumption efficiency is represented using 8 kW load for the FC system).

The zooms presented for the FC net power and the fuel consumption efficiency in Figs. 4.51 and 4.52, respectively, show that values mentioned in Fig. 4.49 (for the FC system under 8 kW load) are accurately found.

The increase in FC net power is about 61 W ($= 6973 - 6912$ W), which means an increase with 0.84% of the electrical energy efficiency for the FC system under an 8 kW load.

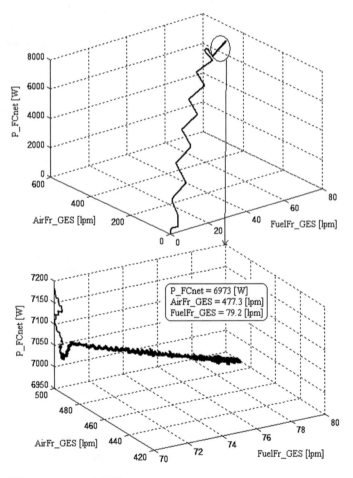

Fig. 4.51 FC net power search [73]

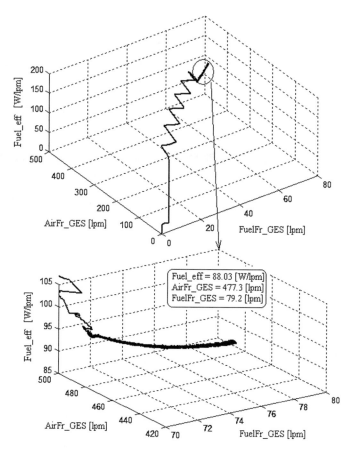

Fig. 4.52 Fuel consumption efficiency search [73]

The increase in fuel consumption efficiency is about 4.28 W/lpm (= 88.03 – 83.75 W/lpm), which means an increase of the fuel economy with 26.6 lpm for the FC system under an 8 kW load.

The performance of the GES-RTO strategy in comparison with the sFF-RTO strategy will be evaluated below using a variable load. A load cycle with average value of $P_{load(AV)}$ will have the levels of $0.75 \cdot P_{load(AV)}$, $1.25 \cdot P_{load(AV)}$, and $P_{load(AV)}$ changed every 4 s. For example, the 5 kW load cycle shown in Fig. 4.53 has the levels of 4, 6, and 5 kW. Because the dithers frequencies are now $f_1 = 5$ Hz and $f_2 = 10$ Hz, the search time will be about 0.2 s (see these search regimes in Fig. 4.53, during changing of the load level). The power flow balance on the DC bus during changing of the load level will be sustained by the battery and the ultracapacitors (see the third plot of Fig. 4.53). The values of the FuelFr and AirFr will be accurately tracked during stationary regimes. For example, the values of the AirFr and FuelFr will be 40.18 and 242.4 lpm, and 41.1 and 245.1 lpm for the GES-RTO strategy and

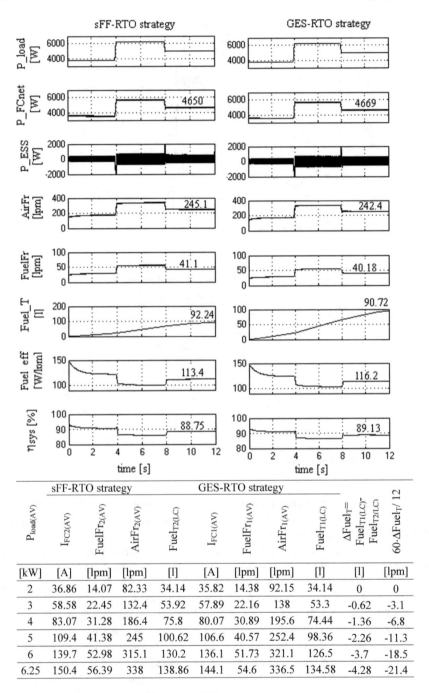

Fig. 4.53 FC system under 5 kW load cycle [73]

the sFF-RTO strategy, respectively (see the plots fourth and fifth of Fig. 4.53 during the 5 kW load level).

The performance indicators (P_{FCnet}, η_{sys}, $Fuel_{eff}$, and $Fuel_T$) are presented in the last plots of Fig. 4.53 and registered using different load cycles for both strategies (the GES-RTO strategy and the sFF-RTO strategy). Better fuel economy across all load cycles can be observed. For example, the differences in performance indicators (ΔP_{FCnet}, $\Delta\eta_{sys}$, $\Delta Fuel_{eff}$, and $\Delta Fuel_T$) are presented in Fig. 4.54 using the sFF-RTO strategy as reference.

It can be observed the fuel economy of about 2.26 L (see Fig. 4.53) obtained at the end of the 5 kW load cycle and the improvements in other performance indicators for different load levels (see the differences in performance indicators presented in Fig. 4.54).

It worth mentioning that the transitory regimes duration is shorter than about 0.2 s (which is the search time).

Fig. 4.54 Differences in performance indicators of the FC system under a 5 kW load cycle [73]

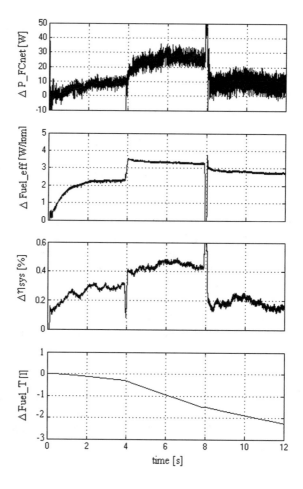

The capacity of the GaPESC algorithm to find the global maximum has been shown in the previous sections using different multimodal patterns in the case studies analyzed. The performance of the GaPESC algorithm is to find a global maximum on other multimodal patterns that are usually used as benchmark tests [72].

4.8.4 GMPP Search on Different Multimodal Patterns

Except the normalization gains (which will be specifically designed considering the maximum value and the scanning range requested for each multimodal pattern), the aforementioned parameters of the GaPESC scheme ($f_d = 100$ Hz, $\beta_h = 0.5$ and $\beta_l = 5.5$, $A_m = 0.001$, $k_1 = 200\pi$ ($\gamma_d = 1$) and $k_2 = 2$) will be used. The startup point (p_0) will be mentioned in each case.

4.8.4.1 Multimodal Patterns with Many LMPPs in the Left and Right Sides of the GMPP

The normalization gains are $k_{Ny} = 1/400$ and $k_{Np} = 5$ for the multimodal pattern with many LMPPs in the left and right sides of the GMPP, which is given by (4.62) (see Fig. 4.55):

$$y = -p^4 + 64 \cdot \sin^2(p^3) + 12 \cdot p^2 + 4 \cdot p + 300 \tag{4.62}$$

It is worth mentioning that this pattern has many LMPPs on the left and right of the GMPP located at about (7.585, 409.8525). To test the capacity of the GaPESC algorithm to find the GMPP, the startup point has been selected at $p(0) = 5$, between the LMPPs located in the left and right sides of this pattern. Because the LMPP located at (7.4185, 409.645) is the nearest to the GMPP (see the zooms presented in Fig. 4.55), the search resolution (R_S) given by (4.58) is about 0.0488% ($\cong$ (409.8525 − 409.645)/409.8525) and the tracking accuracy given by (4.56) is about 99.997% ($\cong$ 409.84/409.8525).

The normalization gains are $k_{Ny} = 1/10$ and $k_{Np} = 10$ for the second example of multimodal pattern with many LMPPs located symmetrically in the left and right sides of the GMPP, which is given by (4.63) (see Fig. 4.56):

$$y = -|p + 0.3 \cdot \sin(10 \cdot p)| \tag{4.63}$$

Furthermore, the derivative is discontinuous in origin (where the GMPP is located). If $|p(0)| > 1$, then the search algorithm must pass over many LMPPs until localization of the GMPP. For example, the search is presented in Fig. 4.56 for startup point $p(0) = -10$. The tracking accuracy given by (4.56) is about 99.6%. The search

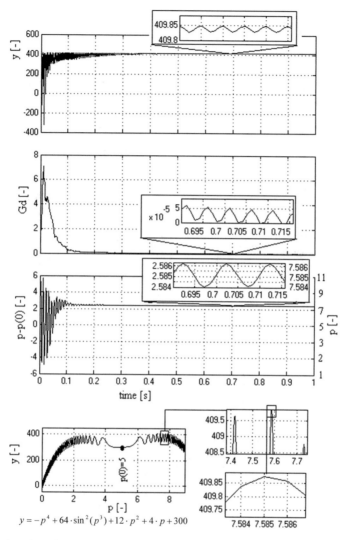

Fig. 4.55 Search on the multimodal pattern (4.62) [28]

resolution (R_S) depends on the chosen scanning range. For a scanning range from -10 to 10, the search resolution is about 0.02%.

It is worth mentioning that the hit count for the GMPP search is less than 100%. Also, the design of the tuning parameters and the normalization gain is more difficult for such patterns with complex shape, providing a more restrictive range for selecting these parameters.

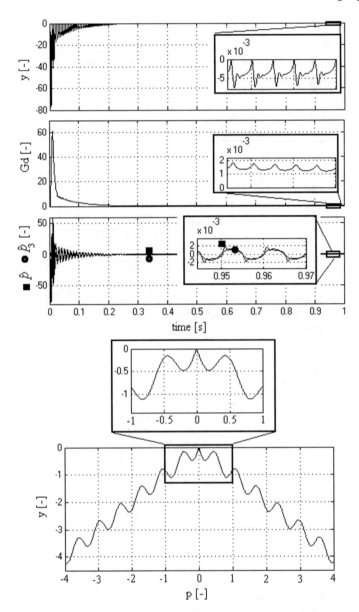

Fig. 4.56 Search on the multimodal pattern (4.63) [30]

4.8.4.2 Multimodal Patterns with the Startup Point Close to a LMPP but Away from GMPP

The normalization gains for a multimodal pattern with many LMPPs in the left and right sides of the GMPP, which is a sixth-order polynomial given by (4.64), are $k_{Ny} = 1/4$ and $k_{Np} = 20$ (see Fig. 4.57):

$$y = \left[-(p-1)^6 + \frac{1}{10} \cdot (p-1)^5 + \frac{623}{400} \cdot (p-1)^4 - \frac{659}{4000} \cdot (p-1)^3 \right.$$
$$\left. - \frac{11287}{20000} \cdot (p-1)^2 + \frac{259}{4000} \cdot (p-1) + \frac{637}{20000} \right] \cdot 100 \tag{4.64}$$

The GMPP and the LMPP that are shown in a zoom of Fig. 4.57 have the coordinates of (1, 3.72) and (16, 3.74). Thus, the resolution and tracking accuracy are of 0.534 and 99.9946%. The startup point has been chosen between the GMPP and the LMPP (e.g., $p(0) = 10$ in Fig. 4.57) or in the left side of the LMPP (e.g., $p(0) = 0$). In the last case, finding the GMPP is quite difficult to achieve without a careful design, especially for the localization parameter k_2 and the gain k_N.

4.8.4.3 Multimodal Patterns with a Plateau for the GMPP and LMPPs

The normalization gains are $k_{Ny} = 5$ and $k_{Np} = 1$ for a multimodal pattern with a plateau for the GMPP and LMPPs, which is given by (4.65), are $k_{Ny} = 1/4$ and $k_{Np} = 20$ (see Fig. 4.58):

$$y = \exp\left(\frac{1}{1 + 0.2 \cdot (p-1)^2} \right) + \exp\left(\frac{1}{1 + 5 \cdot (p-16)^2} \right) \tag{4.65}$$

This multimodal pattern has three maximums located at $u = 1 - 0.8985 = 0.1015$ (the GMPP), $u = 1 + 0.05 = 1.057$ (the first LMPP), and $u = 1 + 0.8951 = 1.8951$ (the second LMPP). The value of the second LMPP differs with 0.5503 by the value of the GMPP ($y_{GMPP} = 6.7947$), so the searching resolution is about 8.1% if the reference is the axis of abscissas (p) and the startup point is on the plateau (e.g., between 0 and 2). The search using the startup point $p(0) = 0$ is represented in Fig. 4.58. The tracking accuracy is about 99.997% ($= 6.7945/6.7947$) considering the zooms presented in Fig. 4.58.

Based on a careful design, the GMPP can be found with high value of the hit count even if the startup point is lower than -0.5, but the hit count will be very low if the startup point is higher than 2.5.

The capacity of the GaPESC algorithm to find the GMPP has been presented in previous section.

So, in the next section, the ability of the GaPESC algorithm to also find a local maximum will be explored using different multimodal patterns [26].

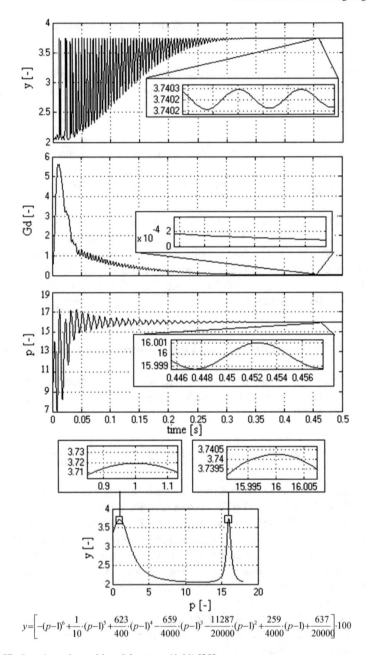

$$y = \left[-(p-1)^6 + \frac{1}{10} \cdot (p-1)^5 + \frac{623}{400} \cdot (p-1)^4 - \frac{659}{4000} \cdot (p-1)^3 - \frac{11287}{20000} \cdot (p-1)^2 + \frac{259}{4000} \cdot (p-1) + \frac{637}{20000} \right] \cdot 100$$

Fig. 4.57 Search on the multimodal pattern (4.64) [28]

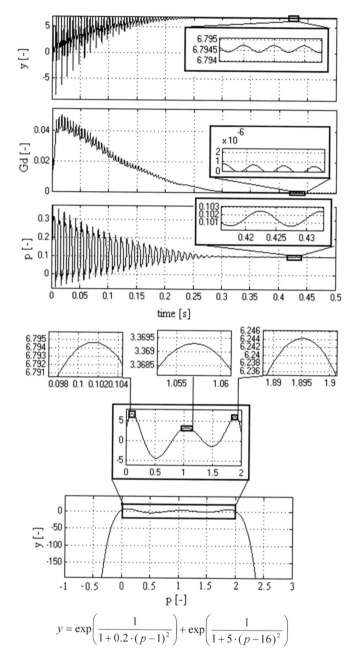

$$y = \exp\left(\frac{1}{1+0.2\cdot(p-1)^2}\right) + \exp\left(\frac{1}{1+5\cdot(p-16)^2}\right)$$

Fig. 4.58   Search on the multimodal pattern (4.65) [30]

4.9 Searching of Local Extreme Based on the GaPESC Scheme

The multimodal patterns represented in Fig. 4.18 will be used in this section

4.9.1 Modeling for Local Search

The quadratic form (4.62) may represents a good approximation of the pattern around a maximum (in this case a LMPP, where $\dot{h}\big|_{P_{LMPP}} \cong 0$, $\ddot{h}\big|_{P_{LMPP}} \cong 0$, and $\dddot{h}\big|_{P_{LMPP}} = 2k_3 < 0$):

$$\tilde{y} \cong \frac{1}{2}\ddot{h} \cdot \tilde{p}^2 = k_3 \cdot \tilde{p}^2 \tag{4.62}$$

where, considering (4.46), $|k_3| = 2$.

But for better evaluation of the gradient, the small-signal model (4.63) will be used in this section. The multimodal pattern can be approximated by (3):

$$\tilde{y} \cong \dot{h} \cdot \tilde{p} + \frac{1}{2}\ddot{h} \cdot \tilde{p}^2 + \frac{1}{6}\dddot{h} \cdot \tilde{p}^3 \tag{4.63}$$

where $\tilde{p} = p - p_{LMPP}$, $\tilde{y} = y - h(p_{LMPP})$, p_{LMMP} the abscissa of the LMPP, and $\dot{h} = \frac{dy}{dp}, \ddot{h} = \frac{d^2y}{dp^2}, \dddot{h} = \frac{d^3y}{dp^3}$ are the derivatives.

If the searching signal p_1 will have three harmonics with initial magnitudes $a_j, j = 1, 2, 3$, and the initial gradient is approximated with $\dot{h}$, then the searching signal $\tilde{p}$ will be (4.64):

$$\tilde{p} \cong \sum_{j=1}^{3} a_j \sin(j\omega t) + k_2|\dot{h}| \cdot \sin(\omega t) \tag{4.64}$$

Thus, (4.63) can be approximated with (4.65):

$$\tilde{y} \cong \dot{h} \cdot \tilde{p} + \frac{1}{2}\ddot{h} \cdot \tilde{p}^2 + \frac{1}{6}\dddot{h} \cdot \tilde{p}^3 \cong \sum_{j=1}^{4}\left(\dot{h} \cdot H_j + \frac{1}{2}\ddot{h} \cdot H_j^2 + \frac{1}{6}\dddot{h} \cdot H_j^3\right) \tag{4.65}$$

where

$$H_j = a_j \sin(j\omega t), \quad j = 1, 2, 3, \quad H_4 = k_2|\dot{h}| \cdot \sin(\omega t) \tag{4.66}$$

After the ideal BPF, the filtered signal y_{BPF} can approximated with (4.67):

$$y_{BPF} = y_{BPF(4)} + \sum_{j=1}^{3} y_{BPF(j)} \tag{4.67}$$

where

$$y_{BPF(j)} = \dot{h} \cdot a_j \cdot \sin(j\omega t) - \frac{1}{2}\ddot{h} \cdot a_j^2 \cdot \frac{1}{2}\cos(2j\omega t)$$
$$+ \frac{1}{6}\dddot{h} \cdot a_j^3 \cdot \frac{1}{4}[3\sin(j\omega t) - \sin(3j\omega t)], \quad j = 1, 2, 3$$
$$y_{BPF(4)} = \dot{h} \cdot k_2|\dot{h}|\sin(\omega t) - \frac{1}{2}\ddot{h} \cdot (k_2|\dot{h}|)^2 \cdot \frac{1}{2}\cos(\omega t)$$
$$+ \frac{1}{6}\dddot{h} \cdot (k_2|\dot{h}|)^3 \cdot \frac{1}{4}[3\sin(\omega t) - \sin(3\omega t)] \tag{4.68}$$

The filtered signal y_{BPF} is demodulated using a sinusoidal dither with unitary amplitude, resulting the signal p_{DM} (4.69):

$$p_{DM} = \sum_{j=1}^{4} \sin(\omega t) \cdot y_{BPF(j)} \tag{4.69}$$

The new components of the searching signal after integration are as follows

$$\tilde{p}_{new} = g_S \cdot t + p_{1LFnew} + k_2|\dot{h}| \cdot \sin(\omega t) + A_m \cdot \sin(\omega t) \tag{4.70}$$

where the searching gradient is now approximated by (4.71):

$$g_S \cong k_1 \frac{\dot{h}}{2} \cdot (a_1 + k_2|\dot{h}|) + k_1 \frac{\dddot{h}}{16} \cdot (a_1^3 + (k_2|\dot{h}|)^3) \tag{4.71}$$

and the harmonics of the low frequency (LF) signal (p_{1LF}):

$$p_{1LFnew} = \sum_{j=1}^{3} (s_j \cdot \sin(j\omega t) + c_j \cdot \cos(j\omega t)) = \sum_{j=1}^{3} b_j \cdot \sin(j\omega t + \phi_j) \tag{4.72}$$

where the coefficients s_j, c_j, and the magnitude b_j and phase ϕ_j of the harmonics are approximated by (4.73):

$$s_1 = \frac{k_1}{\omega} \cdot \left(\frac{\dot{h} \cdot a_2}{2} + \frac{\dddot{h} \cdot a_2^3}{16} \right),$$

$$c_1 = -\frac{k_1}{\omega} \cdot \frac{\ddot{h}}{8} \cdot (a_1^2 + (k_2|\dot{h}|)^2),$$

$$s_2 = \frac{k_1}{2\omega} \cdot \left[-\frac{\dot{h}}{2} \cdot (a_1 + k_2|\dot{h}|) + \frac{\dot{h} \cdot a_3}{2} - \frac{\dddot{h}}{12}(a_1^3 + (k_2|\dot{h}|)^3) + \frac{\dddot{h} \cdot a_3^3}{16} \right],$$

$$c_2 = 0,$$

$$s_3 = -\frac{k_1}{3\omega} \cdot \left[-\frac{\dot{h}}{2} \cdot a_2 - \frac{\ddot{h}}{16} \cdot a_2^3 \right],$$

$$c_3 = \frac{k_1}{3\omega} \cdot \left[-\frac{\ddot{h}}{8} \left(a_1^2 + (k_2|\dot{h}|)^2 \right) + \frac{\ddot{h}}{8} \cdot a_2^2 \right]$$

$$b_j = \sqrt{s_j^2 + c_j^2}, \quad tg\phi_j = \frac{c_j}{s_j} \tag{4.73}$$

Thus, the search gradient (4.71) can be approximated by:

$$g_s \cong g_{s1} + g_{s2} \tag{4.74}$$

where

$$g_{s1} = k_1 \left(\frac{\dot{h}}{2} \cdot a_1 + \frac{\ddddot{h}}{16} \cdot a_1^3 \right) \tag{4.75}$$

$$g_{s2} = k_1 k_2 \frac{\dot{h}}{2} \cdot |\dot{h}| + k_1 k_2^3 \frac{\ddddot{h}}{16} \cdot |\dot{h}|^3 \tag{4.76}$$

The a_1 amplitude of the first harmonic becomes small after the LMP is found and then accurately tracked, so $g_s \cong g_{s2}$. Thus, this gradient will be used to locate the LMPP based on appropriate design of the gain k_2. Then, the gain k_1 will be given by (4.18) to ensure the stability of the search loop.

Some cases studies and design guideline will be presented in the next section.

4.9.2   Designing to Search for Local Maximums

The simulation parameters are as follows: $f_d = 100$ Hz, $A_m = 0.001$, $k_{Ny} = 1$, $k_{Np} = 3$, $p_0 = 0$, $\gamma_d = 1$, and $\beta_h = 0.5$ and $\beta_l = 3.5$ (which defines the cutoff frequencies for the BPF (4.77), $f_h = \beta_h \cdot f_d$ and $f_l = \beta_l \cdot f_d$):

$$H_{BPF}(s) = H_{HPF}(s) \cdot H_{LPF}(s) = \frac{2\pi \beta_l f_d \cdot s}{(s + 2\pi \beta_l f_d) \cdot (s + 2\pi \beta_h f_d)} \tag{4.77}$$

The design of the tuning gain k_2 will be shown specifically for each case study in range of $k_{2(min)}$ to $k_{2(LMPP)}$,

$$k_{2(min)} < k_2 < k_{2(LMPP)} \tag{4.78}$$

where $\Delta p_{LMPP} = k_{Np} k_{2(LMPP)}$.

4.9.2.1   Search for LMPPs Located on the Same Side as the Starting Point

The search for LMPPs located on the same side as the starting point is presented in Fig. 4.59 using the pattern R, where the LMPPs and GMPP (having the magnitude of $LMPP_1 = 1$, $LPM_2 = 1.875$, $LMPP_3 = 2$, $GMPP = 2.375$, and $LMPP_4 = 1$) are situated at $p_1 = 1$, $p_2 = 1.75$, $p_3 = 2.5$, $p_{GMPP} = 3.25$, and $p_4 = 4$.

Thus, the design based on relationship (4.78) is as follows:

$$\Delta p_{LMPP1} = k_{Np}k_{2(R-LMPP1)} \cong p_1 = 1$$
$$\Delta p_{LMPP2} = k_{Np}k_{2(R-LMPP2)} \cong p_2 = 1.75$$
$$\Delta p_{LMPP3} = k_{Np}k_{2(R-LMPP3)} \cong p_3 = 2.5 \qquad (4.79)$$

So, the tuning gain k_2 will be given by (4.80):

$$k_{2(R-LMPP1)} = 0.2 < 1/3 \cong 0.33$$
$$k_{2(R-LMPP2)} = 0.4 \cong 1.75/3 \cong 0.58$$
$$k_{2(R-LMPP3)} = 0.6 < 2.5/3 \cong 0.83 \qquad (4.80)$$

Using the design relationship (4.78) for GMPP search will result:

$$\Delta p_{GMPP} = k_{Np}k_{2(R-GMPP)} \cong p_{GMPP} = 3.25 \Rightarrow k_{2(R-GMPP)} = 1 < 3.35/3 \cong 1.08$$
$$(4.81)$$

The search for LMPPs located on the same side as the starting point is presented in Fig. 4.60 using the pattern M, where the LMPPs and GMPP (having the magnitude of $LMPP_1 = 1$, $LMPP_2 = 1.875$, $GMPP = 2$, $LMPP_3 = 1.875$, and $LMPP_4 = 1$ are situated at $p_1 = 1$, $p_2 = 1.75$, $p_3 = 2.5$, $p_{GMPP} = 3.25$, and $p_4 = 4$.

Thus, the design based on relationship (4.78) is as follows:

$$\Delta p_{LMPP1} = k_{Np}k_{2(M-LMPP1)} \cong p_1 = 1$$
$$\Delta p_{LMPP2} = k_{Np}k_{2(M-LMPP2)} \cong p_2 = 1.75 \qquad (4.82)$$

So, the tuning gain k_2 will be given by (4.84):

$$k_{2(M-LMPP1)} = 0.3 < 1/3 \cong 0.33, \ k_{2(M-LMPP2)} = 0.6 \cong 1.75/3 \cong 0.58 \quad (4.83)$$

Using the design relationship (4.78) for GMPP search will result:

$$\Delta p_{GMPP} = k_{Np}k_{2(M-GMPP)} \cong p_{GMPP} = 2.5 \Rightarrow k_{2(M-GMPP)} = 0.9 \cong 2.5/3 \cong 0.83$$
$$(4.84)$$

Fig. 4.59 Search for LMPPs
located on the pattern R
using $p(0) = 0$ [26]

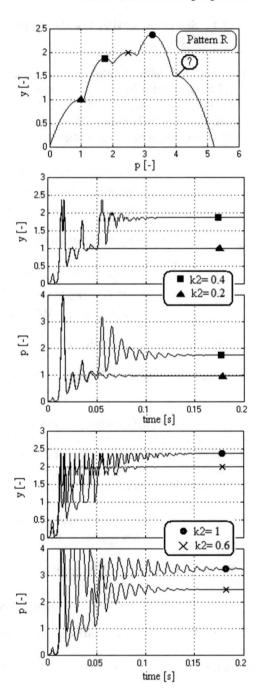

Fig. 4.60 Search for LMPPs located on the pattern R using $p(0) = 0$ [26]

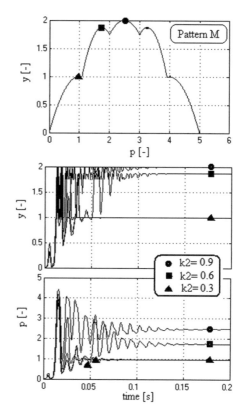

The simulation values for tuning gain k_2 are mentioned in Figs. 4.59 and 4.60 as well.

4.9.2.2 Search for LMPPs Located Opposite the Starting Point

The LMPP$_3$ is located in the opposite side of the starting point, $p_0 = 0$ (across the GMPP), being located at $p_3 = 3.25$ and marked with **x** on the pattern M shown in Fig. 4.61.

Thus, the design based on relationship (4.78) is as follows:

$$\Delta p_{\text{LMPP3}} = k_{\text{Np}} k_{2(\text{M}-\text{LMPP3})} \cong p_3 = 2.5 \tag{4.85}$$

So, if $k_{\text{Np}} = 6$, then the tuning gain k_2 will be given by (4.86):

$$k_{2(\text{M}-\text{LMPP3})} \cong 0.55 \tag{4.86}$$

Using the design relationship (4.78) for GMPP search, the following results:

Fig. 4.61 Search on the
pattern M for LMPPs located
opposite the starting point
$p(0) = 0$ [26]

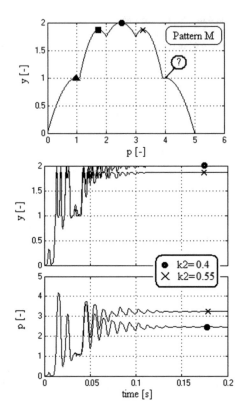

$$\Delta p_{\text{GMPP}} = k_{\text{Np}} k_{2(\text{M}-\text{GMPP})} \cong p_{\text{GMPP}} = 2.5 \Rightarrow k_{2(\text{R}-\text{GMPP})} = 0.4 < 2.5/6 \quad (4.87)$$

The LMPP$_2$ and LMPP$_3$ are located in the opposite side of the starting point, p_0 = 0 (across the GMPP), being located at $p_2 = 2.5$ and $p_3 = 3.25$, and marked with ■ and ▲ on pattern L shown in Fig. 4.62.

Thus, the design based on relationship (4.78) is as follows:

$$\Delta p_{\text{LMPP2}} = k_{\text{Np}} k_{2(L-\text{LMPP2})} \cong p_2 = 2.5, \text{ and}$$
$$\Delta p_{\text{LMPP3}} = k_{\text{Np}} k_{2(L-\text{LMPP3})} \cong p_3 = 3.25 \quad (4.88)$$

So, if $k_{\text{Np}} = 6$, then the tuning gain k_2 will be given by (4.89):

$$k_{2(L-\text{LMPP3})} = 0.4 < 2.5/6 \cong 0.41, \quad k_{3(L-\text{LMPP3})} = 0.5 < 3.25/6 \cong 0.54 \quad (4.89)$$

The value $k_{3(L-\text{LMPP3})} = 0.55 \cong 0.54$ will be also used in simulation to show that this choice is not so restrictive.

Using the design relationship (4.78) for the GMPP search will result:

Fig. 4.62 Search on the
pattern L for LMPPs located
opposite the starting point
$p(0) = 0$ [26]

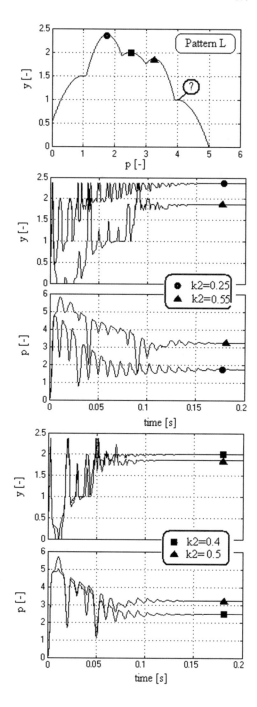

$$\Delta p_{GMPP} = k_{Np}k_{2(L-GMPP)} \cong p_{GMPP} = 1.75 \Rightarrow k_{2(L-GMPP)} = 0.25 < 1.75/6 \cong 0.29 \tag{4.90}$$

So, this design of the tuning gain k_2 can be performed for any patterns using (4.78).

It is worth mentioning that range of the tuning gain k_2 is more restrictive for LMPPs located in the opposite side of the starting point relative to the GMP.

Anyway, the LMP marked with **?** in both Figs. 4.61 and 4.62 cannot be found for any pair of values (k_2, k_1).

The searching resolution (R_{Sj}) for a LMPP$_j$ compared to other LMP$_i$, $i = 1, 2, ...,$ can be estimated using (4.58). For example, the searching resolution ($R_{S(x)}$) for the LMPP$_x$ that is marked with **x** on the pattern R presented in Fig. 4.5 is:

$$R_{S(x)} = \frac{2 - 1.875}{2} \cdot 100 = 6.25\,[\%] \tag{4.91}$$

It is worth mentioning that the design of the tuning and gain parameters for local search is more difficult for an 100% hit count objective.

4.10   Conclusion

The GaPESC scheme has been analyzed in this chapter in order to highlight its capacity for global and local search using different multimodal patterns and case studies such as PV/WT/FC hybrid power system. The performance is measured for 100% hit count and is as follows: The searching resolution is less than 1%, the stationary tracking accuracy is higher than 99.99%, tracking speed is less than 10 periods of the sinusoidal dither, of maximum 15 iterations, and tracking efficiency is higher than 99.96%.

The design for stability has also been presented considering the aforementioned case studies.

The robustness has been tested for variable irradiance, partial shading conditions, noise added in the system under test, different frequencies and shapes of the dither, etc.

So, the GaPESC scheme will be considered as GMPP tracking algorithm to optimize the systems analyzed in the next chapters.

References

1. Zhang C, Ordóñez R (2012) Extremum-seeking control and applications: a numerical optimization-based approach. Springer, London Limited
2. Ariyur KB, Krstic M (2003) Real-time optimization by extremum-seeking control. Wiley, Hoboken

3. Fu L, Özgüner Ü (2009) Variable structure extremum seeking control based on sliding mode gradient estimation for a class of nonlinear systems. In: Proceedings of the American control conference, pp 8–13
4. Ghaffari A, Seshagiri S, Krstic M (2012) Power optimization for photovoltaic micro-converters using multivariable gradient-based extremum-seeking. In: Proceedings of the American control conference, pp 3383–3388
5. Dochain D, Perrier M, Guay M (2011) Extremum seeking control and its application to process and reaction systems: a survey. Math Comput Simulat 82:369–380
6. Zhang C, Ordóñez R (2009) Robust and adaptive design of numerical optimization-based extremum seeking control. Automatica 45:634–646
7. Yang XS (2010) Engineering optimization: an introduction with metaheuristic applications. Wiley, Hoboken
8. Bizon N, Tabatabaei NM, Blaabjerg F, Kurt E (2017) Energy harvesting and energy efficiency: technology, methods and applications. Springer, Available on http://www.springer.com/us/book/9783319498744
9. Zhang C, Siranosian A, Krstic M (2007) Extremum seeking for moderately unstable systems and for autonomous target tracking without position measurements. Automatica 3:1832–1839
10. Ghods N, Krstic M (2011) Source seeking with very slow or drifting sensors. ASME J Dyn Syst Meas Control 133(4):8
11. Sugihira S, Ichikawa K, Ohmori H (2007) Starting speed control of SI engine based on online extremum control. In: SICE annual conference, pp 2569–2573
12. Tanelli M, Astolfi A, Savaresi SM (2006) Non-local extremum seeing control for active braking control systems. In: Proceedings of the American Control Conference, pp 891–96
13. Schneider G, Ariyur KB, Krstic M (2000) Tuning of a combustion controller by extremum seeking: a simulation study. In: Proceedings of the 39th conference on decision and control, vol 5, pp 5219–5223
14. Tan Y, Nešic D, Mareels I (2006) On non-local stability properties of extremum seeking control. Automatica 42(6):889–903
15. Beaudoin JF, Cadot O, Aider JL, Wesfreid JE (2006) Bluff-body drag reduction by extremum-seeking control. J Fluids Struct 22(6–7):973–978
16. Carnevale D, Astolfi A, Centioli C, Podda S, Vitale V, Zaccarian L (2009) A new extremum seeking technique and its application to maximize RF heating on FTU. Fusion Eng Des 84(2–6):554–558
17. Schuster E, Romero C, Yao Z, Si F (2010) Integrated real-time optimization of boiler and post combustion system in coal-based power plants via extremum seeking. In: IEEE international conference on control applications (CCA), pp 2184–2189
18. Bizon N (2020) Energy efficiency of fuel cell/renewable energy sources hybrid power source with controlled fuelling flows. Mathematics 8(2):151–173. Available on https://www.mdpi.com/2227-7390/8/2/151
19. Wang H-H, Yeung S, Krstic M (2000) Experimental application of extremum seeking on an axial-flow compressor. IEEE Trans Control Syst Technol 8(2):300–309
20. Bizon N (2014) Improving the PEMFC energy efficiency by optimizing the fuelling rates based on extremum seeking algorithm. Int J Hydr Energ 39(20):10641–10654
21. Ishaque K, Salam Z (2013) A review of maximum power point tracking techniques of PV system for uniform insolation and partial shading condition. Renew Sustain Energy Rev 19:475–488
22. Bidram A, Davoudi A, Balog RS (2012) Control and circuit techniques to mitigate partial shading effects in photovoltaic arrays. IEEE J Photovolt 2:532–546
23. Liu Z-H, Chen J-H, Huang J-W (2015) A review of maximum power point tracking techniques for use in partially shaded conditions. Renew Sustain Energy Rev 41:436–453
24. Bizon N (2018) Optimal operation of fuel cell/wind turbine hybrid power system under turbulent wind and variable load. Appl Energ 212:196–209
25. Garcıa M, Maruri JM, Marroyo L, Lorenzo E, Perez M (2008) Partial shadowing, MPPT performance and inverter configurations: observations at tracking PV plants. Prog Photovolt Res Appl 16:529–536

26. Bizon N (2017) Searching of the extreme points on photovoltaic patterns using a new asymptotic perturbed extremum seeking control scheme. Energ Convers Manag 144:286–302
27. Bizon N, Kurt E (2017) Performance analysis of the tracking of global extreme on multimodal patterns using the asymptotic perturbed extremum seeking control scheme. Int J Hydr Energ 42(28):17645–17654
28. Bizon N (2016) Global maximum power point tracking (GMPPT) of photovoltaic array using the extremum seeking control (ESC): a review and a new GMPPT ESC scheme. Renew Sustain Energy Rev 57:524–539
29. Bizon N (2016) Global extremum seeking control of the power generated by a photovoltaic array under partially shaded conditions. Energy Convers Manag 109:71–85
30. Bizon N (2016) Global maximum power point tracking based on new extremum seeking control scheme. Prog Photovolt Res Appl 24(5):600–622
31. Orioli A, Di Gangi A (2014) Review of the energy and economic parameters involved in the effectiveness of grid-connected PV systems installed in multi-storey buildings. Appl Energy 113:955–969
32. Reisi AR, Moradi HM, Jamas S (2013) Classification and comparison of maximum power point tracking techniques for photovoltaic system: a review. Renew Sustain Energy Rev 19:433–443
33. Fortunato M, Giustiniani A, Petrone G, Spagnuolo G, Vitelli M (2008) Maximum power point tracking in a one-cycle-controlled single-stage photovoltaic inverter. IEEE Trans Ind Electron 55:2684–2693
34. Patel H, Agarwal V (2009) MPPT scheme for a PV-fed single-phase single-stage grid-connected inverter operating in CCM with only one current sensor. IEEE Trans Energy Convers 24:256–263
35. Harada K, Zhao G (1993) Controlled power interface between solar cells and AC source. IEEE Trans Power Electron 8:654–662
36. Qiang M, Mingwei S, Liying L, Guerrero JMA (2011) Novel improved variable step-size incremental-resistance MPPT method for PV systems. IEEE Trans Ind Electron 58:2427–2434
37. Koutroulis E, Kalaitzakis K, Voulgaris NC (2001) Development of a microcontroller-based, photovoltaic maximum power point tracking control system. IEEE Trans Power Electron 16:46–54
38. Safari A, Mekhilef S (2011) Simulation and hardware implementation of incremental conductance MPPT with direct control method using Cuk converter. IEEE Trans Ind Electron 58:1154–1161
39. Kimball JW, Krein PT (2008) Discrete-time ripple correlation control for maximum power point tracking. IEEE Trans Power Electron 23:2353–2362
40. Yan Hong L, Hamill DC (2000) Simple maximum power point tracker for photo-voltaic arrays. Electron Lett 36:997–999
41. Bodur M, Ermis M (1994) Maximum power point tracking for low power photo-voltaic solar panels. In: Proceedings of 7th Mediterranean electrotechnical conference, vol 752, pp 758–61
42. Shmilovitz D (2005) On the control of photovoltaic maximum power point tracker via output parameters. IEE Proc Electr Power Appl 152:239–248
43. Hart GW, Branz HM, Cox Iii CH (1984) Experimental tests of open-loop maximum-power-point tracking techniques for photovoltaic arrays. Solar Cells 13:185–195
44. HJ Noh, DY Lee, DS Hyun (2002) An improved MPPT converter with current compensation method for small scaled PV-applications. In: Proceedings of IECON conference, vol 1112, pp 1113–1118
45. Chiang SJ, Chang KT, Yen CY (1998) Residential photovoltaic energy storage system. IEEE Trans Ind Electron 45:385–394
46. Bleijs JAM, Gow A (2001) Fast maximum power point control of current-fed DC–DC converter for photovoltaic arrays. Electr Lett 37:5–6
47. Salas V, Olías E, Barrado A, Lázaro A (2006) Review of the maximum power point tracking algorithms for stand-alone photovoltaic systems. Sol Energ Mat Sol C 90(11):1555–1578
48. Esram T, Chapman PL (2007) Comparison of photovoltaic array maximum power point tracking techniques. IEEE Trans Energy Convers 22:439–449

49. Subudhi B, Pradhan R (2013) A comparative study on maximum power point tracking techniques for photovoltaic power systems. IEEE Trans Sustain Energy 4:89–98
50. de Brito MAG, Galotto L, Sampaio LP, de Azevedo e Melo G, Canesin CA (2013) Evaluation of the main MPPT techniques for photovoltaic applications. IEEE Trans Ind Electr 60:1156–1167
51. Bhatnagar P, Nema RK (2013) Maximum power point tracking control techniques: state-of-the-art in photovoltaic applications. Renew Sustain Energy Rev 23:224–241
52. Eltawil MA, Zhao Z (2013) MPPT techniques for photovoltaic applications. Renew Sustain Energy Rev 25:793–813
53. Zhou L, Chen Y, Guo K, Jia F (2011) New approach for MPPT control of photovoltaic system with mutative-scale dual-carrier chaotics search. IEEE Trans Power Electr 26:1038–1048
54. Syafaruddin, Karatepe E, Hiyama T (2009) Polar coordinated fuzzy controller based real-time maximum power point control of photovoltaic system. Renew Energy 34:10
55. Karlis AD, Kottas TL, Boutalis YS (2007) A novel maximum power point tracking method for PV systems using fuzzy cognitive networks (FCN). Electr Pow Syst Res 77(3–4):315–327
56. Liao C-C (2010) Genetic k-means algorithm based RBF network for photovoltaic MPP prediction. Energy 35(2):529–536
57. Shaiek Y, Smida MB, Sakly A, Mimouni MF (2013) Comparison between conventional methods and GA approach for maximum power point tracking of shaded solar PV generators. Sol Energy 90:107
58. Tajuddin MFN, Ayob SM, Salam Z (2012) Tracking of maximum power point in partial shading condition using differential evolution (DE). In: Proceedings of the 2012 IEEE international conference on power and energy (PECon), p 384
59. Ishaque K, Salam Z (2013) A deterministic particle swarm optimization maximum power point tracker for photovoltaic system under partial shading condition. IEEE Trans Ind Electron 60:3195
60. Jianga LL, Maskell DL, Patra JC (2013) A novel ant colony optimization-based maximum power point tracking for photovoltaic systems under partially shaded conditions. Energy Build 58:227–236
61. Sher HA, Addoweesh KE (2012) Micro-inverters—promising solutions in solar photovoltaics. Energy Sustain Dev 16:389–400
62. Tan Y, Nešic D, Mareels I, Astolfi A (2009) On global extremum seeking in the presence of local extrema. Automatica 45(1):245–251
63. Ji YH, Jung DY, Kim JG, Kim JH, Lee TW, Won CY (2011) A real maximum power point tracking method for mismatching compensation in PV array under partially shaded conditions. IEEE Trans Power Electron 26:1001–1009
64. Bizon N, Thounthong P, Raducu M, Constantinescu LM (2017) Designing and modelling of the asymptotic perturbed extremum seeking control scheme for tracking the global extreme. Int J Hydr Energy 42(28):17632–17644
65. Krstic M, Wang H-H (2000) Design and stability analysis of extremum seeking feedback for general nonlinear systems. Automatica 36(2):595–601
66. Krstic M (2000) Performance improvement and limitations in extremum seeking control. Syst Contr Lett 39(5):313–326
67. Bizon N (2013) Energy harvesting from the FC stack that operates using the MPP tracking based on modified extremum seeking control. Appl Energ 104:326–336
68. Bizon N (2014) Tracking the maximum efficiency point for the FC system based on extremum seeking scheme to control the air flow. Appl Energ 129:147–157
69. Bizon N (2013) FC energy harvesting using the MPP tracking based on advanced extremum seeking control. Int J Hydr Energ 38(4):1952–1966
70. Bizon N, Tabatabaei NM, Shayeghi H (2013) Analysis, control and optimal operations in hybrid power systems—advanced techniques and applications for linear and nonlinear systems. Springer. Available on http://dx.doi.org/10.1007/978-1-4471-5538-6
71. Leyva R, Alonso C, Queinnec I, Cid-Pastor A, Lagrange D, Martínez-Salamero L (2006) MPPT of photovoltaic systems using extremum-seeking control. IEEE Trans Aerosp Electron Syst 42(1):249–258

72. Esmaeilzadeh Azar F, Perrier M, Srinivasan B (2011) A global optimization method based on multi-unit extremum-seeking for scalar nonlinear systems. Comput Chem Eng 35:456–463
73. Bizon N, Thounthong P (2018) Real-time strategies to optimize the fueling of the fuel cell hybrid power source: a review of issues, challenges and a new approach. Renew Sustain Energy Rev 91:1089–1102
74. Pukrushpan J, Stefanopoulou A, Peng H (2004) Control of fuel cell power systems: principles, modeling and analysis and feedback design. Springer
75. Sarvi M, Ahmadi S, Abdi S (2015) A PSO-based maximum power point tracking for photovoltaic systems under environmental and partially shaded conditions. Prog Photovolt Res Appl 23(2):201–214
76. Tan T, Dragan N, Iven M (2008) On the choice of dither in extremum seeking systems: a case study. Automatica 44:1446–1450
77. SimPowerSystems TM Reference (2010) Hydro-Québec and the MathWorks, Inc. Natick, MA
78. Rezaeiha A, Kalkman I, Blocken B (2017) Effect of pitch angle on power performance and aerodynamics of a vertical axis wind turbine. Appl Energy 197:132–150
79. Njiri JG, Dirk Söffker D (2016) State-of-the-art in wind turbine control: trends and challenges. Renew Sustain Energy Rev 60:377–393
80. Govind B (2017) Increasing the operational capability of a horizontal axis wind turbine by its integration with a vertical axis wind turbine. Appl Energy 199:479–494

Chapter 5
Fuel Cell Net Power Maximization Strategies

5.1 Introduction

Polymer electrolyte membrane fuel cells (PEMFCs) produce electrical energy based on the chemical reaction of hydrogen and oxygen (air) [1, 2]. PEMFC has become a promising alternative for mobile applications [3, 4], but also for distributed generation that uses PEMFC as a backup source in hybrid power systems (HPS) [5, 6].

The energy efficiency of PEMFC is in the range of 40–50% (reaching almost 85% in cogeneration mode) [7, 8], being competitive compared to competing technologies (which have an energy efficiency in the range of 30–35%) [9, 10]. An increase in energy efficiency can be achieved by optimizing PEMFC operation using advanced control and optimization loops [11, 12]. In general, the design should focus on subsystems that significantly influence the efficiency and safety of the PEMFC system, namely [13, 14]: (1) the air and fuel supply subsystems, (2) the water supply subsystem, and (3) the heat management subsystem. This chapter approaches the optimization of the air and fuel supply subsystems in order to increase the electrical energy efficiency of the PEMFC system.

The PEMFC system can function safely if all auxiliary equipment (such as the air compressor, humidifier, solenoid valves, water and fuel circulation pumps, and the measuring and control equipment) are optimally designed and in good working order [15, 16]. The air compressor consumes more than 80% of the PEMFC power allocated to supply auxiliary equipment, which represents up to 20% of the nominal power of the PEMFC system [17, 18]. Consequently, maximizing the net power of PEMFC is a challenging objective that can be achieved by optimizing the performance parameter, called electrical energy efficiency [19, 20]. The PEMFC system will operate near the maximum efficiency point (MEP) by real-time optimization of energy efficiency [21, 22]. This chapter looks at seven energy efficiency optimization strategies that will operate the PEMFC system near MEP.

© The Editor(s) (if applicable) and The Author(s), under exclusive license to Springer Nature Switzerland AG 2020
N. Bizon, *Optimization of the Fuel Cell Renewable Hybrid Power Systems*,
Green Energy and Technology, https://doi.org/10.1007/978-3-030-40241-9_5

It is noteworthy that the MEP is difficult to track compared to the maximum power point (MPP) because it depends on a larger number of PEMFC parameters, such as supply, humidification, cooling control, and electrical interface subsystems, besides load [23, 24]. The MPP tracking control schemes are simpler than the MPP tracking control schemes and are still being implemented [25, 26] even though the energy efficiency of a PEMFC system operating at MPP is lower than that of the MEP.

This chapter analyzes the tracking techniques of MEP who uses an extremum seeking (ES) scheme [27] that is improved in comparison with the ES basic control schemes [28], ensuring greater search speed and improved accuracy in tracking [30]. The static Feed-Forward (sFF) strategy and the sFF strategy combined with PI control [31, 32] will be used in this study as references.

The MEP strategies use tracking control algorithms based on different control techniques, such as the feedback linearization [33], dynamic feed-forward feedback [34], perturb and observe algorithm [25, 35], LQR/LRS strategies [25], sliding mode [36, 37, 39–41], supper twisting algorithm [42, 43], ES control schemes [43–47], predictive model [48–52], nonlinear differential flatness-based control [53–55], time delay control [56], adaptive control strategies [57–62], and intelligent algorithms using neural networks [63], fuzzy logic [64–66], and differential evolution algorithm [67]. With the exception of ES and adaptive techniques, the above-mentioned studies require search tables, the value of PEMFC status variables that must be acquired using sensors or estimated by observers [68, 69]. In addition, the design of the controller is not robust to the uncertainties of the PEMFC system to variable renewable power and dynamic load [70–72]. Furthermore, improper control may damage the PEMFC system [73–75], so fault-tolerant strategies have been proposed recently [76]. Fuel and oxygen starvation can be avoided by maintaining the oxygen excess ratio (OER) higher than 1.5 [77–79] and limiting the fueling ratio by 100 A/s slope limiters included in both the fueling regulators [80]. The OER-based control is not a real-time solution due to a response time higher than 1 s, the OER measurement is complex and has low accuracy (with tolerance in the range of 1–10%) [81], the OER sensor lifetime is up to 5 years, and its price is still high [35, 65, 82]. The hydrogen and oxygen stoichiometric ratio is recommended to be controlled for an OER in the range of 1.5–3 [81] using low power and speed air compressors [83].

Four MEP-tracking strategies analyzed in this chapter regulate the air or fuel pressure based on the load-following [65] or MEP-tracking [35] modes. The electrical energy efficiency using the best strategy increases with up to 2.6% in comparison with the sFF strategy [86]. Another three MEP-tracking strategies analyzed in this chapter regulate the FC current based on the load-following and the air or fuel pressure based on the MEP-tracking modes [81]. Also, the electrical energy efficiency using the best strategy of this class increases with up to 2.6% in comparison with the sFF strategy. But major differences between the MEP-tracking strategies have been highlighted in fuel economy [88]. A systematic analysis of these MEP-tracking strategies will be performed in this chapter.

The schematic diagram of the MEP-tracking strategy using the load-following for fuel regulator and the MEP-tracking for both the air regulator and the FC current is

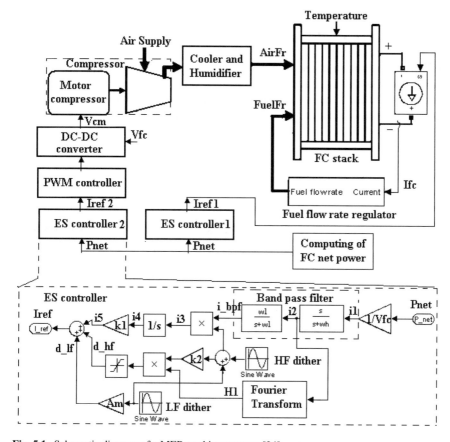

Fig. 5.1 Schematic diagram of a MEP-tracking strategy [24]

presented in Fig. 5.1. Detailed simulation diagrams for each MEP-tracking strategy will be shown in the chapter.

The rest of the chapter is structured as follows: Sect. 5.2 presents the FC net power control strategies based on the load-following mode control and the GES algorithms, highlighting some performance requested to operate in real time; Sect. 5.3 details the behavior of the FC system under control strategies based on load-following mode using the performance indicators such as the electrical efficiency of the FC system, fuel consumption efficiency, and fuel economy; the improvements in FC net power through the ESC-based control of the air regulator are analyzed in Sect. 5.4 for two air compressors; Sect. 5.5 analyzes the improvements in FC net power through the ESC-based control of both the air regulator and the FC current, highlighting the searching advantage with two variables of the maximum efficiency point (MEP); the improvements in FC net power through the ESC-based control of both the fuel regulator and the FC current are analyzed in Sect. 5.6. Performance analysis of all seven strategies is systematically performed in Sect. 5.7 considering the aforementioned indicators. The strategies have been classified based on load-following mode in three classes and the best strategy in each class has been identified. The behavior of the FC system under the best control strategies is further explored in Sect. 5.8 using different dither's frequency. A DC-DC boost converter is used to extract the maximum FC net power using a GES controller with 99.9% tracking accuracy for a dither's frequency in the range of 1–100 Hz. The last section concludes the chapter.

5.2 FC Net Power Control Strategies Based on Load-Following Mode

The load-following (LFW) mode can be used to control the DC-DC boost converter or one of the fueling regulators. Thus, one or two control inputs from the three available inputs to control the FC power can be used to operate the FC system close to MEP or optimum point defined by optimizer function. Consequently, seven strategies may be explored as performance considering the simulation diagram from Fig. 5.2.

The first strategy (called as Fuel-GES-Boost-LFW-RTO strategy [89]) uses the load-following and GES modes to control the DC-DC boost converter and the fuel regulator (see Fig. 5.3). The air regulator is controlled by FC current, which, due to small variation of the FC voltage with FC power level, will approximately follow the loading power.

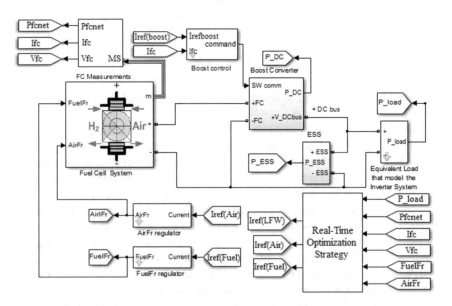

Fig. 5.2 Diagram of the FC system [88]

Fig. 5.3 Diagram of the first strategy

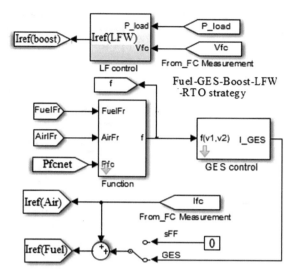

The second strategy (called Air-GES-Boost-LFW-RTO strategy [90]) uses the load-following and GES modes to control the DC-DC boost converter and the air regulator (see Fig. 5.4). The fuel regulator is controlled by the FC current, which, due to small variation of the FC voltage with FC power level, will approximately follow the loading power.

The third strategy (called Boost-GES-Air-LFW-RTO strategy [47]) uses the load-following and GES modes to control the air regulator and the DC-DC boost converter (see Fig. 5.5). The fuel regulator is controlled by FC current, which, due to small

Fig. 5.4 Diagram of the second strategy

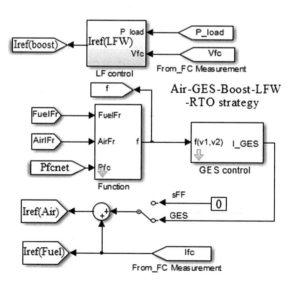

Fig. 5.5 Diagram of the
third strategy

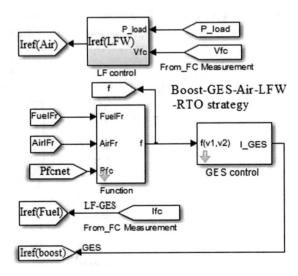

variation of the FC voltage with FC power level, will approximately follow the
loading power.

The fourth strategy (called Boost-GES-Fuel-LFW-RTO strategy [86]) uses the
load-following and GES modes to control the air regulator and the DC-DC boost
converter (see Fig. 5.6). The fuel regulator is controlled by FC current, which, due
to small variation of the FC voltage with FC power level, will approximately follow
the loading power.

The fifth strategy (called Boost-GES1-Fuel-GES2-Air-LFW-RTO strategy [91])
uses the load-following-based control for the air regulator and GES-based control
for the DC-DC boost converter and the fuel regulator (see Fig. 5.7).

Fig. 5.6 Diagram of the
fourth strategy

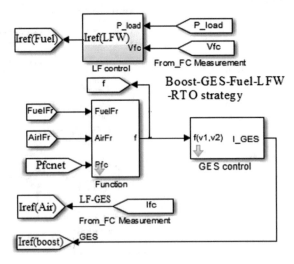

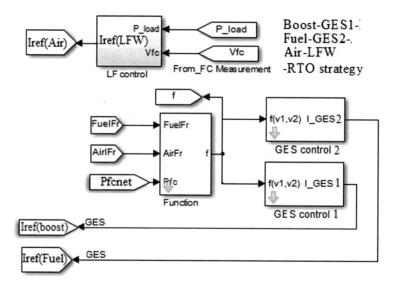

Fig. 5.7 Diagram of the fifth strategy

The sixth strategy (called Boost-GES1-Air-GES2-Fuel-LFW-RTO strategy [92])
uses the load-following-based control for the fuel regulator and GES-based control
for the DC-DC boost converter and the air regulator (see Fig. 5.8).

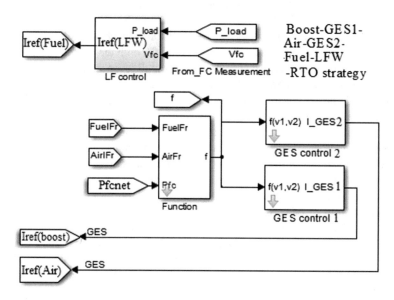

Fig. 5.8 Diagram of the sixth strategy

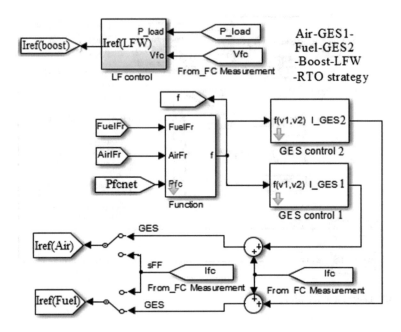

Fig. 5.9 Diagram of the seventh strategy

The seventh strategy (called Air-GES1-Fuel-GES2-Boost-LFW-RTO strategy [22]) uses the load-following-based control for the DC-DC boost converter and GES-based control for both the air and fuel regulators (see Fig. 5.9).

Note that the sFF-based strategy uses the load-following-based control for the DC-DC boost converter and FC current-based control for both the air and fuel regulators [31]. In this manner, the performance evaluation for all seven strategies defined above will be fairly made compared to the sFF-based strategy.

In general, the MEP is located close to the FC current value set by the sFF control for a load demand. So, the searching ranges for the AirFr and FuelFr regulators are limited around the FC current, where the MEP will be found in real time by a GES control. If not, other variants of the aforementioned strategies can be considered. For example, the variants of the strategies 5 and 6 with direct GES control of the AirFr and FuelFr regulators are presented in Figs. 5.10 and 5.11.

It is worth mentioning the disadvantage of the variants under pulsed load related to lower tracking speed compared to original strategies.

Since the FC system will operate under variable load, the load-following (LFW) control is used for the DC-DC converter or the fueling regulators of the FC system to ensure the needed power on the DC bus where the batteries' stack is directly connected and operated in charge-sustained mode. If the FC system will operate in constant-power mode at maximum efficiency point (MEP), many charge-discharge cycles will stress the battery due to load variation. Avoiding these charge-discharge cycles by using the LFW control for the FC system, the battery lifetime will increase

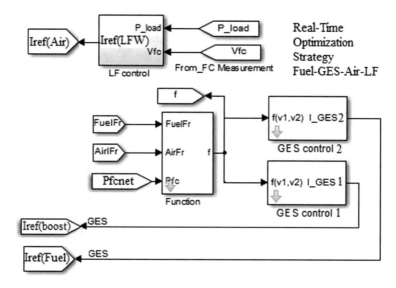

Fig. 5.10 Diagram of the variant for the strategy 5

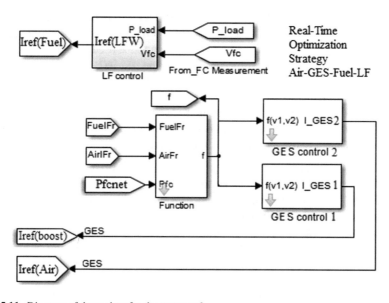

Fig. 5.11 Diagram of the variant for the strategy 6

significantly. Due to optimization loops, the FC power can be a bit different to the load power requested on the DC bus. These energy differences will be supplied by the batteries. The supercapacitors' stack will ensure the dynamic balance of power to a dynamic load profile, assuring a safe operation of the FC system.

The real-time optimization (RTO) strategies based on the advanced algorithm to search the global maximum (GM) could be an alternative to solving the main problems identified in optimizing the operation of the FC system [22]:

- The $P_{FCnet} = f(AirFf, FuelFr)$ surface, or $P_{FCnet} = f(AirF2)$ and $P_{FCnet} = f(FuelFr)$ curves have a unique MEP and many other peaks around the MEP [93]. So, an RTO strategy based on a global maximum power point tracking algorithm (GMPPT) is required;
- The MEP position on the optimizer function plateau varies depending on the complex operating conditions of the FC system [22, 31]. So, an RTO strategy based on an adaptive GMPPT algorithm is recommended, such as one based on global extremum seeking (GES) algorithms proposed for optimizing photovoltaic systems [94–96];
- The MEP position may be close to the local peaks, so the GMPPT algorithm must have a tracking precision of more than 99% to ensure maximum efficiency of available FC power extraction;
- The MEP position must be quickly located (in real time) on the plateau, so an RTO strategy based on a GMPPT algorithm with high tracking speed is required, but within the safe operating limits of the FC system (e.g., a transient search less than 0.1 s, but with slopes limited to 100 A/s by using current slope limiters in air and fuel regulators [80]). GES algorithms need less than 10 periods of perturbation to locate GM, which means less than 0.1 s for a 100 Hz sinusoidal disturbance [94–96]. Note that GMPPT firmware algorithms cannot be used because they work in two stages and cannot search for GM in real time [97, 98].
- The implementation complexity for RTO based on GES algorithms is low, but practical implementations for the FC system have yet to be reported; some hardware-in-the-loop (HIL) implementations were reported in [99];
- RTO strategies do not require periodic adjustment of tuning parameters [100, 101]. Again, the need to use an adaptive algorithm arises.

Following simulation tests (see Fig. 5.12), variant 2 with two band filters (called as GaPESCbpf) and variant 3 using the fast Fourier transform (FFT) (called as GaPESCH1) were chosen in this study because they ensure stable localization (k_2) and tracking (k_1), where k_1 and k_2 are the control parameters to be projected.

Examples of GM search on a multimodal curve and surface are presented in Figs. 5.13 and 5.14. The sweeping of the GM is detailed in the top plot of Fig. 5.13. The evolution of the perturbation amplitude (100 Hz sinusoidal signal) is shown in the bottom plot of Fig. 5.13.

Searching for GM_{M1} starting from different points chosen in a search space near other local peaks is presented in Fig. 5.14. It is worth to mention that the search always converges to GM_{M1}.

Note the search convergence to the GM_{M1} for whatever point initially selected in the search range (Fig. 5.14). It can also be seen on the zooms shown in Fig. 5.15 that the oscillations during stationary mode are negligible and the tracking accuracy is greater than 99.99%. The reader interested in the design and performance of the GES algorithms can find design examples in Chap. 4 of this book.

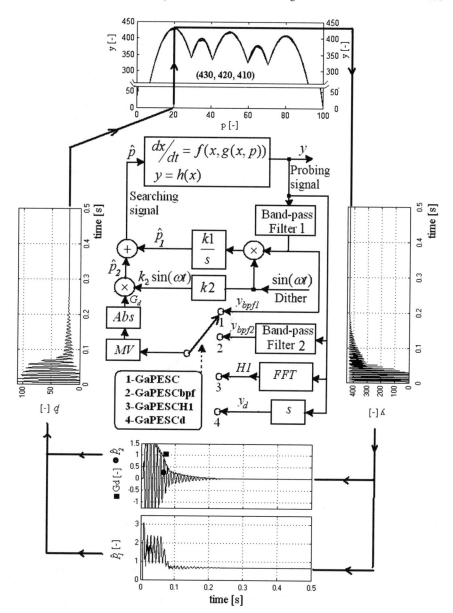

Fig. 5.12 Diagram of the GES algorithms [94]

Fig. 5.13 GM search on a
multimodal curve [94]

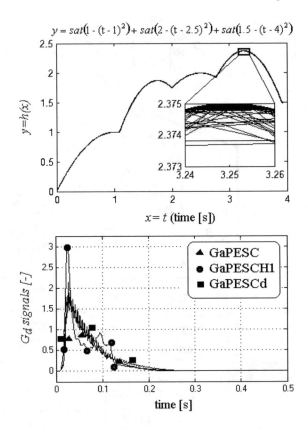

5.3 Behavior of the FC System Under Control Strategies Based on Load-Following Mode

The FC system using the control strategies based on load-following mode will behave in almost the same mode and the signals' shapes will look similar for any strategy. Thus, only to exemplify the behavior of the FC system under control strategies based on load-following mode, the main signals for the FC system using the first strategy (Boost-GES-Fuel-LFW-RTO) and sFF-based strategy are presented in Fig. 5.16.

It is noticed that the fuel flow has a load-like profile due to the LFW control, so the power generated by the FC system ensures the variable load demand, and $p_{ESS} \cong 0$ (see the second plot in Fig. 5.16). It follows that the batteries will be operated in charge-sustained mode regardless of the load demand profile. So, to ensure the balance of powers on the DC bus (see relationship 1), it is necessary to have a low battery capacity or even a lack thereof, dynamic compensation being achieved via a DC-DC bidirectional converter by using an appropriately designed supercapacitors' stack, besides the capacitor C_{dc} directly connected on the DC bus:

$$C_{dc}u_{dc}\mathrm{d}u_{dc}/\mathrm{d}t = h_{boost}p_{FCnet} + p_{ESS} - p_{Load} \qquad (5.1)$$

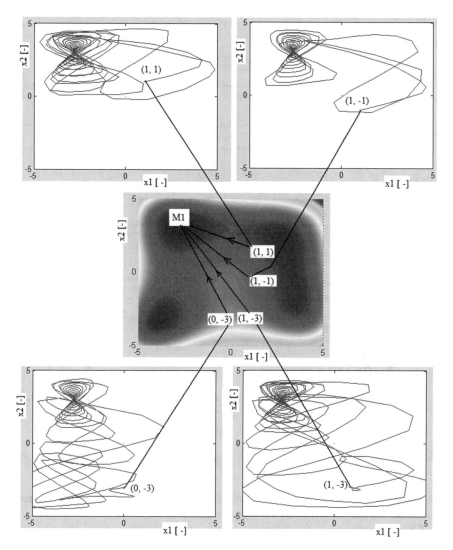

Fig. 5.14 GM search on a multimodal surface [47]

The reference current I_{refLFW} (which is generated by the LFW controller) is estimated as follows, considering the average value (AV) in relation (2) and the energy storage system (ESS) operating in charge-sustaining mode ($P_{ESS(AV)} \cong 0$):

$$I_{ref(LFW)} = I_{FC(AV)} = P_{Load}/(V_{FC} \times h_{boost(AV)}) \tag{5.2}$$

The boost converter is controlled by the GES control to optimize the performance of the FC system. Modifying the power generated by the FC system (so the FC current

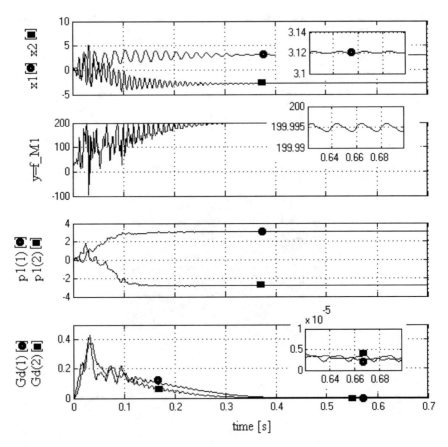

Fig. 5.15 Searching of the GM$_{M1}$ for GMM1 using dithers frequencies $\omega_2 = 2\omega_1 = 100$ Hz [47]

as well), the airflow in the optimization loop will change by default. Performance indicators are shown in the last three graphs in Fig. 5.16.

Although apparently the differences are minor in performance indicators, the representation of fuel consumption in Fig. 5.16 shows a fuel economy of over 4 L for a 12-second cycle (i.e., about 20 lpm) for the Boost-GES-Fuel-LFW-RTO strategy (fuel consumption of 85.2 L), compared to a static Feed-Forward (sFF) commercial algorithm (fuel consumption of 85.24 L).

The differences in performance indicators will be highlighted below. Obviously, the load demand profile is the same in all cases.

Figure 5.17 shows the performance of the strategy 1 (Fuel-GES-Boost-LFW-RTO) compared to the sFF strategy. It is noticed that the fuel economy is over 4 L for a cycle of 12 s (i.e., about 20 lpm).

The performance of the strategy 2 (Air-GES-Boost-LFW-RTO) compared to the sFF strategy is illustrated in Fig. 5.18. For the same load demand profile, fuel economy is only 1.25 L per cycle of 12 s (i.e., about 5 lpm).

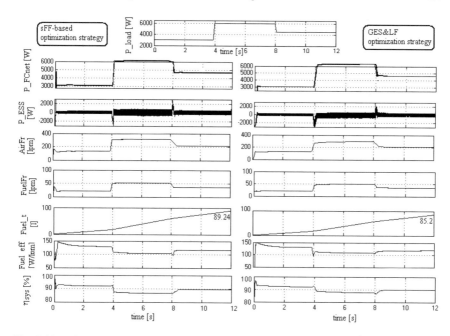

Fig. 5.16 Behavior of the FC system using the Boost-GES-Fuel-LFW-RTO strategy (right) and sFF-based strategy (left) [47]

The performance of the Boost-GES-Air-LFW-RTO strategy compared to the sFF strategy is illustrated in Fig. 5.19. For the same load demand profile, fuel savings are now around 10 L per cycle of 12 s (i.e., approx. 50 lpm).

As it was mentioned before, Fig. 5.20 shows the fuel economy of over 4 L for a 12-second cycle (i.e., about 20 lpm) for the strategy 4 (Boost-GES-Fuel-LFW-RTO), compared to a commercial algorithm based on sFF control.

The strategies analyzed above (strategies 1–4) use a single GES controller to optimize the FC system operation. Considering the aforementioned results, for the 6 kW (AV) load it is advisable to control LFW for airflow and GES optimization by control of the boost converter.

Next, the case of strategies using 2 GES controllers is analyzed, i.e., strategies 5–7, in order to highlight improvements in fuel economy.

The performance of the strategy 5 (Boost-GES1-Air-GES2-Fuel-LFW-RTO) compared to the sFF strategy is illustrated in Fig. 5.21, and compared to strategy 4 (Boost-GES-Fuel-LFW-RTO), a double fuel economy (8 l/12 s = 40 lpm) is observed. Both strategies use LFW fuel flow control.

To analyze the effect of using two GES controllers and LFW control for airflow, Fig. 5.22 illustrates the performance of the Boost-GES1-Fuel-GES2-Air-LFW-RTO strategy compared to the sFF strategy. It is observed that fuel savings are tripled (30 l/12 s = 120 lpm) compared to the Boost-GES-Air-LFW-RTO strategy (Fig. 5.19). So, the same advantage of better fuel economy results in case of using two GES controllers instead of one.

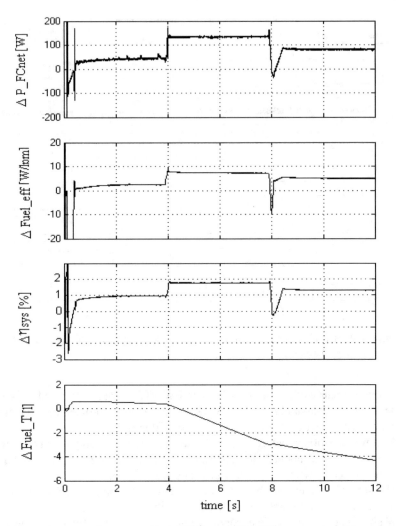

Fig. 5.17 Performance of strategy 1 (Fuel-GES-Boost-LFW-RTO) compared to sFF strategy

However, it is not the case for strategy 7 (Air-GES1-Fuel-GES2-Boost-LFW-RTO) due to use of the LFW control for the DC-DC boost converter (as in the strategies 1 and 2). The strategy 7 will be separately analyzed to highlight its advantage (better performance in full range of load) over sFF control.

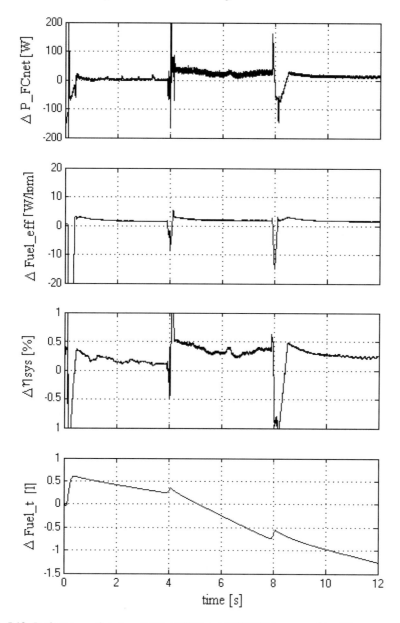

Fig. 5.18 Performance of strategy 2 (Air-GES-Boost-LFW-RTO) compared to sFF strategy

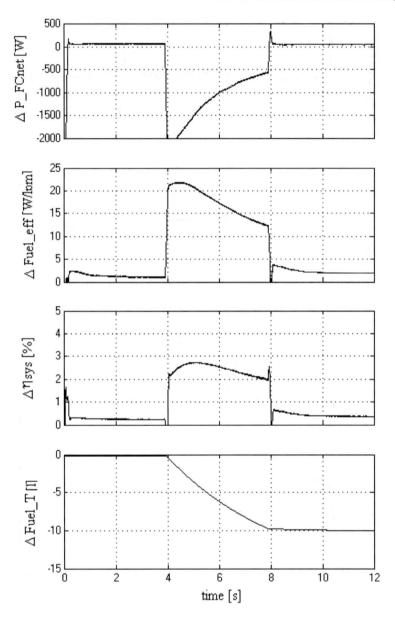

Fig. 5.19 Performance of strategy 3 (Boost-GES-Air-LFW-RTO) compared to sFF strategy

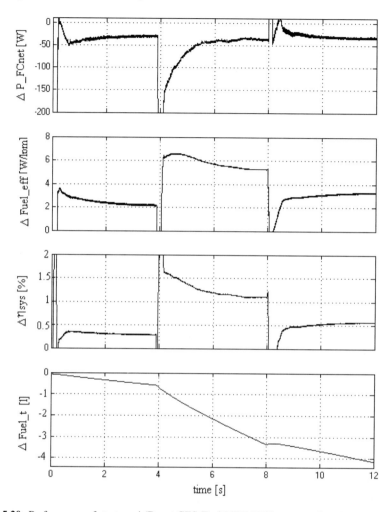

Fig. 5.20 Performance of strategy 4 (Boost-GES-Fuel-LFW-RTO) compared to sFF strategy [47]

5.4   Improve FC Net Power Through the ESC-Based Control of the Air Regulator

The simulation diagram is presented in Fig. 5.23. The FC system 2 (see the bottom of Fig. 5.23) uses the sFF control for fueling regulators.

The sFF control of the FuelFr and AirFr will be performed via the FC current using (Eqs. 5.3 and 5.4):

$$\text{FuelFr} = \frac{60000 \cdot R \cdot (273 + \theta) \cdot N_C \cdot i_{FC}}{2F \cdot (101325 \cdot P_{f(H_2)}) \cdot (U_{f(H_2)}/100) \cdot (x_{H_2}/100)} \tag{5.3}$$

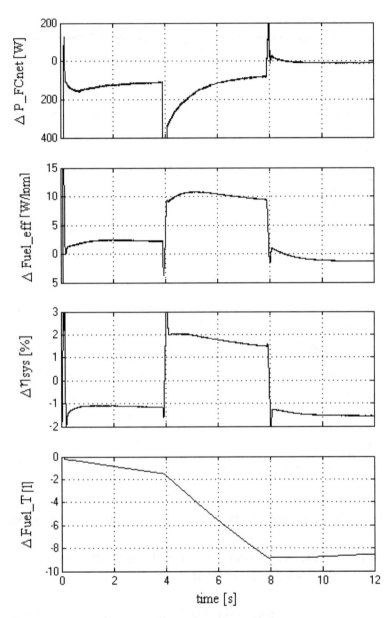

Fig. 5.21 Performance of strategy 5 (Boost-GES1-Air-GES2-Fuel-LFW-RTO) compared to sFF strategy

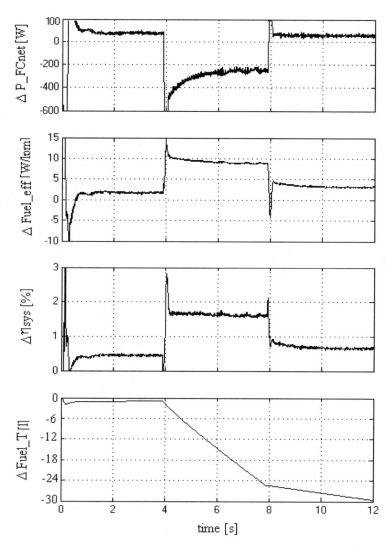

Fig. 5.22 Performance of strategy 6 (Boost-GES1- Fuel-GES2-Air-LFW-RTO) compared to sFF strategy

$$\text{AirFr} = \frac{60000 \cdot R \cdot (273 + \theta) \cdot N_C \cdot i_{\text{FC}}}{4F \cdot (101325 \cdot P_{f(O_2)}) \cdot (U_{f(O_2)}/100) \cdot (y_{O_2}/100)} \qquad (5.4)$$

where $R = 8.3145$ J/(mol K) and $F = 96485$ As/mol. The other parameters are specific to the two FC systems used in this study with nominal power of 6 kW and 1.26 kW (see Fig. 5.24). The parameters and power characteristics for the two FC systems used in this study with nominal power of 6 and 1.26 kW are presented in bottom and top of Fig. 5.24. A time constant of 0.2 s has been set for both FC systems.

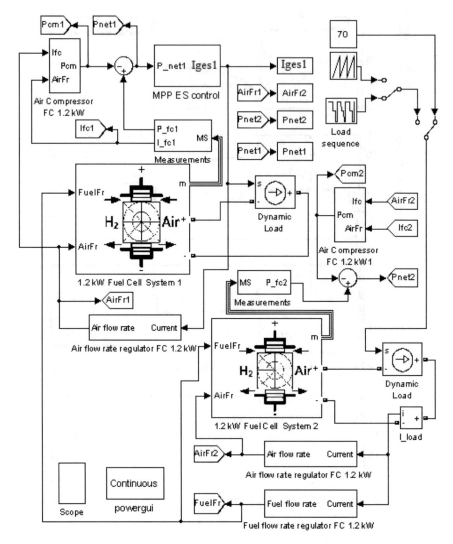

Fig. 5.23 Simulation diagram using ESC-based control of the air regulator [102]

For a fair comparison, the FC system 1 (see the top of Fig. 5.23) uses the same fuel regulator as the FC system 2, but the AirFr is optimally set by the air regulator under the ESC control.

$$\text{AirFr} = \frac{60000 \cdot R \cdot (273 + \theta) \cdot N_C \cdot i_{\text{GES1}}}{4F \cdot (101325 \cdot P_{f(O_2)}) \cdot (U_{f(O_2)}/100) \cdot (y_{O_2}/100)} \tag{5.5}$$

The air compressor is the main consumer of the FC system, being powered from the FC output power. Thus, neglecting other losses, the FC net power, P_{FCnet}, can be approximated using (Eq. 5.6):

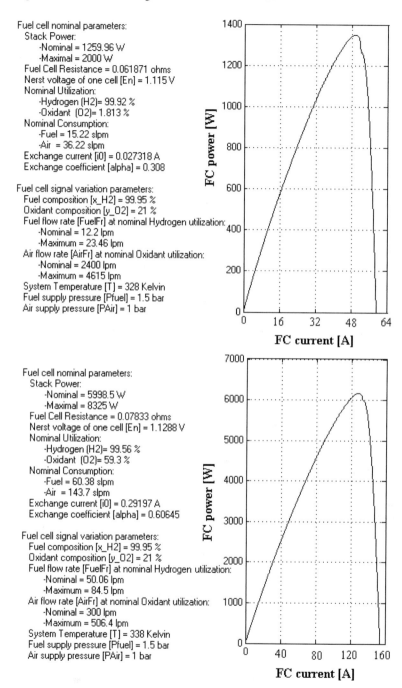

Fig. 5.24 Power characteristics for the FC systems of 6 kW (bottom) and 1.26 kW (top) rated power [102]

$$P_{FCnet} \cong P_{FC} - P_{cm} = V_{FC} \cdot I_{FC} - V_{cm} \cdot I_{cm} \tag{5.6}$$

by measuring the voltage and the current of the FC system (V_{FC} and I_{FC}) and the air compressor (V_{mc} and I_{mc}). By measuring the output FC current, I_{FCnet}, the FC net power can be evaluated using (Eq. 5.7):

$$P_{FCnet} = V_{FC} \cdot I_{FCnet} \tag{5.7}$$

The power consumed by the air compressor, P_{cm}, can be estimated using (Eq. 5.8) [32].

$$P_{cm} = I_{cm} \cdot V_{cm} = k_{cm} \cdot \left(a_2 \cdot AirFr^2 + a_1 \cdot AirFr + a_0\right) \cdot (b_1 \cdot I_{FC} + b_0) \tag{5.8}$$

where $a_0 = 0.6$, $a_1 = 0.04$, $a_2 = -0.00003231$, $b_0 = 0.9987$, and $b_1 = 46.02$.

The parameter k_{cm} is set at 1 and 2 for two compressors powered by about 10% and 20% of the rated FC power, respectively, but this can be set using a lookup table to include the nonlinearities in modeling the air compressor [103].

The FC net power of the FC system 2, P_{FCnet2}, will be used as reference to estimate the availability in increasing the FC net power, $\Delta P_{FCnet} = P_{net(max)} - P_{FCnet2}$, where the $P_{net(max)}$ is the FC net power measured at the MEP on the FC net power characteristics (see Figs. 5.25 and 5.26 for the FC systems of 6 and 1.26 kW rated power).

The increase of the FC net power 1, P_{net1}, considering the FC net power 2, P_{FCnet2}, as reference will be given using the performance indicator:

$$Eff_{FCnet} = \Delta P_{FCnet} / P_{net(max)} \tag{5.9}$$

where $\Delta P_{FCnet} = P_{FCnet1} - P_{FCnet2}$ (see Figs. 5.27 and 5.28 for the FC systems of 6 and 1.26 kW rated power).

The tracking accuracy (T_{acc}) and the oxygen excess ratio (OER) are also recorded in Figs. 5.27 and 5.28.

The tracking accuracy (T_{acc}) has been estimated using (Eq. 5.10):

$$T_{acc} = \frac{P_{FCnet}}{P_{FCnet(max)}} \cdot 100[\%] \tag{5.10}$$

The OER has been estimated using (Eq. 5.11) [32]:

$$OER \equiv \lambda_{O_2} = \frac{a_3 \cdot I_{FC}^3 + a_2 \cdot I_{FC}^2 + a_1 \cdot I_{FC} + a_0}{b_1 \cdot I_{FC} + b_0} \tag{5.11}$$

where $a_0 = 402.4$, $a_1 = 402.4$, $a_2 = -0.8387$, $a_3 = 0.027$, $b_0 = 1$, and $b_1 = 61.4$.

I_{FC} [A]	FuelFr2 [lpm]	AirFr2 [lpm]	P_{FCnet2} [W]	$AirFr_{(max)}$ [lpm]	$P_{net(max)}$ [W]	$P_{net(max)} - P_{FCnet2}$ [W]
115	43.19	259	4730	255.6	4865	135
120	45	270	4870	265.6	5014	144
125	46.94	281	5004	276.5	5158	154
130	48.82	293	5134	287.4	5298	164
135	50.7	304	5259	298.3	5434	175
140	52.58	315	5380	309.2	5565	185
145	54.45	337	5498	320.1	5692	196

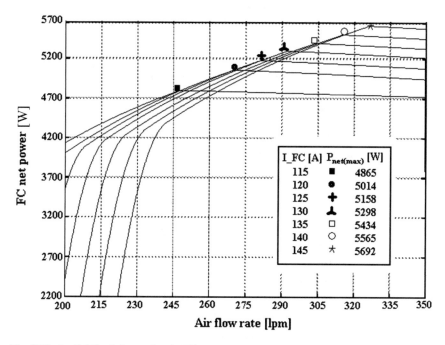

Fig. 5.25 Availability in increasing the FC net power, $P_{net(max)} - P_{FCnet2}$, for the 6 kW FC system [102]

It can be observed that the tracking accuracy and OER are higher than 99.98% and 1.5, respectively. Both fueling regulators of the FC system 1 include a 50 A/s rate limiter for the input to avoid gases starvation, so the values of $AirFr_1$ and P_{FCnet1} increase with limited slope during the start-up phase. The air regulator of the FC system 2 is without a rate limiter to highlight the behavior of the FC systems during the start-up phase.

Also, during the start-up phase, the dither amplitude ($G_d \cong H1$, the first harmonic of the P_{FCnet1}) is high to speed-up MEP search. During the stationary phase, the dither amplitude is very small, so the FC power ripple (see the zoom in the third plot of Fig. 5.28) is negligible (which means a reduced mechanical stress for the proton

I_{FC}	FuelFr2	AirFr2	P_{FCnet2}	$AirFr_{(max)}$	$P_{net(max)}$	$P_{net(max)} - P_{FCnet2}$
[A]	[lpm]	[lpm]	[W]	[lpm]	[W]	[W]
40	9.4	1846	1021	1840	1045.5	24.5
45	10.6	2077	1023	2039	1155.5	32.5
50	11.8	2308	1221	2263	1258.5	37.5
55	13	2538	1314	2487	1357	43
60	14.1	2770	1403	2711	1452	49
65	15.3	3000	1488	2934	1543.5	55.5
70	16.5	3231	1569	3156	1630	61

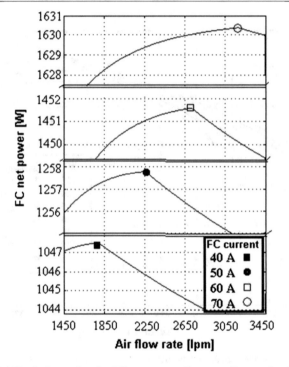

Fig. 5.26 Availability in increasing the FC net power, $P_{net(max)} - P_{FCnet2}$, for the 1.26 kW FC system [102]

exchange membrane (PEM) membrane). This behavior (high and low value during the transitory and stationary regimes) is highlighted for a pulsed load (see Figs. 5.29 and 5.30 for the FC systems of 6 and 1.26 kW rated power).

The performance indicator $Eff_{FCnet} = \Delta P_{FCnet}/P_{net(max)}$ (the normalized increase in FC net power) increases with 2.3–3.7% if both 1.26 kW FC systems use a compressor which is powered by about 10% of rated FC power. A comparable value of 3.5% more power is reported in [56] for the 1.26 kW FC system (but the power consumed by the air compressor is not mentioned).

Fig. 5.27 Increase of the FC net power using the GES control for the air regulator of the 6 kW FC system [102]

I_{FC}	$\Delta P_{FCnet} =$ $P_{FCnet1}\text{-}P_{FCnet2}$	$Eff_{FCnet}=$ $\Delta P_{FCnet}\,/\,P_{net(max)}$	T_{acc}	OER
[A]	[W]	[%]	[%]	[-]
115	134.8	2.77	99.996	4.3
120	143.7	2.87	99.994	4.7
125	153.6	2.98	99.992	5.2
130	163.6	3.09	99.992	5.7
135	174.4	3.21	99.988	6.2
140	184.3	3.31	99.986	6.7
145	195.3	3.43	99.986	7.3

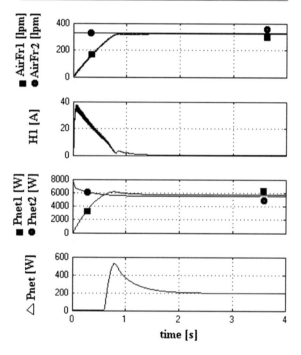

In case of the 6 kW FC systems, a narrower range (2.8–3.4%) is obtained for the Eff_{FCnet} performance indicator. It is worth mentioning that the average increase in performance indicator Eff_{FCnet} is less dependent on FC power level, being around 3%. So, only the 6 kW FC systems will be considered in next sections to explore other ways to improve the FC net power.

If the 6 kW FC systems use a compressor which is powered by about 20% of rated FC power (see Fig. 5.31), then the Eff_{FCnet} performance indicator increases with 10–14%. For the example presented in Fig. 5.32, the energy efficiency increases with about 12.34% (525 W/4254 W $\cong$ 0.1234). A comparable value of 10.5% more power is reported in [25] for a 100 kW FC system using the sFF + PI feedback control. It is worth mentioning that the average increase in performance indicator Eff_{FCnet} is

Fig. 5.28 Increase of the FC net power using the GES control for the air regulator of the 1.26 kW FC system [102]

I_{FC}	$\Delta P_{FCnet} = P_{FCnet1} - P_{FCnet2}$	$Eff_{FCnet} = \Delta P_{FCnet} / P_{net(max)}$	T_{acc}	OER
[A]	[W]	[%]	[%]	[-]
40	24.43	2.337	99.999	3.3
45	32.41	2.805	99.998	4.3
50	37.41	2.973	99.998	5.5
55	42.73	3.149	99.994	6.8
60	48.41	3.334	99.988	8.3
65	54.44	3.527	99.978	9.9
70	60.88	3.735	99.998	11.7

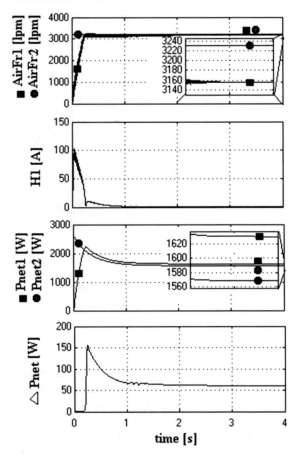

Fig. 5.29 Behavior of the 6 kW FC systems under pulsed load using the sequence for the FC current of 115–145–115 A [102]

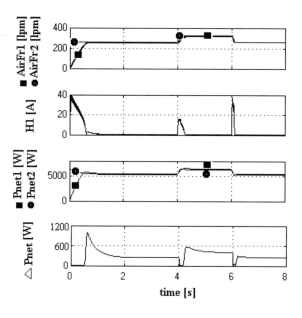

Fig. 5.30 Behavior of the 1.26 kW FC systems under pulsed load using the sequence for the FC current of 40–70–40 A [102]

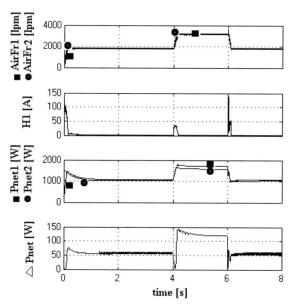

about four times higher (12%/3% = 4) in case of a compressor powered by about 20% of rated FC power compared to one powered by about 10% of rated FC power. This means that the optimization is more efficient for FC systems with higher power losses. This finding must also be validated for other strategies, so the same two air compressors will be used in next sections.

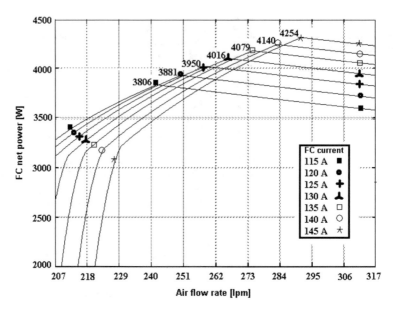

Fig. 5.31 Power characteristics for the 6 kW FC system using a compressor powered by about 20% of rated FC power [102]

The results presented in this section highlight that it is possible to improve the FC net power available to supply the DC bus of the FC hybrid power systems (HPS) by searching the MEP using the air regulator. Thus, the MEP is the maximum value on the curve $P_{FCnet} = f(AirFr)$.

The next section will show that better performance can be obtained by searching the MEP using both the air regulator and the FC current. Thus, the MEP is the maximum value on the surface $P_{FCnet} = f(AirFr, I_{FC})$.

5.5   Improve FC Net Power Through the ESC-Based Control of Both the Air Regulator and the FC Current

The simulation diagram using ESC-based control of the air regulator is represented in Fig. 5.33. The FuelFr value is set to values obtained with the sFF control. The values of the FC current (I_{FC}), AirFr, FC net power (P_{FCnet}), power consumed by air compressor (P_{cm}), and FC power (P_{FC}) are recorded in tables to compute the electrical energy efficiency of the FC system (η_{sys}).

The electrical energy efficiency of the FC system shows the improvement in the FC net power, being given in Eq. (5.11) [104]:

$$\eta_{sys} = P_{FCnet}/P_{FC} \tag{5.12}$$

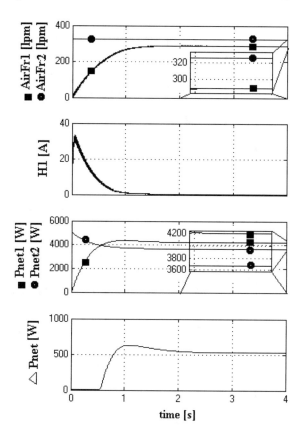

Fig. 5.32 Increase of the FC net power for the 6 kW FC system using a compressor powered by about 20% of rated FC power [102]

The values of the performance indicator η_{sys} are given in Figs. 5.34 and 5.35 in case of a compressor powered by about 10% (subscript 1) and 20% (subscript 2) of rated FC power. Subscript "a" denotes the case of the air regulator controlled using ESC-based algorithm. Subscript "f" will be used in case of the fuel regulator controlled using ESC-based algorithm, which will be analyzed in next section. Subscript "aFF" denotes the case of the fueling regulators controlled using the FC current (the sFF control case).

The sections of the FC net power surfaces in case of a compressor powered by about 10 and 20% of rated FC power are represented in Figs. 5.34 and 5.35, where the MEPs are marked.

The simulation diagram and obtained results using the sFF control are shown in Fig. 5.36.

The improvements in performance indicator η_{sys} are observed compared to sFF control for both air compressors (see Table 5.1, where $\Delta\eta_{sys1a} = \eta_{sys1a} - \eta_{sys1}$ and $\Delta\eta_{sys2a} = \eta_{sys2a} - \eta_{sys2}$). Note that the ratio $\Delta\eta_{sys2a}/\Delta\eta_{sys1a}$ is about 2 (but not 4 as in the previous section).

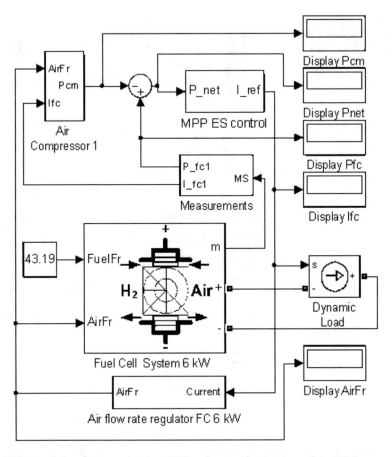

Fig. 5.33 Simulation diagram using the ESC-based control for the air regulator [104]

To highlight the advantages of searching with two variables (using ESC-based control for both the air regulator and the FC current) instead of one (using ESC-based control of the air regulator), the simulation diagram represented in Fig. 5.37 will be used. The reference will be the FC system with ESC-based control for the air regulator, which was analyzed in the previous section. The FC system under test will use the ESC-based control for both the air regulator and FC current. For a fair comparison, both FC systems use the same fuel regulator.

The values of the FC net power estimated with the simulation diagram represented in Fig. 5.37 are recorded in Table 5.2 for both air compressors.

The improvements in FC net power by searching with two variables (using ESC-based control for the air regulator and the FC current) instead of one have been highlighted, but further improvements can be obtained if the search will be performed using ESC-based control for both the fuel regulator and the FC current.

FuelFr$_{sFF}$	I$_{FC1a}$	AirFr$_{1a}$	P$_{net1a}$	P$_{cm1a}$	P$_{FC1a}$	η$_{sys1a}$
[lpm]	[A]	[lpm]	[W]	[W]	[W]	[%]
43.19	113.1	254.6	4866	621	5487	88.7
45	117.8	266.2	5009	657	5666	88.4
46.94	122.8	276.5	5158	697	5855	88.1
48.82	127.7	287.4	5298	735	6034	87.8
50.7	132.6	298.3	5434	774	6208	87.5
52.58	137.4	309.3	5566	813	6379	87.3
54.45	142.2	320.1	5693	853	6546	87.0

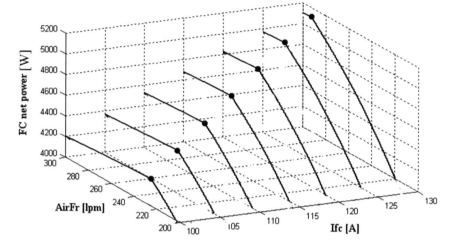

Fig. 5.34 MEPs on the FC net power surfaces in case of a compressor powered by about 10% of rated FC power [104]

5.6 Improve FC Net Power Through the ESC-Based Control of Both the Fuel Regulator and the FC Current

The simulation diagram using ESC-based control of the fuel regulator is represented in Fig. 5.38.

The AirFr value is set to values obtained with the sFF control. The values of the FC current (I_{FC}), FuelFr, FC net power (P_{net}), power consumed by air compressor (P_{cm}), and FC power (P_{FC}) are recorded in tables shown in Fig. 5.38 to compute the electrical energy efficiency of the FC system (η_{sys}).

The improvements in performance indicator η_{sys} are observed compared to sFF control, but also with the ESC-based control for both the fuel regulator and the FC current (see Table 5.3, where the following differences are computed: $\Delta\eta_{sys1f} = \eta_{sys1f}$ − η_{sys1}, $\Delta\eta_{sys2f} = \eta_{sys2f}$ − η_{sys2}, $\Delta\eta_{sys1fa} = \eta_{sys1f}$ − η_{sys1a}, and $\Delta\eta_{sys2fa} = \eta_{sys2f}$ − η_{sys2a}). Note that the ratio $\Delta\eta_{sys2f}/\Delta\eta_{sys1f}$ is about 2.

The improvements in performance indicator η_{sys}, if searching uses two variables, are registered in the last three columns. Note the improvements ($\Delta\eta_{sys1fa}$ and $\Delta\eta_{sys2fa}$ registered in the seventh and eighth columns) if the ESC-based control is used for the fuel regulator instead of the air regulator (of about 8 and 16 for the first and the

FuelFr$_{sFF}$	I$_{FC2a}$	AirFr$_{2a}$	P$_{net2a}$	P$_{cm2a}$	P$_{FC2a}$	η$_{sys2a}$
[lpm]	[A]	[lpm]	[W]	[W]	[W]	[%]
43.19	113.1	254.6	4245	1243	5487	77.4
45	117.8	266.2	4353	1314	5666	76.8
46.94	122.8	276.5	4463	1392	5855	76.2
48.82	127.7	287.4	4564	1470	6034	75.6
50.7	132.6	298.3	4661	1548	6208	75.1
52.58	137.4	309.3	4753	1626	6379	74.5
54.45	142.2	320.1	4840	1706	6546	73.9

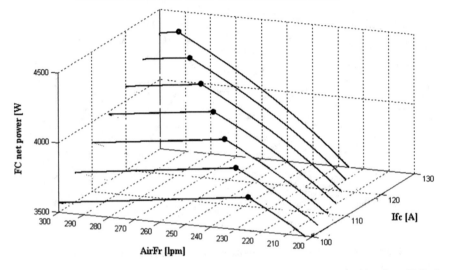

Fig. 5.35 MEPs on the FC net power surfaces in case of a compressor powered by about 20% of rated FC power [104]

second air compressor used in simulation). Also, the ratio $\Delta\eta_{sys2fa}/\Delta\eta_{sys1fa}$ is about 2 (see the last column in Table 5.3).

The advantages of searching with two variables can also be highlighted in this case (using ESC-based control for both the air regulator and the FC current) compared with one searching variable (using ESC-based control of the fuel regulator) by using the simulation diagram represented in Fig. 5.39. The reference will be the FC system with ESC-based control for the fuel regulator, and the FC system under test will use the ESC-based control for both the fuel regulator and the FC current. For a fair comparison, both FC systems use the same air regulator.

The values of the FC net power estimated with the simulation diagram represented in Fig. 5.39 are recorded in Table 5.4 for both air compressors.

The differences ($\Delta P_{netjfa} = \Delta P_{netjf} - \Delta P_{netja}, j = 1,2$) in FC net power registered in Tables 5.2 and 5.4 are shown in Table 5.5, where the ratio $\Delta P_{net2fa}/\Delta P_{net1fa}$ is also computed. The advantage of using ESC-based control for the fuel regulator compared to the air regulator is highlighted again by improvements in FC net power considering both air compressors. Also, the ratio $\Delta P_{net2fa}/\Delta P_{net1fa}$ is around 2 (see the last column in Table 5.5).

I_{FC} [A]	FuelFr$_{sFF}$ [lpm]	AirFr$_{sFF}$ [lpm]	P_{net1} [W]	P_{cm1} [W]	P_{net2} [W]	P_{cm2} [W]	P_{FC} [W]	η_{sys1} [%]	η_{sys2} [%]
115	43.19	258.8	4730	636.2	4094	1272	5366	88.1	76.3
120	45	270.1	4870	675	4194	1350	5545	87.8	75.6
125	46.94	281.3	5004	714.7	4289	1429	5719	87.5	75.0
130	48.82	300.7	5134	754.7	4379	1509	5889	87.2	74.4
135	50.7	303.8	5259	795.1	4463	1590	6054	86.9	73.7
140	52.58	315.1	5380	825.9	4544	1672	6216	86.6	73.1
145	54.45	326.3	5498	877	4619	1754	6373	86.3	72.5

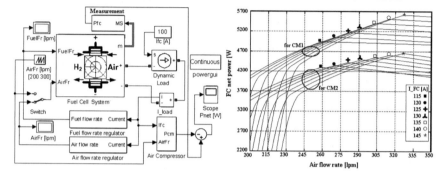

Fig. 5.36 Simulation diagram and results using the sFF control [104]

Table 5.1 Improvements in performance indicator η_{sys}

η_{sys1} [%]	η_{sys2} [%]	η_{sys1a} [%]	η_{sys2a} [%]	$\Delta\eta_{sys1a}$ [%]	$\Delta\eta_{sys2a}$ [%]
88.1	76.3	88.7	77.4	0.6	1.1
87.8	75.6	88.4	76.8	0.6	1.2
87.5	75	88.1	76.2	0.6	1.2
87.2	74.4	87.8	75.6	0.6	1.2
86.9	73.7	87.5	75.1	0.6	1.4
86.6	73.1	87.3	74.5	0.7	1.4
86.3	72.5	87	73.9	0.7	1.4

The performance analysis of the FC power control strategies based on load-following mode will be systematically presented in next section.

5.7 Performance Analysis of the FC Power Control Strategies Based on Load-Following Mode

Three classes can be defined after the place where the LFW reference ($I_{ref(LFW)}$) is used to control the FC power (see Fig. 5.40) [88, 105]: at boost controller for the strategies sFF, S1, S2, and S7 of class C1 ($I_{ref(boost)} = I_{ref(LF)}$), at air regulator for

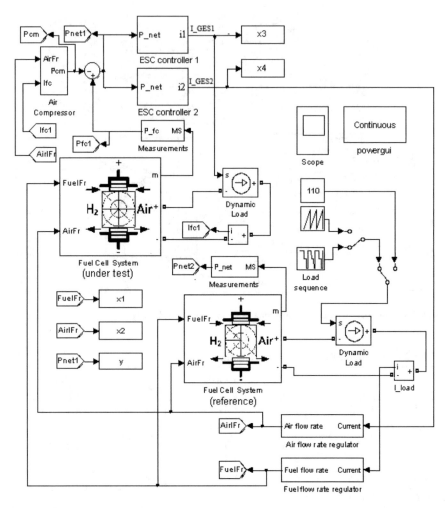

Fig. 5.37 Simulation diagram using the ESC-based control for both the air regulator and the FC current [104]

Table 5.2 Improvements in FC net power using the simulation diagram shown in Fig. 5.37

I_{FC} [A]	ΔP_{net1a} [W]	ΔP_{net2a} [W]
115	111	121
120	118	129
125	126	137
130	134	146
135	142	155
140	150	164
145	159	173

AirFr$_{SFF}$	I$_{FC1f}$	FuelFr$_{1f}$	P$_{net1f}$	P$_{cm1f}$	P$_{FC1f}$	η$_{sys1f}$	AirFr$_{SFF}$	I$_{FC2f}$	FuelFr$_{2f}$	P$_{net2f}$	P$_{cm2f}$	P$_{FC2f}$	η$_{sys2f}$
[lpm]	[A]	[lpm]	[W]	[W]	[W]	[%]	[lpm]	[A]	[lpm]	[W]	[W]	[W]	[%]
258.8	128.3	48.18	5214	192	5406	96.4	258.8	128.3	48.18	5022	384	5406	92.9
270.1	132.4	49.74	5376	201	5577	96.4	270.1	132.4	49.74	5175	402	5577	92.8
281.3	136.4	51.25	5533	210	5743	96.3	281.3	136.4	51.25	5323	420	5743	92.7
300.7	143.2	53.79	5797	225	6022	96.3	300.7	143.2	53.79	5572	450	6022	92.5
303.8	144.3	54.19	5838	228	6066	96.2	303.8	144.3	54.19	5610	456	6066	92.5
315.1	148.1	55.62	5986	237	6223	96.2	315.1	148.1	55.62	5749	474	6223	92.4
326.3	151.8	57.01	6129	247	6376	96.1	326.3	151.8	57.01	5882	494	6376	92.3

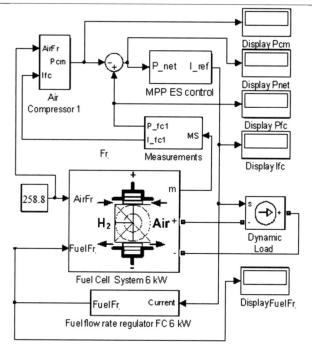

Fig. 5.38 Simulation diagram and results using the ESC-based control for the fuel regulator

Table 5.3 Improvements in performance indicator η_{sys}

η_{sys1} [%]	η_{sys2} [%]	η_{sys1f} [%]	η_{sys2f} [%]	$\Delta\eta_{sys1f}$ [%]	$\Delta\eta_{sys2f}$ [%]	$\Delta\eta_{sys1fa}$ [%]	$\Delta\eta_{sys2fa}$ [%]	$\Delta\eta_{sys2fa}/\Delta\eta_{sys1fa}$ [−]
88.1	76.3	96.4	92.9	8.3	16.6	7.7	15.5	2.01
87.8	75.6	96.4	92.8	8.6	17.2	8	16	2.00
87.5	75	96.3	92.7	8.8	17.7	8.2	16.5	2.01
87.2	74.4	96.3	92.5	9.1	18.1	8.5	16.9	1.99
86.9	73.7	96.2	92.5	9.3	18.8	8.7	17.4	2.00
86.6	73.1	96.2	92.4	9.6	19.3	8.9	17.9	2.01
86.3	72.5	96.1	92.3	9.8	19.8	9.1	18.4	2.02

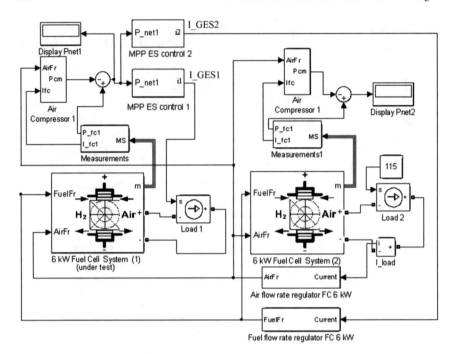

Fig. 5.39 Simulation diagram using the ESC-based control for the fuel regulator and the FC current [104]

Table 5.4 Improvements in FC net power using the simulation diagram shown in Fig. 5.39	I_{FC} [A]	ΔP_{net1f} [W]	ΔP_{net2f} [W]
	115	333	516
	120	337	539
	125	342	563
	130	348	589
	135	355	616
	140	363	645
	145	373	675

the strategies S3 and S5 of class C2 ($I_{ref(Air)} = I_{ref(LF)}$), or at fuel regulator for the strategies S4 and S6 of class C3 ($I_{ref(Fuel)} = I_{ref(LF)}$).

The values of the improvements in electrical energy efficiency ($\Delta \eta_{sysj}, j = 1–7$) of the FC system are estimated for each strategy $S_j, j = 1–7$ for different load levels and represented in Fig. 5.41.

In addition to the best strategy in each class (strategies S1, S5, and S4 of class C1, C2, and C3, respectively), the S6-strategy of class C3 is also represented in Fig. 5.41 due to good values in another performance indicators related to fuel economy, such

Table 5.5 Improvements in FC net power using ESC-based control for the fuel regulator compared to the air regulator

I_{FC} [A]	ΔP_{net1fa} [W]	ΔP_{net2fa} [W]	$\Delta P_{net2fa}/\Delta P_{net1fa}$ [−]
115	222	395	1.78
120	219	410	1.87
125	216	426	1.97
130	214	443	2.07
135	213	461	2.16
140	213	481	2.26
145	214	502	2.35

as the fuel consumption efficiency (Fuel$_{eff}$ = P_{FCnet}/FuelFr) and the total fuel consumption (Fuel$_T$ = $\int$ FuelFr(t)dt) measured in W/lpm (where lpm means liters per minute) and liters.

The values of the improvements in fuel consumption efficiency (ΔFuel$_{eff\,j}$, j = 1–7 are estimated for each strategy S_j, j = 1–7, for different load levels and represented in Fig. 5.42. The same strategies (strategies S1, S5, and S4 of class C1, C2, and C3, respectively) are the best in their class.

The values of the improvements in total fuel consumption (named fuel savings, ΔFuel$_{T\,j}$, j = 1–7 are estimated using different load levels for each strategy S_j, j = 1–7, and represented in Fig. 5.43. The S5-strategy is the best of class C2, but the S6-strategy is better than the S4-strategy of class C3, and the strategies S1 and S7 of class C1 are better for high- and low-load levels, respectively.

These results suggest the use of a strategy for high-load levels (e.g., the strategy S5 or S3) and another for low-load levels (e.g., strategy S4 or S6). It is worth mentioning that the fuel economy is only a bit higher for the S6-strategy compared to the S4-strategy, so any strategy can be used for low-load levels.

The behavior of the strategies S3, S5, and S6 is presented in next section to highlight the operating modes of the battery's stack: the charge-sustaining and charging modes if P_{load} > 0 and P_{load} < 0 (e.g., during accelerating and braking regimes of the FC vehicle, respectively).

5.8 Behavior of the FC System Under Control Strategies S3, S5, and S6

The behavior of the strategies S3, S5, and S6 will be analyzed using the simulation diagram of the FC HPS shown in Fig. 5.44. The FC HPS uses (see also in Fig. 5.44 the numbers mentioned between the brackets): (1) a 6 kW FC system; (2) an ESS represented by the battery's symbol; (3) a DC load with the profile set by the drive sequence; (4) a boost converter using a hysteresis controller; (5) two ES controllers; (6) the LFW controller.

$I_{ref(Boost)}$	$I_{ref(Air)}$	$I_{ref(Fuel)}$	Strategy	Class
$I_{ref(LF)}$	I_{FC}	I_{FC}	sFF	reference
$I_{ref(LF)}$	I_{FC}	$I_{GES1}+I_{FC}$	S1	C1
$I_{ref(LF)}$	$I_{ref(GES1)}+I_{FC}$	I_{FC}	S2	C1
$I_{ref(GES1)}$	$I_{ref(LF)}$	I_{FC}	S3	C2
$I_{ref(GES1)}$	I_{FC}	$I_{ref(LF)}$	S4	C3
$I_{ref(GES1)}$	$I_{ref(LF)}$	$I_{ref(GES2)}+I_{FC}$	S5	C2
$I_{ref(GES1)}$	$I_{ref(GES2)}+I_{FC}$	$I_{ref(LF)}$	S6	C3
$I_{ref(LF)}$	$I_{ref(GES1)}+I_{FC}$	$I_{ref(GES2)}+I_{FC}$	S7	C1

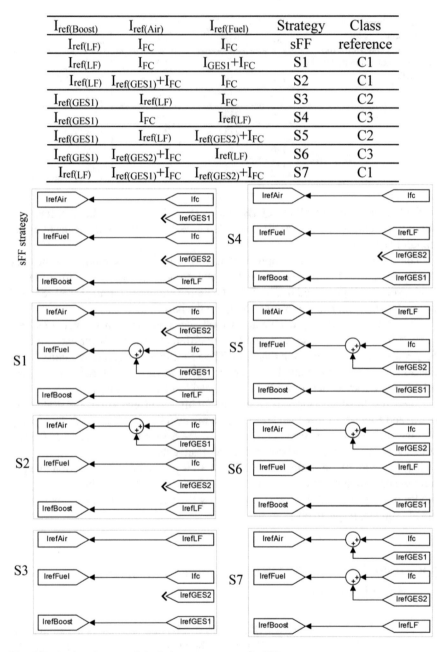

Fig. 5.40 Setting diagram of the fuel economy strategies [88]

P_{load}	$\Delta\eta_{sys1}$	$\Delta\eta_{sys2}$	$\Delta\eta_{sys3}$	$\Delta\eta_{sys4}$	$\Delta\eta_{sys5}$	$\Delta\eta_{sys6}$	$\Delta\eta_{sys7}$
[kW]	[%]	[%]	[%]	[%]	[%]	[%]	[%]
2	0.27	-0.62	-0.35	0.14	0.09	-3.26	0
3	0.42	-0.51	-0.01	0.14	0.23	-2.06	0.1
4	0.53	-0.48	0.06	0.3	0.28	-1.46	0.29
5	0.61	-0.31	0.13	0.43	0.67	-0.79	0.38
6	0.69	-0.15	0.27	0.94	0.94	0.24	0.47
7	0.91	0.18	0.63	2.4	1.52	1.09	0.57
8	2.65	1.61	1.61	2.65	2.13	1.88	0.84
Class	C1	C1	C2	C3	C2	C3	C1

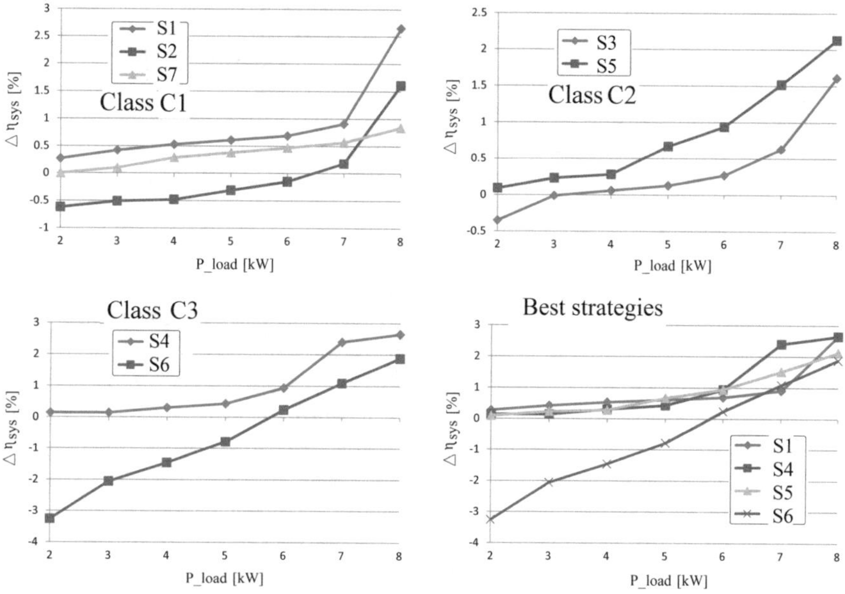

Fig. 5.41 Improvements in electrical energy efficiency

The ESS simulation diagram is represented in Fig. 5.45 using a semi-active ESS topology: battery connected directly to the DC bus and the ultracapacitors' stack exchanging power with the DC bus via a bidirectional DC-DC power converter of buck–boost type.

The DC voltage will be regulated to the reference of 200 V using the buck-boost controller.

The air compressor model (see Fig. 5.46) uses (5.8) to compute the air compressor power (implemented as a function), the second-order system to model the dynamic part, and a static gain $k_{cm} = 0.45$ to set the air compressor power at about 1 kW if the FC system operates at 6 kW nominal conditions (1 kW means about 16.6% from 6 kW).

P_{load}	$\Delta Fuel_{eff1}$	$\Delta Fuel_{eff2}$	$\Delta Fuel_{eff3}$	$\Delta Fuel_{eff4}$	$\Delta Fuel_{eff5}$	$\Delta Fuel_{eff6}$	$\Delta Fuel_{eff7}$
[kW]	[W/lpm]	[W/lpm]	[W/lpm]	[W/lpm]	[W/lpm]	[W/lpm]	[W/lpm]
2	-1	-1.8	-15.3	1.2	-15.3	-2.3	-0.1
3	-0.7	-1.5	-3	1.8	-9	-0.2	1.4
4	0.7	-0.7	-0.7	2.4	-3.4	1.5	2.3
5	1.8	-0.5	0.4	3	3.4	3.1	2.8
6	2.5	-0.4	1.4	4.7	4.9	5.3	3.1
7	3.91	0.62	3.31	7.6	8.84	8.14	3.52
8	10.35	11.2	11.2	10.35	11.47	11.26	4.28
Class	C1	C1	C2	C3	C2	C3	C1

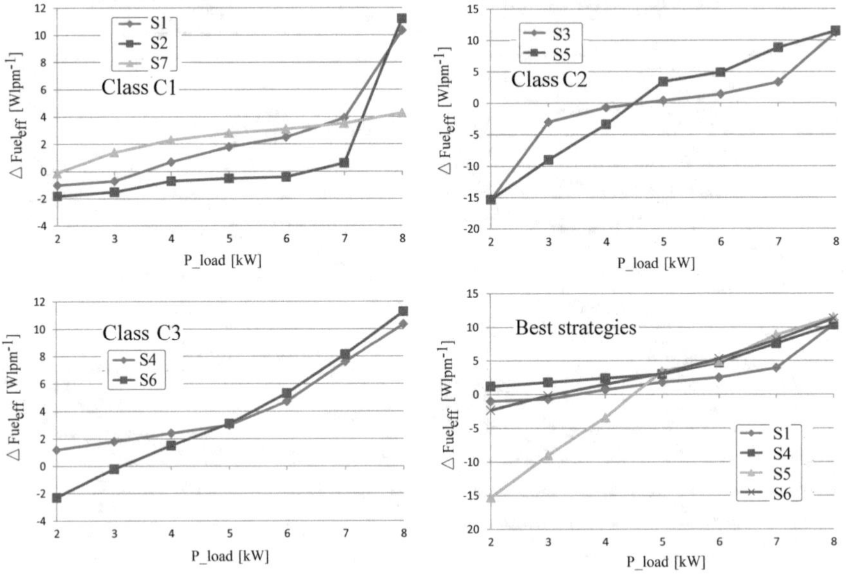

Fig. 5.42 Improvements in fuel consumption efficiency

The FC net power is computed using (5.6) and the FC net power characteristics are represented in Fig. 5.47.

The LFW controller is implemented in Fig. 5.48 based on (5.2).

Searching for and tracking the maximum on the FC net power characteristics (named MEP) can be performed using one searching variable (the case of the S3-strategy) or two searching variables (the case of the strategies S5 and S6).

The S3-strategy uses the reference $I_{ref(LFW)}$ for the air regulator and performs the fueling optimization via the boost controller using the $I_{ref(GES1)}$ reference (see the setting presented in Fig. 5.40) [47]:

$$I_{ref(Fuel)} = I_{FC}, \ I_{ref(Air)} = I_{ref(LF)}, \text{ and } I_{ref(boost)} = I_{ref(GES1)} \tag{5.13}$$

The simulation diagram of the S3-strategy is shown in Fig. 5.49.

P_{load}	$\Delta Fuel_{T1}$	$\Delta Fuel_{T2}$	$\Delta Fuel_{T3}$	$\Delta Fuel_{T4}$	$\Delta Fuel_{T5}$	$\Delta Fuel_{T6}$	$\Delta Fuel_{T7}$
[kW]	[liters]	[liters]	[liters]	[liters]	[liters]	[liters]	[liters]
2	1.24	1.2	11.26	-0.46	8	-0.42	0.22
3	0.13	0.79	4.14	-1.22	6.16	-1.7	-0.32
4	-0.13	0.77	2.08	-2.28	1.94	-3.1	-0.88
5	-0.38	0.55	-0.08	-5.6	-5.18	-5.24	-1.52
6	-1.38	0.42	-2.28	-7.66	-11.56	-8.48	-2.2
7	-4.34	-0.14	-12.16	-13.56	-24.48	-14.04	-3.48
8	-11.8	-4	-28.48	-22.92	-43.34	-27.36	-5.32
Class	C1	C1	C2	C3	C2	C3	C1

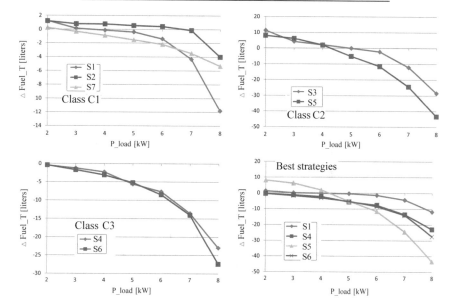

Fig. 5.43 Improvements in total fuel consumption

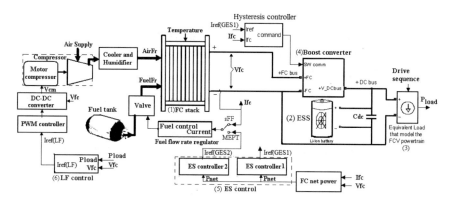

Fig. 5.44 FC HPS system architecture [24]

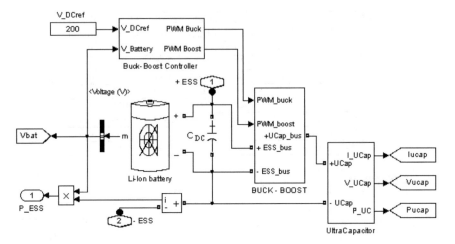

Fig. 5.45 ESS simulation diagram [24]

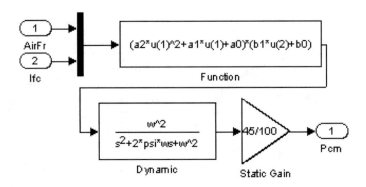

Fig. 5.46 Air compressor model [24]

The S5-strategy uses the reference $I_{ref(LFW)}$ for the air regulator and performs the fueling optimization via both the boost controller and the fuel regulators by using the references $I_{ref(GES1)}$ and $I_{ref(GES2)}$ [91]:

$$I_{ref(Fuel)} = I_{FC} + I_{ref(GES2)}, \; I_{ref(Air)} = I_{ref(LF)}, \text{ and } I_{ref(boost)} = I_{ref(GES1)} \qquad (5.14)$$

The simulation diagram of the S5-strategy is shown in Fig. 5.50.

The S6-strategy uses the reference $I_{ref(LFW)}$ for the fuel regulator and performs the fueling optimization via both the boost controller and air regulators by using the references $I_{ref(GES1)}$ and $I_{ref(GES2)}$ [92]:

$$I_{ref(Fuel)} = I_{ref(LF)}, \; I_{ref(Air)} = I_{FC} + I_{ref(GES2)}, \text{ and } I_{ref(boost)} = I_{ref(GES1)} \qquad (5.15)$$

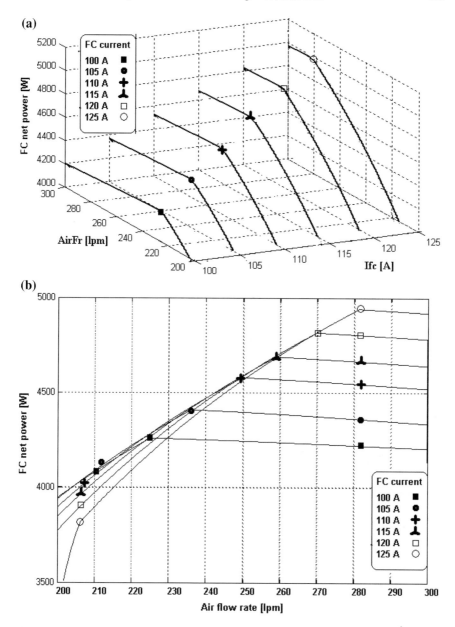

Fig. 5.47 FC net power characteristics. **a** The FC net power surface [24]. **b** The FC net power versus AirFr [24]

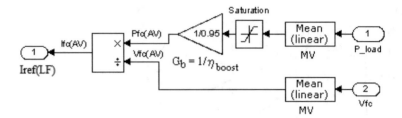

Fig. 5.48 Diagram of the LFW controller [24]

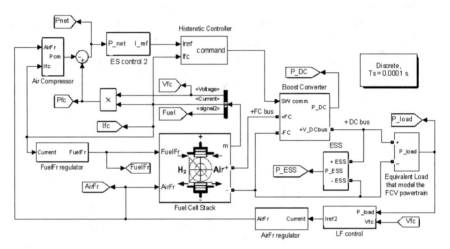

Fig. 5.49 Simulation diagram of the S3-strategy [24]

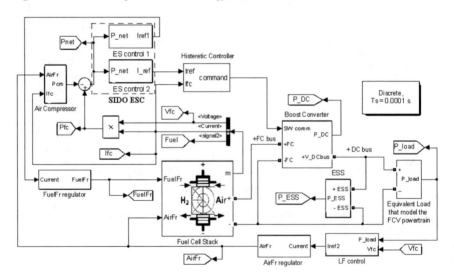

Fig. 5.50 Simulation diagram of the S5-strategy [24]

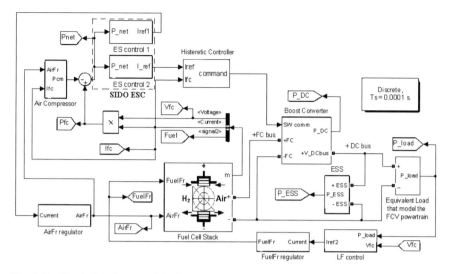

Fig. 5.51 Simulation diagram of the S6-strategy [24]

The simulation diagram of the S6-strategy is shown in Fig. 5.51.

The simulation results using the strategies S3, S5, and S6 are represented in Fig. 5.50. The plots' structure of Fig. 5.50 is the following: the first plot represents the load profile, which contains positive pulses (acceleration phases) and negative pulses (braking phases); the second plot represents the profiles of the FC power and FC net power, which are trying to follow the load profile with a delay given by the FC time constant (which is of 0.2 s) and the 100 A/s slope limiters included in both the fueling regulators; the third plot represents the ESS power exchanged with the DC bus to compensate the power flow balance (1); the fourth plot represents the current fuel consumption for the strategies S3 and S5 and the FuelFr for the S6-strategy; the fifth plot represents the current air consumption (AirFr) for the strategies S3 and S5 and the fuel consumption for the S6-strategy.

If $P_{\text{load}} > 0$, then the battery operates in charge-sustaining mode for all strategies S3, S5, and S6 ($P_{ESS} \cong 0$). If $P_{\text{load}} < 0$, then the battery will operate in charging mode ($P_{ESS} \cong P_{\text{load}}$), sustaining the power flow balance instead of the FC system, which will operate in standby mode using a low FC power level ($P_{FC} \cong 0$). During transitory regimes, the lack of power or the excess power on the DC bus will be compensated by the ultracapacitors' stack (see the positive and negative pulses on the ESS power due to sharp changes in load).

The current fuel consumption under a 3 kW load is about 38 lpm for both the strategies S3 and S5 (see Figs. 5.52 and 5.53), which means the almost same fuel economy compared to the sFF strategy (see Fig. 5.43).

The current fuel consumption under a 6 kW load is a bit lower for the S5-strategy compared to the S3-strategy (see Figs. 5.52 and 5.53), which means a better fuel economy for S5-strategy compared to the S3-strategy (see Fig. 5.43).

Fig. 5.52 Simulation results for S3-strategy [24]

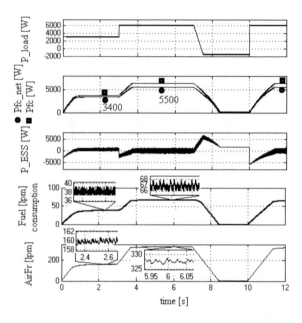

Fig. 5.53 Simulation results for S5-strategy [24]

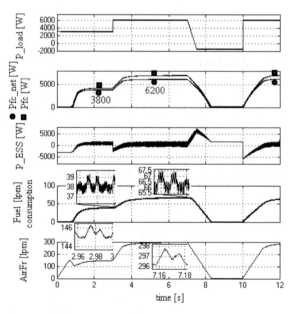

The current fuel consumption under a 3 kW load is lower for the S6-strategy (of about 26 lpm; see Fig. 5.54) compared to the S5-strategy (of about 38 lpm; see Fig. 5.53), which means a better fuel economy for S6-strategy compared to the S5-strategy (see Fig. 5.43).

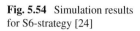

Fig. 5.54 Simulation results for S6-strategy [24]

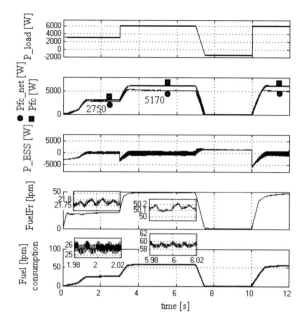

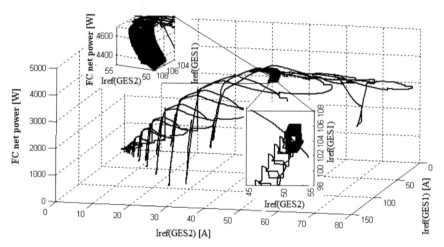

Fig. 5.55 MEP searching process on the FC net power surface

The 100 Hz oscillations on the searching signals (FuelFr and AirFr) appear due to the 100 Hz sinusoidal dither used by the ES controller (see the zooms presented in Figs. 5.52, 5.53 and 5.54).

The MEP searching process under the S5-strategy is shown in Fig. 5.55 on the FC net power surface if the FC current is 110 A. The projection of the MEP searching process on the phase plane of the searching variables is presented in Fig. 5.56. Zooms

Fig. 5.56 MEP searching process on the phase plane of the searching variables

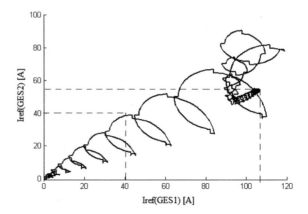

reveal that the MEP is obtained for $I_{ref(GES1)} \cong 105$ and $I_{ref(GES2)} \cong 52$. It is worth mentioning that $I_{ref(GES1)} \cong I_{FC}$.

The MEP searching process under the S6-strategy using different dither's frequencies is shown in Fig. 5.57 on the FC net power surface if the FC current is of 145 A. The projection of the MEP searching process on the phase plane of the searching variables is presented in Fig. 5.58. The MEP tracked for FC current of 145 A using the dither's frequency of 100, 10, and 1 Hz is located at $I_{ref(GES1)} \cong I_{FC}$ and different values of the $I_{ref(GES2)} = I_{ref(Air)}$.

It is worth mentioning that the initial searching gradient and the tracking accuracy is almost the same for different dither's frequency. The MEP searching process in time using the S6-strategy is presented in Fig. 5.59. The tracking accuracy is higher than 99.9% for different dither's frequency.

The MEP searching process in time using the S5-strategy is also presented in Fig. 5.60 to validate the same 99.9% tracking accuracy for different dither's frequency.

Fig. 5.57 MEP searching process using different dither's frequencies [104]

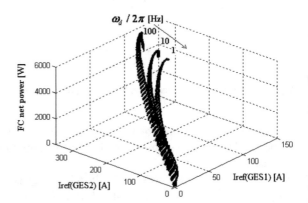

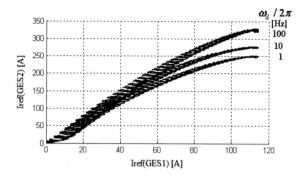

Fig. 5.58 Projection on phase plane of the MEP searching process using different dither's frequencies [104]

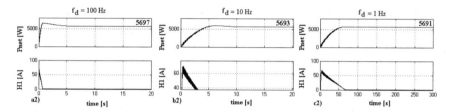

Fig. 5.59 MEP searching process in time using the S6-strategy [104]

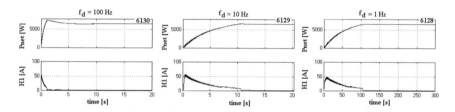

Fig. 5.60 MEP searching process in time using the S5-strategy [104]

5.9 Conclusion

The performance analysis of seven strategies is systematically performed in this chapter based on performance indicators such as the electrical efficiency of the FC system, fuel consumption efficiency, and fuel economy. The strategies have been classified based on load-following mode in three classes, and the best strategy in each class has been identified for each indicator. The best strategies to improve the electrical efficiency of the FC system are the S1, S5, and S4 of class C1, C2, and C3, respectively. These strategies provide the best improvement to the fuel consumption efficiency, but the S6-strategy gives comparable improvements as well. The improvements in total fuel consumption are estimated for all strategies, and the results are as follows: the S5-strategy is the best of class C2; in class C3, the S6-strategy is better than the S4-strategy; in class C1, the strategies S1 and S7 are better for high- and low-load levels, respectively.

These results suggest the design of a fuel economy strategy using the strategies S5 or S3 for high-load levels and strategy S4 or S6 for low-load levels. It is worth mentioning that the fuel economy is only a bit higher for the strategy S6 compared to the strategy S4, so any strategy can be used for low-load levels. But if the S3-strategy is chosen for high-load levels, then it is recommended for low-load levels to use a strategy of the same type (with only one GES controller), this being the S4-strategy. If the S5-strategy will be chosen for high-load levels, then the S6-strategy is recommended because both strategies use two GES controllers.

If the strategy is focused only on improving the electrical efficiency of the FC system, then any of the strategies S1, S5, and S4 of class C1, C2, and C3 may be used because minor differences appear for high-load levels. For example, the improvements in electrical efficiency of the FC system can be about 8 and 16% for the first and the second air compressor if the S5-strategy is used compared to S6-strategy. These mean the increase of the FC net power with 213 and 461 W in case of a 6 kW FC system using the first and the second air compressor with power consumed of about 10 and 20% of rated FC power.

Finally, it is worth mentioning that none of the strategies offer the best results in all performance indicators. Thus, a decision unit must choose the best strategy to current optimization objective, as it will be shown in next chapter.

References

1. Teichmann D, Arlt W, Wasserscheid P (2012) Liquid organic hydrogen carriers as an efficient vector for the transport and storage of renewable energy. Int J Hydrogen Energ 27(23):18118–18132
2. Sulaiman N, Hannan MA, Mohamed A, Ker PJ, Majlan EH, Wan Daud WR (2018) Optimization of energy management system for fuel-cell hybrid electric vehicles: issues and recommendations. Appl Energ 228:2061–2079
3. Sorrentino M, Cirillo V, Nappi L (2019) Development of flexible procedures for co-optimizing design and control of fuel cell hybrid vehicles. Energ Convers Manage 185:537–551
4. Bizon N (2019) Hybrid power sources (HPSs) for space applications: analysis of PEMFC/Battery/SMES HPS under unknown load containing pulses. Renew Sustain Energy Rev 105:14–37. https://doi.org/10.1016/j.rser.2019.01.044
5. Lawan Bukar AL, Wei Tan CW. A review on stand-alone photovoltaic-wind energy system with fuel cell: system optimization and energy management strategy. J Clean Prod 2019 (in press) https://doi.org/10.1016/j.jclepro.2019.02.228
6. Gielen D, Boshell F, Saygin D, Bazilian MD, Wagner N, Gorini R (2019) The role of renewable energy in the global energy transformation. Energy Strateg Rev 24:38–50
7. Ghribi D, Khelifa A, Diaf S, Belhamel M (2013) Study of hydrogen production system by using PV solar energy and PEM electrolyser in Algeria. Int J Hydrogen Energ 38(20):8480–8490
8. Petrov K, Baykara SZ, Ebrasu D, Gulin M, Veziroglu A (2011) An assessment of electrolytic hydrogen production from H2S in Black Sea waters. Int J Hydrogen Energ 36(15):8936–8942

9. Hesselink LXW, Chappin EJL (2019) Adoption of energy efficient technologies by households—Barriers, policies and agent-based modelling studies. Renew Sustain Energy Rev 99:29–41

10. Wang F-C, Yi-Shao Hsiao Y-S, Yi-Zhe Yang Y-Z (2018) The optimization of hybrid power systems with renewable energy and hydrogen generation. Energies 11(8):1948. https://doi.org/10.3390/en11081948

11. Bizon N, Tabatabaei NM, Blaabjerg F, Kurt E (2017) Energy harvesting and energy efficiency: technology, methods and applications. Springer, Berlin, available on http://www.springer.com/us/book/9783319498744

12. Das V, Padmanaban S, Venkitusamy K, Selvamuthukumaran R, Siano P (2017) Recent advances and challenges of fuel cell based power system architectures and control—a review. Renew Sustain Energy Rev 73:10–18

13. Priya K, Sathishkumar K, Rajasekar N (2018) A comprehensive review on parameter estimation techniques for Proton Exchange Membrane fuel cell modelling. Renew Sustain Energy Rev 93:121–144

14. Yue M, Jemei S, Gouriveau R, Zerhouni N (2019) Review on health-conscious energy management strategies for fuel cell hybrid electric vehicles: degradation models and strategies. Int J Hydrogen Energy 44(13):6844–6861

15. Dafalla AM, Jiang J (2018) Stresses and their impacts on proton exchange membrane fuel cells: a review. Int J Hydrogen Energy 43(4):2327–2348

16. Chen H (2019) The reactant starvation of the proton exchange membrane fuel cells for vehicular applications: a review. Energ Convers Manage 182:282–298

17. Vielstich W, Gasteiger H, Lamm A (2003) Handbook of fuel cells-fundamentals, technology, applications. Wiley, New York

18. Larminie J, Dicks A (2000) Fuel cell system explained. Wiley, Chichester

19. Bizon N, Tabatabaei NM, Shayeghi H (2013) Analysis, control and optimal operations in hybrid power systems—advanced techniques and applications for linear and nonlinear systems. Springer, Berlin. Available on http://dx.doi.org/10.1007/978-1-4471-5538-6

20. Daud WRW, Rosli RE, Majlan EH, Hamid SAA, Mohamed R, Husaini T (2017) PEM fuel cell system control: a review. Renew Energ 113:620–638

21. Olatomiwa L, Mekhilef S, Ismail MS, Moghavvemi M (2016) Energy management strategies in hybrid renewable energy systems: a review. Renew Sustain Energy Rev 62:821–835

22. Bizon N, Thounthong P (2018) Real-time strategies to optimize the fueling of the fuel cell hybrid power source: a review of issues, challenges and a new approach. Renew Sustain Energy Rev 91:1089–1102

23. Wang F-C, Lin K-M (2019) Impacts of load profiles on the optimization of power management of a green building employing fuel cells. Energies 12(1):57. https://doi.org/10.3390/en12010057

24. Bizon N, Radut M, Oproescu M (2015) Energy control strategies for the fuel cell hybrid power source under unknown load profile. Energy 86:31–41

25. Becherif M, Hissel D (2010) MPPT of a PEMFC based on air supply control of the motocompressor group. Int J Hydrogen Energ 35(22):12521–12530

26. Li Q, Chen W, Liu Z, Guo A, Liu S (2013) Control of proton exchange membrane fuel cell system breathing based on maximum net power control strategy. J Power Sources 241(1):212–218

27. Bizon N (2010) On tracking robustness in adaptive extremum seeking control of the fuel cell power plants. Appl Energ 87(10):3115–3130

28. Krstiç M, Wang H–H (2000) Design and stability analysis of extremum seeking feedback for general nonlinear systems. Automatica 36(2):595–601

29. Ariyur KB, Krstic M (2003) Real–time optimization by extremum–seeking Control (Wiley–Interscience edn). Wiley, Chichester

30. Kunusch C, Puleston PF, Mayosky MA, Riera J (2009) Sliding mode strategy for PEM fuel cells stacks breathing control using a super–twisting algorithm. IEEE T Contr Syst T 17(1):167–173

31. Pukrushpan J, Stefanopoulou A, Peng H (2004) Control of fuel cell power systems: principles, modeling and analysis and feedback design. Springer, Berlin
32. Ramos-Paja CA, Spagnuolo G, Petrone G, Mamarelis E (2014) A perturbation strategy for fuel consumption minimization in polymer electrolyte membrane fuel cells: analysis, design and FPGA implementation. Appl Energ 119:21–32
33. Na W, Gou B (2008) Feedback-linearization-based nonlinear control for PEM fuel cells. IEEE Trans Energy Convers 23:179–190
34. Grujicic M, Chittajallu KM, Law EH, Pukrushpan JT (2004) Model-based control strategies in the dynamic interactions of air supply and fuel cell. Proc Inst Mech Eng Part A: J Power Energy 218:487–99
35. Jang M-H, Lee J-M, Kim J-H, Park J-H, Cho B-H (2011) Maximum efficiency point tracking algorithm using oxygen access ratio control for fuel cell systems. J Power Electronics 11(2):194–201
36. Niknezhadi A, AlluT-Fantova M, Kunusch C, Ocampo-Martfnez C (2011) Design and implementation of LQR/LQG strategies for oxygen stoichiometry control in PEM fuel cells based systems. J Pow Sour 196(9):4277–4282
37. Rodatz P, Paganelli G, Sciarretta A, Guzzella L (2005) Optimal power management of an experimental fuel cell/supercapacitor-powered hybrid vehicle. Control Eng Pract 13:41–53
38. Garcia-Gabin W, Dorado F, Bordons C (2010) Real-time implementation of a sliding mode controller for air supply on a PEM fuel cell. J Process Contr 20:325–336
39. Deng H, Li Q, Chen W, Zhang G (2018) High order sliding mode observer-based OER control for PEM fuel cell air-feed system. IEEE Trans Energy Convers 33(1):232–244
40. Sankar K, Jana AK (2018) Nonlinear multivariable sliding mode control of a reversible PEM fuel cell integrated system. Energy Convers Manage 171:541–565
41. Deng H, Li Q, Cui Y, Zhu Y, Chen W. Nonlinear controller design based on cascade adaptive sliding mode control for PEM fuel cell air supply systems. Int J Hydrogen Energy 2019 (in press). https://doi.org/10.1016/j.ijhydene.2018.10.180
42. Kunusch C, Puleston PF, Mayosky MA, Davila A (2009) Sliding mode strategy for PEM fuel cells stacks breathing control using a super-twisting algorithm. IEEE Trans Control Syst Technol 17:167–174
43. Derbeli M, Farhat M, Barambones O, Sbita L (2017) Control of PEM fuel cell power system using sliding mode and supertwisting algorithms. Int J Hydrogen Energy 42(13):8833–8844
44. Methekar RN, Patwardhan SC, Gudi RD, Prasad V (2010) Adaptive peak seeking control of a proton exchange membrane fuel cell. J Process Contr 20:73–82
45. Chang YA, Moura SJ (2009) Air flow control in fuel cell systems: an extremum seeking approach. American Control Conference pp 1052–1059
46. Dalvi A, Guay M (2019) Control and real-time optimization of an automotive hybrid fuel cell powers system. Control Eng Pract 17:924–938
47. Bizon N (2017) Energy optimization of fuel cell system by using global extremum seeking algorithm. Appl Energ 206:458–474
48. Gruber JK, Doll M, Bordons C (2009) Design and experimental validation of a constrained MPC for the air feed of a fuel cell. Control Eng Pract 17:874–885
49. Vahidi A, Stefanopoulou A, Peng H (2006) Current management in a hybrid fuel cell power system: a model-predictive control approach. IEEE Trans Control Syst Technol 14:1047–1057
50. Li G-P, Zhan J-L, He H-W (2017) Battery SOC constraint comparison for predictive energy management of plug-in hybrid electric bus. Appl Energy 194:578–587
51. Ziogou C, Papadopoulou S, Pistikopoulos E et al (2017) Model-based predictive control of integrated fuel cell systems—from design to implementation. Advances in energy systems engineering. Springer, Berlin, pp 387–430
52. Ziogou C, Voutetakis S, Georgiadis MC, Papadopoulou S (2018) Model predictive control (MPC) strategies for PEM fuel cell systems—a comparative experimental demonstration. Chem Eng Res Des 131:656–670
53. da Fonseca R, Bideaux E, Gerard M, Jeanneret B, Desbois-Renaudin M, Sari A (2014) Control of PEMFC system air group using differential flatness approach: validation by a dynamic fuel cell system model. Appl Energ 113:219–229

54. Thounthong P, Tricoli P, Davat B (2014) Performance investigation of linear and nonlinear controls for a fuel cell/supercapacitor hybrid power plan. Int J Elec Power 54:454–464

55. Saadi R, Kraa O, Ayad MY, Becherif M, Ghodbane H, Bahri M, Aboubou A (2016) Dual loop controllers using PI, sliding mode and flatness controls applied to low voltage converters for fuel cell applications. Int J Hydrogen Energy 41(42):19154–19163

56. Wanga Y-X, Xuan D-J, Kim Y-B (2013) Design and experimental implementation of time delay control for air supply in a polymer electrolyte membrane fuel cell system. Int J Hydrogen Energ 38(1):3381–3392

57. Onori S, Tribioli L (2015) Adaptive Pontryagin's Minimum Principle supervisory controller design for the plug-in hybrid GM Chevrolet Volt. Appl Energy 147:224–234

58. Zhang J, Liu G, Yu W, Ouyanga M (2008) Adaptive control of the airflow of a PEM fuel cell system. J Pow Sour 179:649–659

59. Jiang Z-H, Gao LJ, Dougal RA (2007) Adaptive control strategy for active power sharing in hybrid fuel cell/battery power sources. IEEE Trans Energ Convers 22(2):507–515

60. Jannelli E, Minutillo M, Perna A (2013) Analyzing microcogeneration systems based on LT-PEMFC and HT-PEMFC by energy balances. Appl Energ 108:82–91

61. Han J, Yu S, Yi S (2017) Adaptive control for robust air flow management in an automotive fuel cell system. Appl Energ 190:73–83

62. Li Q, Wang T-H, Dai C-H, Chen W-R, Ma L (2018) Power management strategy based on adaptive droop control for a fuel cell-battery-supercapacitor hybrid tramway. IEEE Trans Veh Technol 67(7):5658–5670

63. Almeida PEM, Simoes MG (2005) Neural optimal control of PEM fuel cells with parametric CMAC networks. IEEE Trans Ind Applic 41(1):237–245

64. Tekin M, Hissel D, Pera M-C, Kauffmann J-M (2006) Energy consumption reduction of a PEM fuel cell motor-compressor group thanks to efficient control laws. J Pow Sour 156:57–63

65. Guo A, Chen W, Li Q, Liu Z, Que H (2013) Air flow control based on optimal oxygen excess ratio in fuel cells for vehicles. J Mod Transport 21(2):79–85

66. Ameur K, Hadjaissa A, Ait Cheikh MS, Cheknane A, Essounbouli N (2017) Fuzzy energy management of hybrid renewable power system with the aim to extend component lifetime. Int J Energy Res 41(13):1867–1879

67. Beirami H, Shabestari AZ, Zerafat MM (2015) Optimal PID plus fuzzy controller design for a PEM fuel cell air feed system using the self-adaptive differential evolution algorithm. Int J Hydrogen Energy 40(30):9422–9434

68. Pilloni A, Pisano A, Usai E (2015) Observer-based air excess ratio control of a PEM fuel cell system via high-order sliding mode. IEEE T Ind Electron 62(8):5236–5246

69. Piffard M, Gerard M, Bideaux E, Da Fonseca R, Massioni P (2015) Control by state observer of PEMFC anodic purges in dead-end operating mode. IFAC-Papers Online 48(15):237–243. https://doi.org/10.1016/j.ifacol.2015.10.034

70. Bizon N, Lopez-Guede JM, Kurt E, Thounthong P, Mazare AG, Ionescu LM, Iana VG (2019) Hydrogen economy of the fuel cell hybrid power system optimized by air flow control to mitigate the effect of the uncertainty about available renewable power and load dynamics. Energ Convers Manage 179:152–165. https://doi.org/10.1016/j.enconman.2018.10.058

71. Bizon N (2018) Optimal operation of fuel cell/wind turbine hybrid power system under turbulent wind and variable load. Appl Energ 212:196–209

72. Bizon N, Lopez-Guede JM, Hoarca IC, Culcer M, Iliescu M (2018) Fuel cell (FC) hybrid power system with mitigation of the load power variability by the FC fuel flow control information. ECAI—10th edition of international conference on electronics, computers and artificial intelligence. Iasi, ROMÂNIA, 27 June–30 June

73. Ahmadi P, Torabi SH, Afsaneh H, Sadegheih Y, Ganjehsarabi H, Ashjaee M (2019) The effects of driving patterns and PEM fuel cell degradation on the lifecycle assessment of hydrogen fuel cell vehicles. Int J Hydrogen Energy (in press). https://doi.org/10.1016/j.ijhydene.2019.01.165

74. Luo Y, Jiao K (2018) Cold start of proton exchange membrane fuel cell. Prog Energy Combust Sci 64:29–61

75. Zhang T, Wang P, Chen H, Pei P (2018) A review of automotive proton exchange membrane fuel cell degradation under start-stop operating condition. Appl Energy 223:249–262
76. Dijoux E, Steiner NY, Benne M, Péra M-C, Pérez BG (2017) A review of fault tolerant control strategies applied to proton exchange membrane fuel cell systems. J Power Sources 359:119–133
77. Laghrouche S, Matraji I, Ahmed FS, Jemei S, Wack M (2013) Load governor based on constrained extremum seeking for PEM fuel cell oxygen starvation and compressor surge protection. Int J Hydrogen Energ 38(33):14314–14322
78. Zhong D, Lin R, Liu D, Cai X (2018) Structure optimization of anode parallel flow field for local starvation of proton exchange membrane fuel cell. J Power Sources 403:1–10
79. Sun J, Kolmanovsky IV (2005) Load governor for fuel cell oxygen starvation protection: a robust nonlinear reference governor approach. IEEE Trans Contr Syst Technol 13(6):911–920
80. Nikiforow K, Koski P, Ihonen J (2017) Discrete ejector control solution design, characterization, and verification in a 5 kW PEMFC system. Int J Hydrogen Energy 42:16760–16772
81. Restrepo C, Ramos-Paja CA, Giral R, Calvente J, Romero A (2012) Fuel cell emulator for oxygen excess ratio estimation on power electronics applications. Comp Elec Eng 38:926–937
82. Bao C, Ouyang M, Yi B (2006) Modeling and control of air stream and hydrogen flow with recirculation in a PEM fuel cell system—II. Int J Hydrogen Energ 31:1897–1913
83. He Y, Xing L, Zhang Y, Zhang J, Cao F, Xing Z (2018) Development and experimental investigation of an oil-free twin-screw air compressor for fuel cell systems. Appl Therm Eng 145:755–762
84. Bizon N (2014) Load-following mode control of a standalone renewable/fuel cell hybrid power source. Energ Convers Manage 77:763–772
85. Bizon N, Hoarcă CI (2019) Hydrogen saving through optimized control of both fueling flows of the Fuel Cell Hybrid Power System under a variable load demand and an unknown renewable power profile. Energ Convers Manage 184:1–14
86. Bizon N (2018) Real-time optimization strategy for fuel cell hybrid power sources with load-following control of the fuel or air flow. Energ Convers Manage 157:13–27
87. Bizon N, Oproescu M (2018) Experimental comparison of three real-time optimization strategies applied to renewable/FC-based hybrid power systems based on load-following control. Energies 11(12):3537–3569. https://doi.org/10.3390/en11123537
88. Bizon N (2019) Real-time optimization strategies of FC Hybrid Power Systems based on Load-following control: a new strategy, and a comparative study of topologies and fuel economy obtained. Appl Energ 241C:444–460
89. Bizon N, Culcer M, Iliescu M, Mazare A, Laurentiu I, Beloiu R (2017) Real-time strategy to optimize the fuel flow rate of fuel cell hybrid power source under variable load cycle. ECAI—9th edition of international conference on electronics, computers and artificial intelligence. Targoviste, ROMÂNIA
90. Bizon N, Culcer M, Oproescu M, Iana G, Laurentiu I, Mazare A, Iliescu M (2017) Real-time strategy to optimize the airflow rate of fuel cell hybrid power source under variable load cycle. The 22rd International Conference on Applied Electronics—APPEL 2017, Pilsen, Czech Republic
91. Bizon N (2018) Optimization of the proton exchange membrane fuel cell hybrid power system for residential buildings. Energy Convers Manag 163:22–37
92. Bizon N, Iana VG, Kurt E, Thounthong P, Oproescu M, Culcer M, Iliescu M (2018) Air flow real-time optimization strategy for fuel cell hybrid power sources with fuel flow based on load-following. Fuel Cell 18(6):809–823. https://doi.org/10.1002/fuce.201700197
93. Bizon N (2017) Searching of the extreme points on photovoltaic patterns using a new asymptotic perturbed extremum seeking control scheme. Energ Convers Manage 144:286–302
94. Bizon N (2016) Global extremum seeking control of the power generated by a photovoltaic array under partially shaded conditions. Energy Convers Manage 109:71–85
95. Bizon N (2016) Global maximum power point tracking based on new extremum seeking control scheme. Prog Photovoltaics Res Appl 24(5):600–622

96. Bizon N (2016) Global maximum power point tracking (GMPPT) of photovoltaic array using the extremum seeking control (ESC): a review and a new GMPPT ESC scheme. Renew Sustain Energy Rev 57:524–539
97. Ishaque K, Salam Z (2013) A review of maximum power point tracking techniques of PV system for uniform insolation and partial shading condition. Renew Sustain Energy Rev 19:475–488
98. Liu Z-H, Chen J-H, Huang J-W (2015) A review of maximum power point tracking techniques for use in partially shaded conditions. Renew Sustain Energy Rev 41:436–453
99. Ettihir K, Boulon L, Agbossou K (2016) Optimization-based energy management strategy for a fuel cell/battery hybrid power system. Appl Energ 163:142–153
100. Bizon N, Kurt E (2017) Performance analysis of tracking of the global extreme on multimodal patterns using the asymptotic perturbed extremum seeking control scheme. Int J Hydrogen Energ 42(28):17645–17654
101. Bizon N, Thounthong P, Raducu M, Constantinescu LM (2017) Designing and modelling of the asymptotic perturbed extremum seeking control scheme for tracking the global extreme. Int J Hydrogen Energy 42(28):17632–17644
102. Bizon N (2014) Tracking the maximum efficiency point for the FC system based on extremum seeking scheme to control the air flow. Appl Energ 129:147–157
103. Tirnovan R, Giurgea S (2012) Efficiency improvement of a PEMFC power source by optimization of the air management. Int J Hydrogen Energ 37:7745–7756
104. Bizon N (2014) Improving the PEMFC energy efficiency by optimizing the fueling rates based on extremum seeking algorithm. Int J Hydrogen Energ 39(20):10641–10654
105. Bizon N, Stan VA, Cormos AC (2019) Optimization of the fuel cell renewable hybrid power system using the control mode of the required load power on the DC bus. Energies 12(10):1889–1904. https://doi.org/10.3390/en12101889

Chapter 6
Fuel Economy Maximization Strategies

6.1 Introduction

The need to reduce air pollution has become stringent in recent years, imposing national emission restrictions [1], which need to be rapidly expanded internationally [2]. Consequently, hybrid power systems (HPS) based on proton exchange membrane fuel cell (PEMFC) and renewable energy sources (RESs) becomes ecological alternatives for energy generation under environmental policy compliance [3]. The PEMFC system is a non-polluting solution for replacing the diesel generator [4] as a source of backup energy in HPS [5], generally using an electrolyzer (ELZ) to store excess energy in hydrogen tanks [6]. Thus, the PEMFC/ELZ combination becomes a more feasible solution for storing surplus energy than conventional energy storage devices, such as batteries [7]. Advantages (such as high electrical efficiency, non-polluting operation), progress (low-temperature operation), and the recent challenges (fuel consumption reduction) for PEMFC/ELZ HPS architectures are systematically analyzed in [8].

In order to dynamically offset the balance of power flow, besides batteries, the energy storage systems (ESS) must also contain a power storage device, such as supercapacitors' stacks [9], superconducting magnetic energy storage (SMES) devices [10], and high-speed flywheels [11]. In addition to the active and semi-active topologies, the multiport architecture for the PEMFC vehicle is proposed to improve the energy efficiency of the ESS hybrid topologies [12, 13].

Therefore, the main objectives of an energy management strategy (EMS) can be defined as follows [3, 4, 14, 15]: (1) HPS must be able to generate power required by the task, regardless of the conditions; (2) HPS performance should be improved by optimizing the operation of the PEMFC system, which according to the chosen objective can operate near the maximum efficiency point (MEP) [16], the point where is getting the best fuel economy [5, 17] or the optimum point defined by a specific optimization function (such as the aggregate objective to obtain low emissions at minimal cost) [18], but also providing a safe HPS control mode [19]; (3) the state

© The Editor(s) (if applicable) and The Author(s), under exclusive license to Springer
Nature Switzerland AG 2020
N. Bizon, *Optimization of the Fuel Cell Renewable Hybrid Power Systems*,
Green Energy and Technology, https://doi.org/10.1007/978-3-030-40241-9_6

of charge (SoC) of the battery should be monitored to vary in an imposed window, regardless of the type of load cycle; (4) the DC voltage must be stabilized and the load impulses attenuated by the anti-control of the DC–DC bidirectional converter associated with the power storage device; and (5) performance required by design must be obtained at a minimum cost [20].

The optimization strategies are based on an optimization function and different type of controllers to implement the optimization loops such as the hysteresis controller [21], the proportional-integral (PI) controller [22], the static Feed-Forward (sFF) control [23], the filtering-based split control [24], the heuristic control [25], the rule-based control [26], the fuzzy logic (FL) control [27], etc.

The most known and efficient strategies are based on the advanced maximization techniques [21, 28], the equivalent consumption minimization [29], the Pontryagin's minimum principle [5], the model predictive control [30], the nonlinear control [31, 32], the dynamic programming [33], but also on other hybrid techniques proposed in [34–37].

Equivalent consumption minimization strategy (ECMS) and rule-based control strategy are the most commonly used strategies to optimally divide power between available energy sources [29, 38, 39].

Recently, other advanced EMSs for FC vehicles have been identified in [40–42]. The peaking power source strategy uses the power storage device during the brake and accelerating phase to compensate the power flow on the DC bus [41]. The operating mode control (OMC) strategy controls the charge–discharge phases of the battery and operation close to MEP for the FC system, ensuring the best fuel economy by this optimal operation of the FCHPS [42].

The proposed EMSs were tested using software tools [43] (such as MAT-LAB/Simulink® [44] or Advisor® [45]), hardware and software techniques (such as Hardware-in-Loop (HIL)), or by experiments [5, 46].

As a conclusion, most of optimization strategies use a weighting function of the performance indicators such as hydrogen consumption, PEMFC performance degradation, HPS efficiency, battery charging state, and lifetime of FC and battery [15, 26, 34].

In this chapter, an optimization function is proposed to optimize the fuel cell (FC) system and analyze fuel economy improvements, but also the influence of optimization on other performance indicators.

The proposed optimization problem is the following [47]:

Maximize

$$k_{net} \cdot P_{FCnet} + k_{fuel} \cdot Fuel_{eff} = f(x, AirFr, FuelFr, P_{Load})$$

considering the dynamic model $\dot{x} = g(x, AirFr, FuelFr, P_{Load}), x \in X$ \quad (6.1)

where k_{net} and k_{fuel} are the weight coefficients, *AirFr* and *FuelFr* are the air and hydrogen flows, x is the state vector, and P_{load} is the load demand (perturbation).

The performance indicators used are as follows [48]: FC net power ($P_{FCnet} = P_{FC} - P_{loss}$), fuel consumption efficiency ($Fuel_{eff} \cong P_{FCnet}/FuelFr$), FC system electrical efficiency ($\eta_{sys} = P_{FCnet}/P_{FC}$), and total fuel consumption ($Fuel_T = \int FuelFr(t)dt$).

The air and fuel flows can be controlled to optimize fuel cell operation considering the optimization function defined above and the global extremum seeking (GES) algorithm presented in Chap. 3.

The FC system's diagram (including the control loops for the fueling regulators) and the searching scheme based on two GES control algorithms are presented in Fig. 6.1 [24].

This searching process is based on a variable multimodal function over time where the global maximum (GM) in real-time optimization (RTO) has to be located. RTO

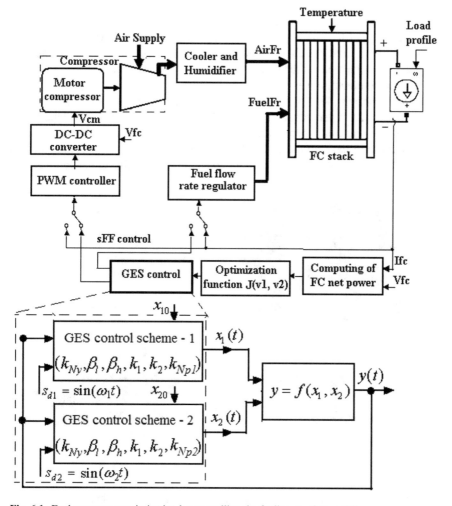

Fig. 6.1 Fuel economy maximization by controlling the fueling regulators [24]

algorithms fall into two categories [4, 49–54]: (1) rule-based and (2) optimization-based. In the literature, a lot of rules-based RTO algorithms are proposed, but they do not always find GM and remain locked in local maxima on the multimodal plateau of the optimization function [55].

To exemplify, the performance of the GES algorithm in case of a multimodal function with four maximum points on the plateau is presented (see Fig. 6.2). The GM is point M_1 for this multimodal function and the local maxima are M_2, M_3, M_4, very close to the GM point as position and magnitude (less than 5%). On the right side of Fig. 6.2 is presented the GM search process starting from the point (0.0) located between the maxima. The frequencies of the sinusoidal dithers s_{d1} and s_{d2} are 50 Hz and 100 (a double value) to increase the dither persistency due to nonlinear response (the harmonics will be at the multiple of 50 Hz and will increase as a mixing result of those).

Time search of the GM point highlights the performance of the GES algorithm (see Fig. 5.15 in Chap. 5): excellent tracking accuracy (less than 0.01%) and speed (less than 0.3 s), and searching resolution about 3% (less than the 5% difference between the magnitudes of the points) [24]. So, this algorithm will be used to search for the maximum of the fuel optimization function.

Recent research [47, 56] has shown that fuel consumption can be reduced by up to 20 lpm (liters per minute) for a 6 kW FC system compared to a static Feed-Forward (sFF) control (where air and fuel controllers depend on FC current), used as a reference in most applications [23]. The innovative solution presented in this chapter is to use a GM RTO algorithm for real-time GM tracking of the optimization

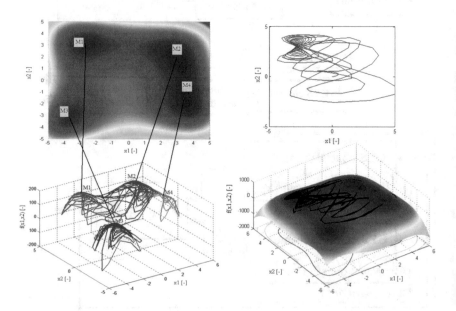

Fig. 6.2 Searching of the GM point (point M_1) on the multimodal function [24]

function by adapting fuel flows with one or two GM RTO-based algorithms that have the function f as input (see Fig. 6.1). Thus, fuel flows will look for GM, ensuring real-time performance of the FC system.

The rest of the chapter is structured as follows: Sect. 6.2 presents briefly the components of the FCHPS and the principle of fuel-saving strategies implemented based on load-following mode for the FC system; Sect. 6.3 details how the fuel economy strategies have been implemented based on optimization loops that use as controlled variables for the FC system the fueling regulators or the FC boost controller; the systematic analysis of the fuel-saving strategies is performed in Sect. 6.4, highlighting the fuel economy obtained using same load cycles for the FCHPS in order to compare the results; fuel economy achieved with algorithms for tracking the global maximum power point (MPP) compared to MPP tracking algorithms is presented in Sect. 6.5; the fuel economy obtained in searching the global MPP with different slope limits for the fueling regulators is highlighted in Sect. 6.6; the behavior of the FC system under the aforementioned fuel economy strategies is shown in Sect. 6.7 in order to sustain the findings presented in above sections and identify other ways to improve fuel economy; the last section shows a comparison of the fuel economy strategies and the final conclusions.

6.2 Fuel Cell Hybrid Power System

The diagram of the fuel cell hybrid power system (FCHPS) is shown in Fig. 6.3. All the components and parameters used in simulation have been explained in Chaps. 2, 3 and 4. The operation of the FC system is controlled by input references of the fueling regulators ($I_{ref(Air)}$ and $I_{ref(Fuel)}$) and the switching command of the boost DC–DC converter (SW$_{command}$).

The Energy Management Unit's (EMU) diagram presented in Fig. 6.4 highlights the optimization and control loops, and involved signals. The switching command is the output signal of the boost controller, having as inputs the reference $I_{ref(Boost)}$ and the FC current (I_{FC}). So, these three references ($I_{ref(Air)}$, $I_{ref(Fuel)}$, and $I_{ref(Boost)}$) will be used to systematically explore the improvements in fuel economy based on load-following (LF) control and GES-based optimization.

As it was mentioned in Chap. 3, the reference $I_{ref(LF)}$ is given by (6.2):

$$I_{ref(LF)} = P_{Load}/(V_{FC} \cdot \eta_{boost}) \tag{6.2}$$

The reference $I_{ref(LF)}$ is the output of the LF controller shown in Fig. 6.4, having as inputs the load demand (P_{load}) and the FC voltage (V_{FC}). η_{boost} is the boost efficiency.

The references used in search of optimal fuel economy ($I_{ref(GES1)}$ and $I_{ref(GES2)}$) are the outputs of two GES controllers, having as inputs the fuel optimization function f given by (6.1).

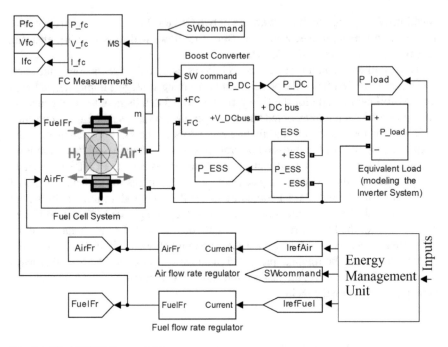

Fig. 6.3 The FCHPS diagram [47]

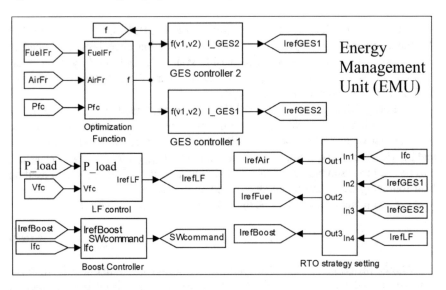

Fig. 6.4 The EMU diagram [47]

Different fuel economy strategies may be obtained by appropriate selection of the four inputs of the strategy setting block, $I_{ref(GES1)}$, $I_{ref(Ges2)}$, I_{FC}, and $I_{ref(LF)}$, to the three aforementioned references ($I_{ref(Air)}$, $I_{ref(Fuel)}$, and $I_{ref(Boost)}$).

A systematic analysis of the performances obtained with the fuel economy strategies that can be potentially used to improve the fuel economy will be performed in this chapter. The fuel economy obtained is compared with those obtained with the sFF control-based strategy proposed in reference [23].

6.3 Fuel Economy Strategies

The sFF control-based strategy of the FC system uses the FC current for the fueling regulators and the reference $I_{ref(LF)}$ for the boost controller [23]:

$$I_{ref(Fuel)} = I_{FC}, I_{ref(Air)} = I_{FC}, \quad \text{and} \quad I_{ref(boost)} = I_{ref(LF)} \tag{6.3}$$

The S1 strategy uses the reference $I_{ref(LF)}$ for the boost controller and performs the fueling optimization via the fuel regulator using the $I_{ref(GES1)}$ reference [57]:

$$I_{ref(Fuel)} = I_{FC} + I_{ref(GES1)}, I_{ref(Air)} = I_{FC}, \quad \text{and} \quad I_{ref(boost)} = I_{ref(LF)} \tag{6.4}$$

The S2 strategy uses the reference $I_{ref(LF)}$ for the boost controller and performs the fueling optimization via the air regulator using the $I_{ref(GES1)}$ reference [58]:

$$I_{ref(Fuel)} = I_{FC}, I_{ref(Air)} = I_{FC} + I_{ref(GES1)}, \quad \text{and} \quad I_{ref(boost)} = I_{ref(LF)} \tag{6.5}$$

The S3 strategy uses the reference $I_{ref(LF)}$ for the air regulator and performs the fueling optimization via the boost controller using the $I_{ref(GES1)}$ reference [24]:

$$I_{ref(Fuel)} = I_{FC}, I_{ref(Air)} = I_{ref(LF)}, \quad \text{and} \quad I_{ref(boost)} = I_{ref(GES1)} \tag{6.6}$$

The S4 strategy uses the reference $I_{ref(LF)}$ for the fuel regulator and performs the fueling optimization via the boost controller using the $I_{ref(GES1)}$ reference [35]:

$$I_{ref(Fuel)} = I_{ref(LF)}, I_{ref(Air)} = I_{FC}, \quad \text{and} \quad I_{ref(boost)} = I_{ref(GES1)} \tag{6.7}$$

The S5 strategy uses the reference $I_{ref(LF)}$ for the air regulator and performs the fueling optimization via the boost controller and fuel regulators by using the references $I_{ref(GES1)}$ and $I_{ref(GES2)}$ [37]:

$$I_{ref(Fuel)} = I_{FC} + I_{ref(GES2)}, I_{ref(Air)} = I_{ref(LF)}, \quad \text{and} \quad I_{ref(boost)} = I_{ref(GES1)} \tag{6.8}$$

The S6 strategy uses the reference $I_{ref(LF)}$ for the fuel regulator and performs the fueling optimization via the boost controller and air regulators by using the references $I_{ref(GES1)}$ and $I_{ref(GES2)}$ [47]:

$$I_{ref(Fuel)} = I_{ref(LF)}, \; I_{ref(Air)} = I_{FC} + I_{ref(GES2)}, \quad \text{and} \quad I_{ref(boost)} = I_{ref(GES1)} \quad (6.9)$$

The S7 strategy uses the reference $I_{ref(LF)}$ for the boost controller and performs the fueling optimization via the air and fuel regulators by using the references $I_{ref(GES1)}$ and $I_{ref(GES2)}$ [3]:

$$I_{ref(Fuel)} = I_{FC} + I_{ref(GES2)}, \; I_{ref(Air)} = I_{FC} + I_{ref(GES1)}, \quad \text{and} \quad I_{ref(boost)} = I_{ref(LF)}$$
$$(6.10)$$

The aforementioned settings for the fuel economy strategies are summarized in Fig. 6.5.

For fair comparison of the fuel economy obtained, it is worth mentioning that load-following mode for the FC power is implemented by each strategy. Thus, the battery operates in charge-sustained mode using each strategy.

Three classes (C1, C2, and C3) can be defined after the setting of reference $I_{ref(LF)}$ to the references $I_{ref(Air)}$, $I_{ref(Fuel)}$, and $I_{ref(Boost)}$, where the load-following mode for the FC power is implemented (see the top of Fig. 6.5). The C1 class uses the reference $I_{ref(LF)}$ for the boost controller and contains the strategies S1, S2, and S7. The C2 class uses the reference $I_{ref(LF)}$ for the air regulator and contains the strategies S3 and S5. The C3 class uses the reference $I_{ref(LF)}$ for the fuel regulator and contains the strategies S4 and S6.

6.4   Fuel Economy Analysis

The fuel economy analysis will be systematically performed in this section using constant and variable load for the FCHPS. The levels of load power must be in the admissible range for which the FC power is lower than the maximum power (8.3 kW). The levels for constant power start from 2 to 8 kW with 1 kW step. The levels for a 12 s load cycle (LC) are $0.75 \cdot P_{load(AV)}$, $1.25 \cdot P_{load(AV)}$, and $1.00 \cdot P_{load(AV)}$ on each 4 s in order to obtain a pulsed load with average value of $P_{load(AV)}$. The fuel economy (measured in liter [l]) is registered in tables and represented in Figs. 6.6, 6.7, 6.8, 6.9, 6.10, 6.11 and 6.12 for strategies S1 to S7 considering $k_{net} = 1 \text{ W}^{-1}$ and three values for k_{fuel}: 0 (case A), 25 (case B), and 50 (case C) lpm/W (the subscripts A, B, and C mention the fuel economy in each case). The values for the k_{fuel} weight coefficient were chosen in the range where the sensitivity analysis has been performed to highlight the best fuel economy for $k_{fuel} = 25$ lpm/W, the fuel economy when the optimization objective is the maximization of the FC net power, and the fuel economy for $k_{fuel} > 25$ lpm/W.

$I_{ref(Boost)}$	$I_{ref(Air)}$	$I_{ref(Fuel)}$	Strategy	Class
I_{LF}	I_{FC}	I_{FC}	sFF	reference
I_{LF}	I_{FC}	$I_{GES1}+I_{FC}$	S1	C1
I_{LF}	$I_{ref(GES1)}+I_{FC}$	I_{FC}	S2	C1
$I_{ref(GES1)}$	I_{LF}	I_{FC}	S3	C2
$I_{ref(GES1)}$	I_{FC}	I_{LF}	S4	C3
$I_{ref(GES1)}$	I_{LF}	$I_{ref(GES2)}+I_{FC}$	S5	C2
$I_{ref(GES1)}$	$I_{ref(GES2)}+I_{FC}$	I_{LF}	S6	C3
I_{LF}	$I_{ref(GES1)}+I_{FC}$	$I_{ref(GES2)}+I_{FC}$	S7	C1

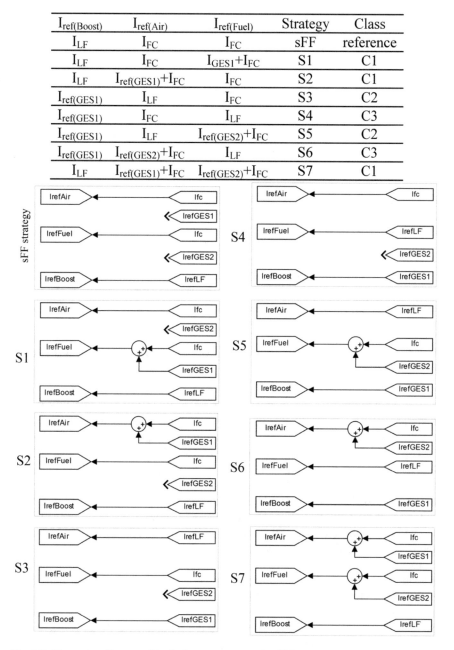

Fig. 6.5 The setting diagram of the fuel economy strategies [47]

(a)

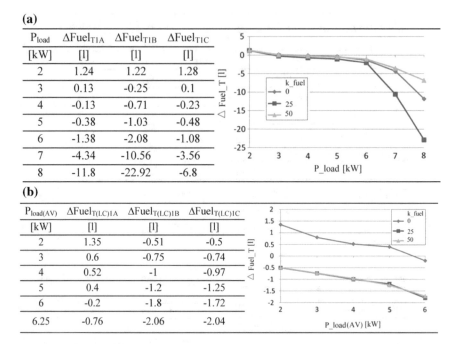

P$_{load}$	ΔFuel$_{T1A}$	ΔFuel$_{T1B}$	ΔFuel$_{T1C}$
[kW]	[l]	[l]	[l]
2	1.24	1.22	1.28
3	0.13	-0.25	0.1
4	-0.13	-0.71	-0.23
5	-0.38	-1.03	-0.48
6	-1.38	-2.08	-1.08
7	-4.34	-10.56	-3.56
8	-11.8	-22.92	-6.8

(b)

P$_{load(AV)}$	ΔFuel$_{T(LC)1A}$	ΔFuel$_{T(LC)1B}$	ΔFuel$_{T(LC)1C}$
[kW]	[l]	[l]	[l]
2	1.35	-0.51	-0.5
3	0.6	-0.75	-0.74
4	0.52	-1	-0.97
5	0.4	-1.2	-1.25
6	-0.2	-1.8	-1.72
6.25	-0.76	-2.06	-2.04

Fig. 6.6 a Fuel economy for S1 strategy under constant load. **b** Fuel economy for S1 strategy under variable load

The main findings of this analysis are as follows: (1) The best fuel economy is obtained for $k_{fuel} = 25$ lpm/W (except the case of S7 strategy (where the fuel economy is anyway lower than that obtained with the other strategies); (2) the fuel economy for $k_{fuel} \neq 0$ is almost the same for strategies S3 and S4 under constant and variable load and strategies S1 and S2 under variable load; (3) the fuel economy increases with increasing demand for the load; (4) the highest fuel economy under constant and variable load is obtained for strategies S5 and S6, and then for strategies S3 and S4 (due to the use of two GES controllers instead of one).

The last conclusion is very interesting, highlighting that the fuel economy can be improved if the load-following mode is not used for the boost controller.

To better see the improvements in fuel economy and the classification of the strategies, the fuel economy for all strategies is presented in the same plot of Figs. 6.13, 6.14, and 6.15 for $k_{fuel} = 0$, 25, 50 lpm/W.

The main findings of this analysis are as follows: (1) Starting with the best strategy, the classification of the strategies at high load is S5, S6, S3, and S4 for constant load (Fig. 6.14a) and variable load (Fig. 6.14b); (2) strategies S5 and S3 are the best at high load for $k_{fuel} = 0$ in case of constant load (Fig. 6.13a) and variable load (Fig. 6.13b); (3) strategies S5 and S3 are the best at high load for $k_{fuel} = 0$ in case of constant load (Fig. 6.15a) and the S5 strategy remains the best for variable load (Fig. 6.15b); and

(a)

P_{load}	$\Delta Fuel_{T2A}$	$\Delta Fuel_{T2B}$	$\Delta Fuel_{T2C}$
[kW]	[l]	[l]	[l]
2	1.2	-0.09	1.22
3	0.79	-0.24	0.56
4	0.77	-0.25	0.42
5	0.55	-0.46	0.28
6	0.42	-1.58	0.22
7	-0.14	-4.24	-1.14
8	-14	-18.48	-8.48

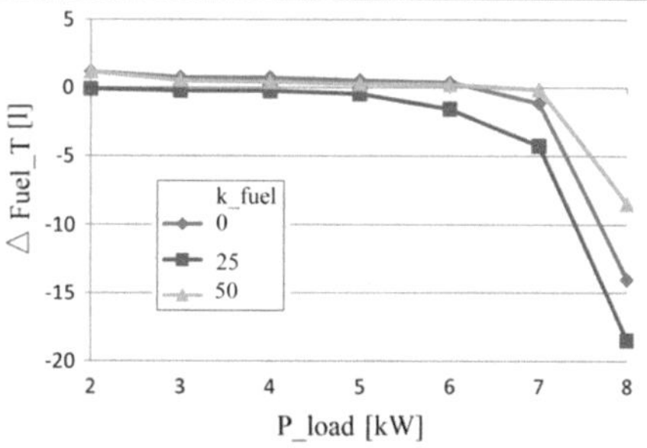

(b)

$P_{load(AV)}$	$\Delta Fuel_{T(LC)2A}$	$\Delta Fuel_{T(LC)2B}$	$\Delta Fuel_{T(LC)2C}$
[kW]	[l]	[l]	[l]
2	1.3	0.5	0.51
3	0.71	-0.48	-0.47
4	0.07	-1.8	-1.58
5	-1.6	-3	-2.99
6	-3.8	-5.3	-5.23
6.25	-4.56	-6.36	-6.21

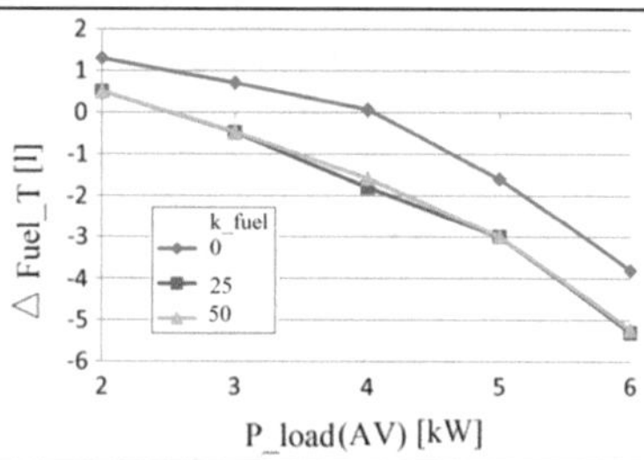

Fig. 6.7 a Fuel economy for S2 strategy under constant load. **b** Fuel economy for S2 strategy under variable load

(4) for any values of k_{fuel}, at reduced load, the strategies S6 and S4 are the best in case of constant load and the S6 strategy remains the best for variable load.

Other results, which may complete the brief discussion shown here, can be found in [47, 56, 59].

An interesting conclusion results from the aforementioned findings: The best strategies are from classes C2 and C3, but must be used in a switching mode (one at low load and the other at high load).

To highlight this finding, the fuel economy is represented in separate plots for strategies from classes C2 and C3, respectively, under constant load (Fig. 6.16a) and variable load (Fig. 6.16b).

It can be observed for strategies from class C2 that the S4 strategy may operate best at low load and the S3 strategy at high load. The threshold between the low and high ranges looks to be between 5 and 6 kW.

Also, for strategies from class C3, it can be noted that the S6 strategy may operate best at low load and the S5 strategy at high load. The threshold between the low and high ranges looks to be between 4 and 5 kW.

(a)

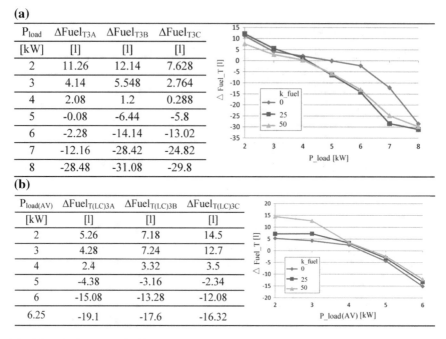

P_{load}	$\Delta Fuel_{T3A}$	$\Delta Fuel_{T3B}$	$\Delta Fuel_{T3C}$
[kW]	[l]	[l]	[l]
2	11.26	12.14	7.628
3	4.14	5.548	2.764
4	2.08	1.2	0.288
5	-0.08	-6.44	-5.8
6	-2.28	-14.14	-13.02
7	-12.16	-28.42	-24.82
8	-28.48	-31.08	-29.8

(b)

$P_{load(AV)}$	$\Delta Fuel_{T(LC)3A}$	$\Delta Fuel_{T(LC)3B}$	$\Delta Fuel_{T(LC)3C}$
[kW]	[l]	[l]	[l]
2	5.26	7.18	14.5
3	4.28	7.24	12.7
4	2.4	3.32	3.5
5	-4.38	-3.16	-2.34
6	-15.08	-13.28	-12.08
6.25	-19.1	-17.6	-16.32

Fig. 6.8 a Fuel economy for S3 strategy under constant load. **b** Fuel economy for S3 strategy under variable load

A more precise threshold value can be obtained using symmetrical stair load profile with small step and performing a sensitivity analysis in each case. It is obvious that these switching strategies will give the best fuel economy compared to basic strategies from the same class. Thus, the fuel economy of S3–S4 switching strategy will be higher compared to S3 strategy or S4 strategy. Also, the fuel economy of S5–S6 switching strategy will be higher compared to S5 strategy or S6 strategy. But switching strategies are still under test [60].

Anyway, it is worth mentioning that the fuel economy of S5–S6 switching strategy may be higher compared to S3–S4 switching strategy, but also the complexity of implementation is slightly higher for S5–S6 switching strategy compared to S3–S4 switching strategy (two GES controllers instead of one). A comparative analysis of the complexity of implementation versus performance is presented in the conclusion section of this chapter.

Also, the complexity of implementation is slightly higher for a global maximum power point (MPP) tracking algorithm compared to an MPP tracking algorithm. Consequently, the performance versus the complexity of implementation will be explored in the next section.

(a)

P_{load}	$\Delta Fuel_{T4A}$	$\Delta Fuel_{T4B}$	$\Delta Fuel_{T4C}$
[kW]	[l]	[l]	[l]
2	-0.46	-0.644	-0.1
3	-1.22	-3.876	-3.7
4	-2.28	-5.176	-5.264
5	-5.6	-8.76	-8.76
6	-7.66	-12.54	-13.98
7	-13.56	-24.26	-20.74
8	-22.92	-26	-25

(b)

$P_{load(AV)}$	$\Delta Fuel_{T(LC)4A}$	$\Delta Fuel_{T(LC)4B}$	$\Delta Fuel_{T(LC)4C}$
[kW]	[l]	[l]	[l]
2	0.7	1.92	3.4
3	-0.56	0.82	2.9
4	-0.82	-0.64	-0.5
5	-4.72	-4.16	-3.98
6	-11.46	-10.08	-9.46
6.25	-14.54	-12.86	-12.06

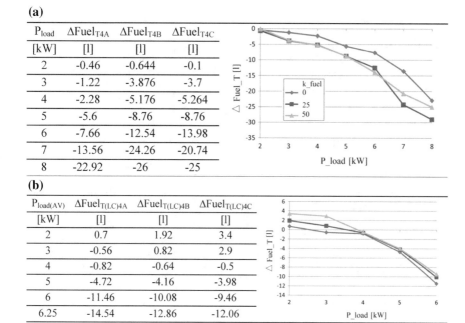

Fig. 6.9 a Fuel economy for S4 strategy under constant load. **b** Fuel economy for S4 strategy under variable load

6.5 Fuel Economy Achieved with Different Tracking Algorithms: Global MPP Versus MPP Tracking

Due to the difficulty of real-time implementation of a global MPP algorithm (search for global maximum), an alternative solution is to implement an MPP algorithm (search for a maximum of the multimodal optimization function).

The GES-based algorithm and the Perturb and Observe (P & O) algorithm are chosen as global MPP algorithm and MPP algorithm [61, 62], respectively. The optimization function and the strategy considered will be the same. The strategy S4 was chosen because it uses only one searching controller and the parameters in optimization function (1) are $k_{net} = 1$ W^{-1} and $k_{fuel} = 25$ lpm/W.

The initialization and implementation routines of the P & O algorithm are shown below:

```
% Initialization
%
global Iref;
global Increment;
global Pold;
Pold = 0; % the initial value of FC power
Iref = 10; % the initial value of searching current
```

(a)

P_{load} [kW]	$\Delta Fuel_{T5A}$ [l]	$\Delta Fuel_{T5B}$ [l]	$\Delta Fuel_{T5C}$ [l]
2	8	6.78	8.56
3	6.16	1.76	4
4	1.94	-3.72	1.1
5	-5.18	-11.42	-6.34
6	-11.56	-17.82	-13
7	-24.48	-30.24	-23.9
8	-43.34	-47.72	-45.52

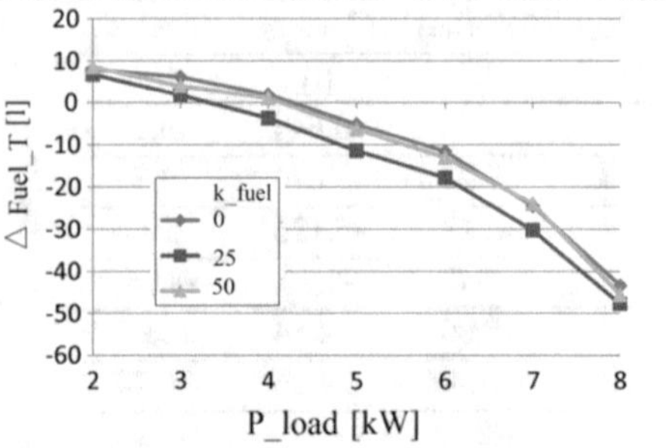

(b)

$P_{load(AV)}$ [kW]	$\Delta Fuel_{T(LC)5A}$ [l]	$\Delta Fuel_{T(LC)5B}$ [l]	$\Delta Fuel_{T(LC)5C}$ [l]
2	8.26	10.8	10.14
3	6.7	8.74	8.82
4	2.84	-0.26	4.92
5	-2.84	-12.96	-9.66
6	-29.08	-42.54	-37.26
6.25	-34.28	-52.7	-42.4

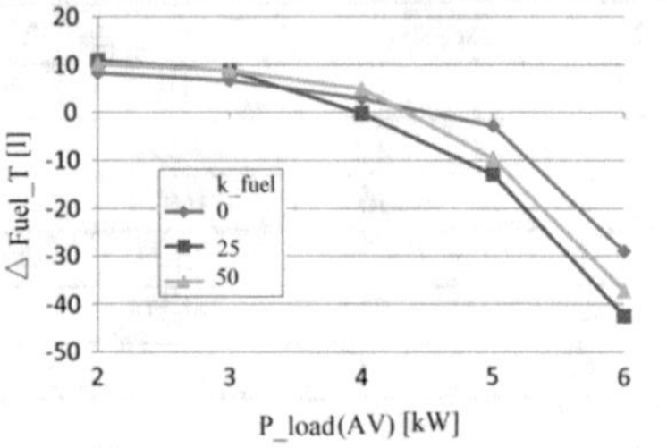

Fig. 6.10 **a** Fuel economy for S5 strategy under constant load. **b** Fuel economy for S5 strategy under variable load

```
Direction = 1; % the initial direction: the searching
current increases

% Input: the FC power, Pfc
% Output: the searching current, Iref
% Objective: the FC power needs to be maximized
function y = MPPtrackIrefFC(Pfc)
global Pold;
global Iref;
global Increment;

IrefH = 240; % the maximum value of the FC current is
240 A
IrefL = 0; % the minimum value of the FC current
DeltaI = 1; % the searching step is 1 A

if (Pfc < Pold)
Direction = -Direction; % the direction is changed: the
searching current will decrease
end

% the searching current will find and follow the MPP
current, I_MPP
```

(a)

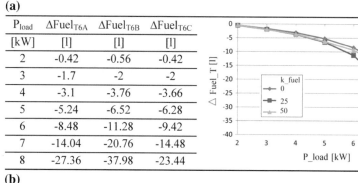

P_{load}	$\Delta Fuel_{T6A}$	$\Delta Fuel_{T6B}$	$\Delta Fuel_{T6C}$
[kW]	[l]	[l]	[l]
2	-0.42	-0.56	-0.42
3	-1.7	-2	-2
4	-3.1	-3.76	-3.66
5	-5.24	-6.52	-6.28
6	-8.48	-11.28	-9.42
7	-14.04	-20.76	-14.48
8	-27.36	-37.98	-23.44

(b)

$P_{load(AV)}$	$\Delta Fuel_{T(LC)6A}$	$\Delta Fuel_{T(LC)6B}$	$\Delta Fuel_{T(LC)6C}$
[kW]	[l]	[l]	[l]
2	-0.18	-0.1	-0.38
3	-1.32	-1.04	-1.18
4	-2.84	-3.84	-2.88
5	-4.52	-9.3	-6.2
6	-8.22	-18.56	-16.38
6.25	-10.4	-21.86	-19.68

Fig. 6.11 a Fuel economy for S6 strategy under constant load. **b** Fuel economy for S6 strategy under variable load

```
Iref=Iref + Direction*DeltaI;

% the upper limit is checked
if (Iref > IrefH)
Iref = IrefH;
end

% the lower limit is checked
if (Iref < IrefL)
Iref = IrefL;
end

% the current values of the FC power and the searching
current are saved
Pold = Pfc;
y = Iref;
```

The hardware-in-the-loop (HIL) implementation for the GES-based algorithm and the Perturb and Observe (P & O) algorithm are presented in Fig. 6.17.

The fuel economy using the P & O algorithm ($\Delta Fuel_{P \& O}$) and the S4 strategy ($\Delta Fuel_{T4B}$) is shown in Fig. 6.18 for constant and variable load, respectively.

The gap in fuel economy ($\Delta Fuel_{T4B} - \Delta Fuel_{P \& O}$) for the S4 strategy compared to the P & O algorithm increases with the increase of the load demand. For an

(a)

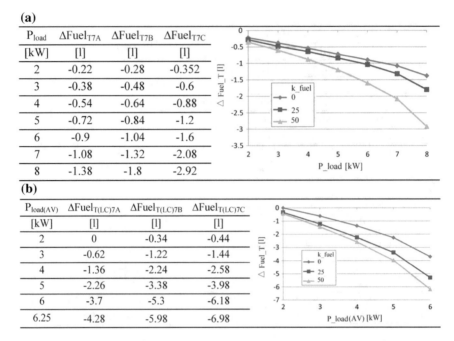

P_{load}	$\Delta Fuel_{T7A}$	$\Delta Fuel_{T7B}$	$\Delta Fuel_{T7C}$
[kW]	[l]	[l]	[l]
2	-0.22	-0.28	-0.352
3	-0.38	-0.48	-0.6
4	-0.54	-0.64	-0.88
5	-0.72	-0.84	-1.2
6	-0.9	-1.04	-1.6
7	-1.08	-1.32	-2.08
8	-1.38	-1.8	-2.92

(b)

$P_{load(AV)}$	$\Delta Fuel_{T(LC)7A}$	$\Delta Fuel_{T(LC)7B}$	$\Delta Fuel_{T(LC)7C}$
[kW]	[l]	[l]	[l]
2	0	-0.34	-0.44
3	-0.62	-1.22	-1.44
4	-1.36	-2.24	-2.58
5	-2.26	-3.38	-3.98
6	-3.7	-5.3	-6.18
6.25	-4.28	-5.98	-6.98

Fig. 6.12 a Fuel economy for S7 strategy under constant load. **b** Fuel economy for S7 strategy under variable load

8 kW load demand, the loss in fuel economy for the P & O algorithm is of 3.59 L compared to the S4 strategy (see Fig. 6.18a). For a 6 kW load cycle (LC), this loss in fuel economy increases to 5 L. This loss is due to the low performance of the P & O algorithm (reduced tracking speed), but first of all because it cannot follow a global maximum (can remain stuck at a local maximum that is far from the global maximum). To increase the tracking speed, it is necessary to increase the adjustment step (*DeltaI*) for the search reference (*Iref*), but the FC power ripple increases as well. This ripple on the DC bus will be partially compensated by the ESS, but the DC voltage of 200 V remains noisy.

The behavior of the FCHPS using the P & O algorithm with an 1 A step and the S4 strategy is presented in Fig. 6.19 for the FCHPS under a 6 kW constant load. Note the noisy power exchanged by the ESS with the DC bus if the P & O algorithm with 1 A step is used (Fig. 6.19a). On average, the power exchanged by the ESS is almost zero. This is more clear in case of using the S4 strategy. The power exchanged by the ESS is almost zero and not noisy (Fig. 6.19b). Thus, the battery operates in both cases in charge-sustained mode.

The battery also operates in charge-sustained mode for variable load due to load-following mode implemented for the fuel regulator in both cases (see Fig. 6.17). The behavior of the FCHPS using the P & O algorithm with 1 A step and the S4 strategy is presented in Fig. 6.20 for the FCHPS under a 6 kW load cycle.

(a)

P_{load}	$Fuel_{T0}$	$\Delta Fuel_{T1}$	$\Delta Fuel_{T2}$	$\Delta Fuel_{T3}$	$\Delta Fuel_{T4}$	$\Delta Fuel_{T5}$	$\Delta Fuel_{T6}$	$\Delta Fuel_{T7}$
[kW]	[l]	[l]	[l]	[l]	[l]	[l]	[l]	[l]
2	34.02	1.22	1.2	11.26	-0.46	8	-0.42	-0.22
3	56.3	0.13	0.79	4.14	-1.22	6.16	-1.7	-0.38
4	74.88	-0.13	0.77	2.08	-2.28	1.94	-3.1	-0.54
5	98.6	-0.38	0.55	-0.08	-5.6	-5.18	-5.24	-0.72
6	125.58	-1.38	0.42	-2.28	-7.66	-11.56	-8.48	-0.9
7	158.34	-4.34	-0.14	-12.16	-13.56	-24.48	-14.04	-1.08
8	176	-11.8	-4	-28.48	-22.92	-43.34	-27.36	-1.38

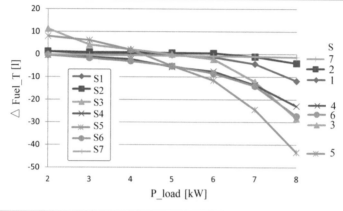

(b)

$P_{load(AV)}$	$Fuel_{T0}$	$\Delta Fuel_{T1}$	$\Delta Fuel_{T2}$	$\Delta Fuel_{T3}$	$\Delta Fuel_{T4}$	$\Delta Fuel_{T5}$	$\Delta Fuel_{T6}$	$\Delta Fuel_{T7}$
[kW]	[l]	[l]	[l]	[l]	[l]	[l]	[l]	[l]
2	34.14	1.3	1.35	5.26	0.7	8.26	-0.18	0
3	53.92	0.71	0.6	4.28	-0.56	6.7	-1.32	-0.62
4	75.8	0.07	0.52	2.4	-0.82	2.84	-2.84	-1.36
5	100.62	-1.6	0.4	-4.38	-4.72	-2.84	-4.52	-2.26
6	130.2	-3.8	-0.2	-15.08	-11.46	-29.08	-8.22	-3.7

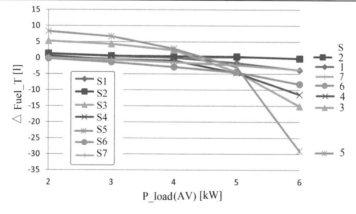

Fig. 6.13 **a** Fuel economy using $k_{fuel} = 0$ for all strategies under constant load. **b** Fuel economy using $k_{fuel} = 0$ for all strategies under variable load

(a)

P_{load}	$Fuel_{T0}$	$\Delta Fuel_{T1}$	$\Delta Fuel_{T2}$	$\Delta Fuel_{T3}$	$\Delta Fuel_{T4}$	$\Delta Fuel_{T5}$	$\Delta Fuel_{T6}$	$\Delta Fuel_{T7}$
[kW]	[l]	[l]	[l]	[l]	[l]	[l]	[l]	[l]
2	34.02	1.22	-0.09	12.14	-0.644	6.78	-0.56	-0.352
3	56.3	-0.25	-0.24	5.548	-3.876	1.76	-2	-0.6
4	74.88	-0.71	-0.25	1.2	-5.176	-3.72	-3.76	-0.88
5	98.6	-1.03	-0.46	-6.44	-8.76	-11.42	-6.52	-1.2
6	125.58	-2.08	-1.58	-14.14	-12.54	-17.82	-11.28	-1.6
7	158.34	-10.56	-4.24	-28.42	-24.26	-30.24	-20.76	-2.08
8	176	-22.92	-18.48	-31.08	-26	-47.72	-37.98	-2.92

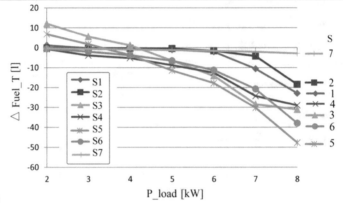

(b)

$P_{load(AV)}$	$Fuel_{T0}$	$\Delta Fuel_{T1}$	$\Delta Fuel_{T2}$	$\Delta Fuel_{T3}$	$\Delta Fuel_{T4}$	$\Delta Fuel_{T5}$	$\Delta Fuel_{T6}$	$\Delta Fuel_{T7}$
[kW]	[l]	[l]	[l]	[l]	[l]	[l]	[l]	[l]
2	34.14	0.5	-0.51	7.18	1.92	10.8	-0.1	-0.44
3	53.92	-0.48	-0.75	7.24	0.82	8.74	-1.04	-1.44
4	75.8	-1.8	-1	3.32	-0.64	-0.26	-3.84	-2.58
5	100.62	-3	-1.2	-3.16	-4.16	-12.96	-9.3	-3.98
6	130.2	-5.3	-1.8	-13.28	-10.08	-42.54	-18.56	-6.18

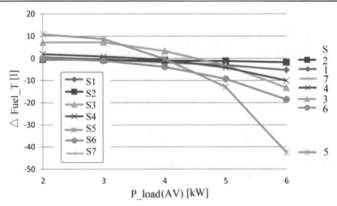

Fig. 6.14 a Fuel economy using $k_{fuel} = 25$ lpm/W for all strategies under constant load. **b** Fuel economy using $k_{fuel} = 25$ lpm/W for all strategies under variable load

(a)

P_{load}	$Fuel_{T0}$	$\Delta Fuel_{T1}$	$\Delta Fuel_{T2}$	$\Delta Fuel_{T3}$	$\Delta Fuel_{T4}$	$\Delta Fuel_{T5}$	$\Delta Fuel_{T6}$	$\Delta Fuel_{T7}$
[kW]	[l]	[l]	[l]	[l]	[l]	[l]	[l]	[l]
2	34.02	1.28	1.22	7.628	-0.1	8.56	-0.42	-0.28
3	56.3	0.1	0.56	2.764	-3.7	4	-2	-0.48
4	74.88	-0.23	0.42	0.288	-5.264	1.1	-3.66	-0.64
5	98.6	-0.48	0.28	-5.8	-8.76	-6.34	-6.28	-0.84
6	125.58	-1.08	0.22	-13.02	-13.98	-13	-9.42	-1.04
7	158.34	-3.56	-1.14	-24.82	-20.74	-23.9	-14.48	-1.32
8	176	-6.8	-8.48	-29.8	-25	-45.52	-23.44	-1.8

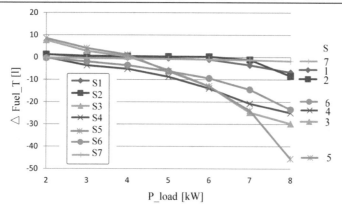

(b)

$P_{load(AV)}$	$Fuel_{T0}$	$\Delta Fuel_{T1}$	$\Delta Fuel_{T2}$	$\Delta Fuel_{T3}$	$\Delta Fuel_{T4}$	$\Delta Fuel_{T5}$	$\Delta Fuel_{T6}$	$\Delta Fuel_{T7}$
[kW]	[l]	[l]	[l]	[l]	[l]	[l]	[l]	[l]
2	34.14	0.51	-0.5	14.5	3.4	10.14	-0.38	-0.34
3	53.92	-0.47	-0.74	12.7	2.9	8.82	-1.18	-1.22
4	75.8	-1.58	-0.97	3.5	-0.5	4.92	-2.88	-2.24
5	100.62	-2.99	-1.25	-2.34	-3.98	-9.66	-6.2	-3.38
6	130.2	-5.23	-1.72	-12.08	-9.46	-37.26	-16.38	-5.3

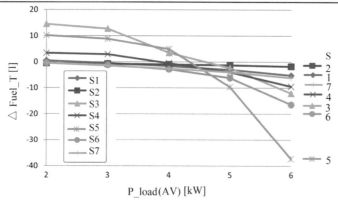

Fig. 6.15 **a** Fuel economy using $k_{fuel} = 50$ lpm/W for all strategies under constant load. **b** Fuel economy using $k_{fuel} = 50$ lpm/W for all strategies under variable load

(a)

P_{load}	$Fuel_{T0}$	$\Delta Fuel_{T1}$	$\Delta Fuel_{T2}$	$\Delta Fuel_{T3}$	$\Delta Fuel_{T4}$	$\Delta Fuel_{T5}$	$\Delta Fuel_{T6}$	$\Delta Fuel_{T7}$
[kW]	[l]	[l]	[l]	[l]	[l]	[l]	[l]	[l]
2	34.02	1.22	-0.09	12.14	-0.644	6.78	-0.56	-0.352
3	56.3	-0.25	-0.24	5.548	-3.876	1.76	-2	-0.6
4	74.88	-0.71	-0.25	1.2	-5.176	-3.72	-3.76	-0.88
5	98.6	-1.03	-0.46	-6.44	-8.76	-11.42	-6.52	-1.2
6	125.58	-2.08	-1.58	-14.14	-12.54	-17.82	-11.28	-1.6
7	158.34	-10.56	-4.24	-28.42	-24.26	-30.24	-20.76	-2.08
8	176	-22.92	-18.48	-31.08	-26	-47.72	-37.98	-2.92
	Class	C1	C1	C2	C3	C2	C3	C1

(b)

$P_{load(AV)}$	$Fuel_{T0}$	$\Delta Fuel_{T1}$	$\Delta Fuel_{T2}$	$\Delta Fuel_{T3}$	$\Delta Fuel_{T4}$	$\Delta Fuel_{T5}$	$\Delta Fuel_{T6}$	$\Delta Fuel_{T7}$
[kW]	[l]	[l]	[l]	[l]	[l]	[l]	[l]	[l]
2	34.14	0.5	-0.51	7.18	1.92	10.8	-0.1	-0.44
3	53.92	-0.48	-0.75	7.24	0.82	8.74	-1.04	-1.44
4	75.8	-1.8	-1	3.32	-0.64	-0.26	-3.84	-2.58
5	100.62	-3	-1.2	-3.16	-4.16	-12.96	-9.3	-3.98
6	130.2	-5.3	-1.8	-13.28	-10.08	-42.54	-18.56	-6.18
	Class	C1	C1	C2	C3	C2	C3	C1

Fig. 6.16 **a** Fuel economy using $k_{fuel} = 25$ lpm/W for strategies S3–S6 under constant load. **b** Fuel economy using $k_{fuel} = 25$ lpm/W for strategies S3–S6 under variable load

(a)

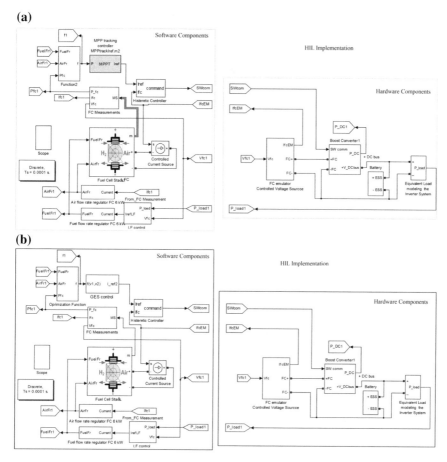

(b)

Fig. 6.17 **a** HIL implementation for the P & O-based algorithm. **b** HIL implementation for the GES-based algorithm

The levels for a 12 s/6 kW load cycle are $0.75 \cdot P_{load(AV)} = 4.5$ kW, $1.25 \cdot P_{load(AV)} = 7.5$ kW, and $1.00 \cdot P_{load(AV)} = 6$ kW on each 4 s (see Fig. 6.19). So, for a step-up in load demand from 4.5 to 7.5 kW, the step-up in FC current is 15 A ($=(7.5 - 4.5)$ kW/200 V). To avoid oxygen and fuel starvation, the fueling regulators will limit the changes of the fueling rates with slope limiters applied to the inputs (the FC current in sFF control and some strategies analyzed here or a reference that follows the FC current in the other strategies analyzed here). The maximum recommended variation of the FC current is 100 A/s [63], so the search delay for the next optimum is about 150 ms ($=15$ A/100 A/s). This is added to the 200 ms time constant of the FC system, resulting on the DC bus a lack of FC power of about 350 ms. The DC power flow balance will be sustained by the ESS (mainly by the supercapacitors' stack). For a step-down in load demand from 7.5 to 6 kW, an excess of FC power on the DC bus

P_{load}	$\Delta Fuel_{T4B}$	$\Delta Fuel_{P\&O}$	$\Delta Fuel_{T4B}-\Delta Fuel_{P\&O}$
[kW]	[l]	[l]	[l]
2	-0.644	-0.124	-0.52
3	-3.876	-2.696	-1.18
4	-5.176	-3.876	-1.3
5	-8.76	-7.04	-1.72
6	-12.54	-10.61	-1.93
7	-24.26	-22.06	-2.2
8	-29	-25.41	-3.59

$P_{load(AV)}$	$\Delta Fuel_{4B(LC)}$	$\Delta Fuel_{P\&OB(LC)}$	$\Delta Fuel_{4B(LC)}-\Delta Fuel_{P\&OB(LC)}$
[kW]	[l]	[l]	[l]
2	-0.08	-0.02	-0.06
3	-1.41	-0.68	-0.73
4	-3.76	-2.48	-1.28
5	-8.44	-5.65	-2.79
6	-17.1	-12.1	-5
6.25	-20.56	-15.46	-5.1

(a) For constant load   (b) For variable load

Fig. 6.18 Fuel economy using the P & O algorithm and the S4 strategy [62]

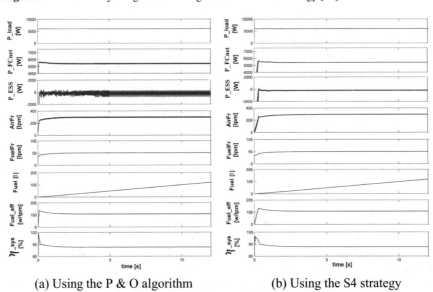

(a) Using the P & O algorithm   (b) Using the S4 strategy

Fig. 6.19 The behavior of the FCHPS under a 6 kW constant load [62]

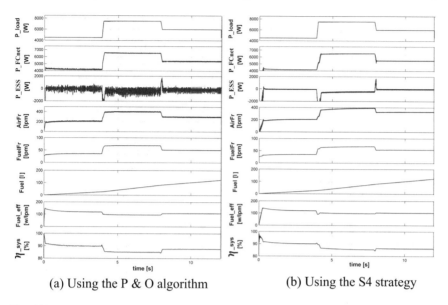

Fig. 6.20 The behavior of the FCHPS under a 6 kW load cycle [62]

will result for about 275 ms (=75 ms + 200 ms). Thus, the ESS will be discharged and charged during the step-up and step-down in load demand (see Fig. 6.19).

As it was mentioned in Chap. 4, but also in this chapter's introduction, the GES-based global MPP tracking algorithm has the tracking time less than 300 ms (about 10–15 periods of sinusoidal dither, so the tracking time can be further reduced by increasing the dither's frequency, but note that the tracking accuracy may be reduced due to dynamical interference of the dither with the FC system if the frequency is higher than 100 Hz [64–70]).

Because the 300 ms tracking time is comparable to the FC system's response delay, the effect of slope reducing to 50 A/s will be analyzed in the next section.

6.6 Fuel Economy Obtained in Searching with Different Slope Limits

The behavior of the FCHPS using the S4 strategy with a 100 A/s slope limiter and the S4* strategy with a 50 A/s slope limiter is analyzed in the same conditions, using the optimization function for three values of k_{fuel} parameter: 0 (case A), 25 (case B), and 50 (case C) lpm/W (the subscripts A, B, and C mention the fuel economy in each case).

The simulation results are recorded and the gap in fuel economy (Fuel$_4$–Fuel*$_4$) is represented in Fig. 6.21 during the aforementioned load cycles (LC).

$P_{load(AV)}$	$Fuel_{4A(LC)}$-$Fuel*_{4A(LC)}$	$Fuel_{4B(LC)}$-$Fuel*_{4B(LC)}$	$Fuel_{4C(LC)}$-$Fuel*_{4C(LC)}$
[kW]	[l]	[l]	[l]
2	-0.44	-2	-1.84
3	-1.03	-2.23	-2.02
4	-2.11	-3.12	-2.53
5	-3.29	-4.28	-4.03
6	-4.44	-7.02	-6.44
6.625	-4.72	-7.7	-7.2

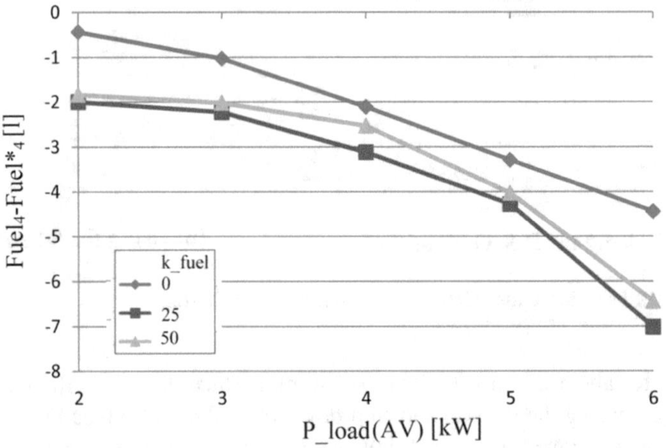

Fig. 6.21 Fuel economy using different slope limits [62]

Note that the greater loss on the fuel economy is in the case of $k_{fuel} = 25$ lpm/W, where the best fuel economy has been achieved as well. A loss of 7 L from the 120 L of total fuel consumption (see Fig. 6.16b) means about 5.8%, so all parameters must be carefully chosen based on the analysis of the FC system's safe operation [63–69].

The behavior of the FC system under other fuel economy strategies will be detailed in next section to highlight the differences between the strategies from different classes in case of $k_{fuel} = 25$ lpm/W.

6.7 Behavior of the FC System Under Fuel Economy Strategies

Because the behavior of the S4 strategy (from class C3) has been highlighted in the previous section, the behavior of the FCHPS using the fuel economy strategies S5 and S6 from class C2 and C3 is presented in Fig. 6.22 and discussed comparatively.

The searching time is less than 200 ms for both S5 and S6 strategies (even if they are from different classes). Note a better fuel consumption efficiency for the

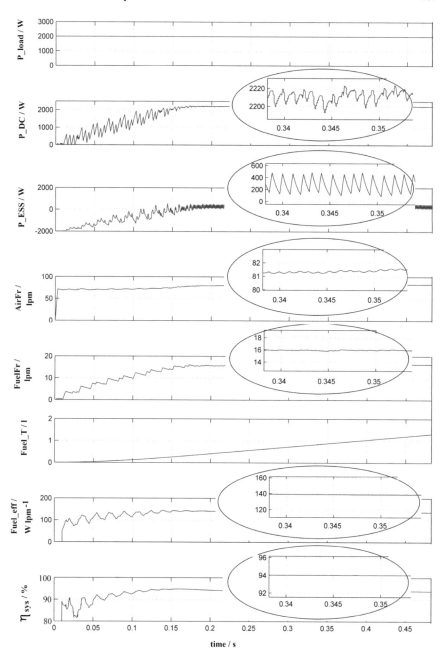

Fig. 6.22   **a** The behavior of the FCHPS using S5 strategy [71]. **b** The behavior of the FCHPS using S6 strategy [71]. **c** The searching of best fuel economy using the strategies S5 and S6

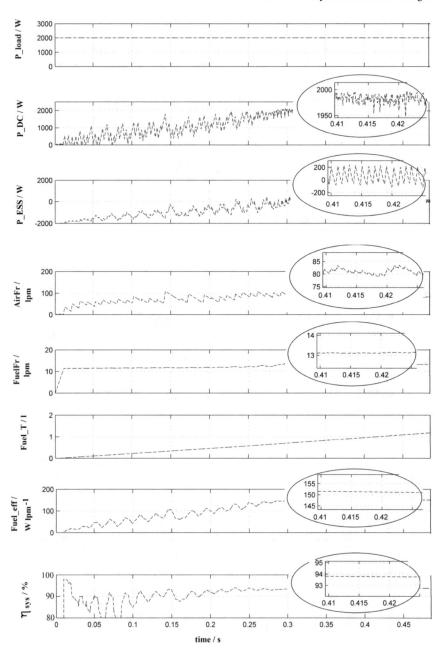

Fig. 6.22 (continued)

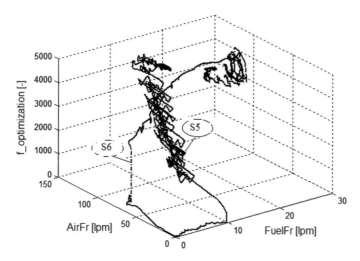

Fig. 6.22 (continued)

S6 strategy (152 W/lpm) compared to the S5 strategy (140 W/lpm), which means a better fuel economy for the S6 strategy compared to the S5 strategy (with about 7.34 L; see Figs. 6.10a and 6.11a). The approximately 94% FC electrical efficiency is obtained for both S5 and S6 strategies. The power ripple of FC power generated on the DC bus is less than 25 W for both S5 and S6 strategies, which means about 1% from the FC power (2220 W/90%).

The searching of the best fuel economy using the strategies S5 and S6 is illustrated in Fig. 6.22c. It can be observed that the shape of multimodal surface depends on the strategy used (being time dependent), so two different maximums have been found (using different searching routes).

The behavior of the FCHPS under variable load is presented in Fig. 6.23 using the fuel economy strategies S1 and S2 from same class C1. Note that only minor differences can be highlighted in the shapes of signals represented in Fig. 6.23.

Note that a better fuel economy for the S2 strategy compared to the S1 strategy (with about 6.3 L; see Figs. 6.6b and 6.7b) because the fuel consumption efficiency for the S2 strategy is higher than that of the S1 strategy for any levels of the pulsed load. But the FC electrical efficiency is higher for the S1 strategy compared to the S2 strategy for any levels of the pulsed load. The ESS power ripple is less than 300 W for both S1 and S2 strategies, which means a ripple less than 1% for the FC power. Also, the searching time is less than 200 ms for both S1 and S2 strategies, because the same GES-based algorithm is used for both strategies to search the optimum on a multimodal curve. This means that the search time is slightly dependent on the number of search variables. Other results, which may fully complete the comparative discussion presented above, can be found in [47, 56, 71].

Next section will highlight the fuel economy obtained with a Hydrogen Mobility Demonstrator developed in the joint research projects between National Research and

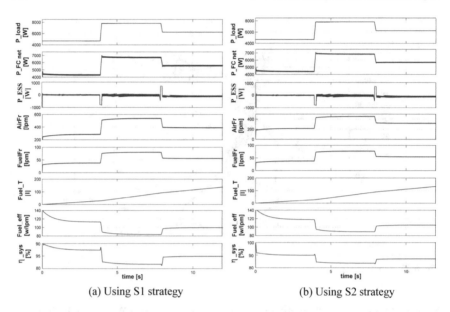

(a) Using S1 strategy   (b) Using S2 strategy

Fig. 6.23 The behavior of the FCHPS under 6.25 kW load cycle

Development Institute for Cryogenics and Isotopic Technologies Ramnicu Valcea
and the University of Pitesti.

6.8 Hydrogen Mobility Demonstrator

Emerging issues on fuel availability and price, but also greenhouse gas emissions
have attracted attention on the alternative energy sources, especially in transportation
sector, which accounts for over 40% of total fuel consumption. The acceptance of
electric vehicles (EVs)—for years championed as the obvious zero emissions solution
and quasi-unanimous accepted as being the future—is still, to a good extent, held
back by consumer concerns about price, convenience and range. To overcome these
problems, hybridization of electric power train was envisaged.

To achieve this, hydrogen fuel cells have proved to be a good candidate.

A fuel cell EV is just an EV that refuels with hydrogen instead of recharging the
battery pack. In this way, it will overcome the range anxiety and slow recharging
rates associated with battery EVs while still delivering the benefits of a zero emission
vehicle. Despite the enthusiasm for fuel cells, which was shared by the large majority
of the specialists it became clear that the future vehicle power trains will be drawn
from a portfolio of alternatives, each optimized for particular applications.

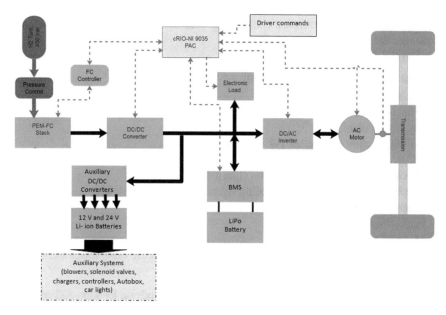

Fig. 6.24 FCHEV block diagram

A R&D project, **_Hy-DeMo—Hydrogen Mobility Demonstrator_**, was proposed by National Research and Development Institute for Cryogenics and Isotopic Technologies Ramnicu Valcea in collaboration with the University of Pitesti, in order to achieve a demonstrator for the mobility application of a hybrid electric power train based on proton exchange membrane fuel cells (PEMFC) and Li-ion polymer (LiPo) battery pack.

In order to highlight the benefits the hydrogen fuel cell stack could bring to the electric power train, tests were carried out in a simulated environment, using an electronic load. The test rig is based on a lightweight electric vehicle, accordingly modified, and is shown schematically in Fig. 6.24.

The hybrid electric vehicle (HEV) in Fig. 6.24 is driven by an AC motor supplied by a DC/AC inverter from a hybrid power assembly with two power streams, one from a LiPo rechargeable battery and another from a PEMFC stack. The PEM fuel cell assembly is connected to the DC bus via a DC/DC converter, and the LiPo rechargeable battery is directly connected to the DC bus. The PEM fuel cell stack supplies the stationary or slow variable load, operating close to the maximum efficiency, and the battery supplies the load transients. The fuel cells stack is also recharging the battery whenever necessary and there is available extra energy.

The AC motor is able to recover the braking energy, and the recovered electricity is also used to recharge the battery, along with the PEMFC.

An auxiliary power system, based on 12 and 24 V LiPo rechargeable batteries, supplies the auxiliary power load, composed of all auxiliary equipments electrical parts self-consumption.

A programmable automation controller provides the interface with the driver and manages the HEV operation.

The HEV features are:

- Total mass of the vehicle [kg]: 540;
- Vehicle cross section [m^2]: 2.05;
- Wheel diameter [m]: 0.3;
- Drag coefficient [–]: 0.3;
- Rolling friction coefficient [–]: 0.01;
- DC bus voltage [V]: 106;
- Maximum hydrogen consumption [lpm]: 40;
- Maximum air consumption (cooling circuit included) [lpm]: 195;
- Rated power [kW]: 10;
- Peak power [kW]: 30;
- Maximum speed [km/h]: 60;
- Minimum drive range [km]: 35, determined in all electric drive mode and battery discharge from 90 to 20%.

The power required for HEV propulsion, P_{prop}, depends on its mass and speed as in Eq. (6.11):

$$P_{\text{prop}} = \vartheta(t)\left(m_v \frac{\text{d}}{\text{d}t}\vartheta(t) + F_{\text{roll}}(t) + F_{\text{aero}}(t) + F_{\text{gra}}(t)\right) \tag{6.11}$$

where $\vartheta(t)$ is the vehicle speed, m_v is its mass, F_{roll} is the rolling force, F_{aero} is the aerodynamic force and F_{gra} is the gravitational force.

The HEV available power is supplied by the battery and the PEMFC, as in Eq. (6.12):

$$P_{\text{vehicle}} = \eta_{\text{BT}} P_{\text{BT}}(t) + \eta_{\text{FC}} P_{\text{FC}}(t) \tag{6.12}$$

where η_{BT} and η_{FC} are, respectively, the battery and the PEM fuel cell efficiency, and P_{BT} and P_{FC} are their power.

The PEMFC used in this HEV is a 3 kW HORIZON fuel cell stack made by 72 in series cells in "open cathode" configuration (see Fig. 6.25). Four blowers on the casing are supplying the required airflow providing the oxygen needed in the PEMFC electrochemical reaction and also are cooling the stack. The fuel used in this reaction, namely hydrogen, is stored in a 38 Nl pressurized tank at max. 400 bar. The total hydrogen amount available in a full tank is of 13 m^3 at a compressibility factor of 1.132, which means approximately 1.2 kg, or 39.9 kW in terms of power. The PEMFC stack I–V characteristic is presented in Fig. 6.26.

The electric energy storage device in HEV is a battery pack made up by 26 in series rechargeable cells of LiPo from LG Chem, having a voltage range between 109 and 80 V, a maximum discharge current of 120 A (3 °C) and a maximum peak discharge current (for 10 s, at 25 °C and SoC > 70%) of 200 A. In Fig. 6.27, the

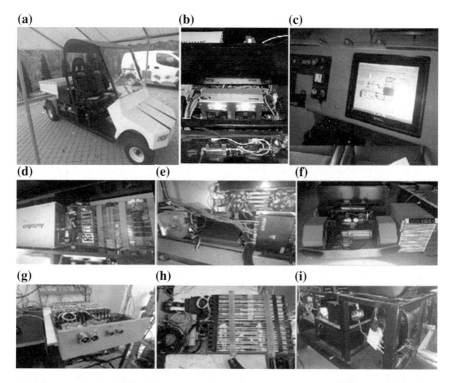

Fig. 6.25 HEV and main subsystems. **a** HEV right side view; **b** PEM fuel cells stack; **c** Dashboard; **d** Autobox and electrical connection panel; **e** Power Distribution Unit and electrical connections; **f** Hydrogen supply control system and electronic load; **g** DC/DC converter without housing; **h** LiPo battery without housing; **i** Hydrogen Tank

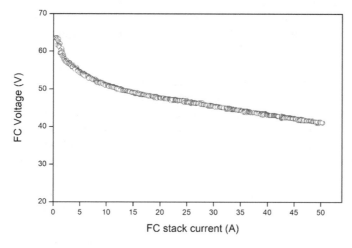

Fig. 6.26 I–V characteristic of the PEMFC

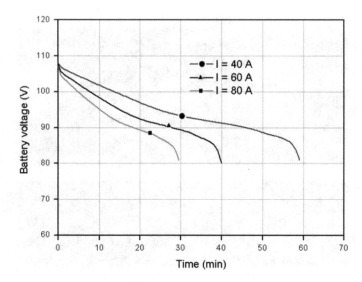

Fig. 6.27 Discharge characteristics of the LiPo battery for different current values

discharge characteristics of the battery from fully charged to fully discharged state for three values of the discharge currents are presented.

The battery energy management is realized by an Orion Battery Management System (BMS), recommended by its increased noise immunity, demanded by the automotive applications. It provides, among others, current, voltage and temperature protection, cells equilibration and state of charge (SoC) accurate monitoring. Another valuable feature of this BMS is the availability of a dual programmable controller, enabling CAN to communicate with a CompactRIO controller.

The tasks of monitoring and controlling the entire drive train system of FCHEV are performed by a specially designed, dedicated software implemented on a NI CompactRIO controller, cRIO-9035, by means of NI LabVIEW 2015 software platform. CompactRIO-9035 is a reliable, high-performance embedded real-time (RT) controller with modular I/O for receiving and transmitting digital or analog signals from/to external devices. Data are measured and analyzed by the RT controller based on custom virtual instruments (VI) realized using LabVIEW graphical programming environment. The role of this hardware and software sub-system is to put in the work the energy management strategy (EMS) of FCHEV, having as objectives the hydrogen consumption minimizing, the drive range increasing and equipments state of health (SoH) preserving.

In order to accomplish these objectives, EMS monitors the battery SoC and the pressure of hydrogen in the tank, controls the PEMFC DC/DC boost converter, starts and stops the fuel cell stack, monitors the throttle pedal position and works by limiting driving when an error occurs. The PEMFC operation at the maximum efficiency point is generally considered. The battery SoC decides the power amount supplied by the PEMFC: For $50\% < SoC < 80\%$, PEMFC will be operated at the maximum efficiency point; when SoC > 80%, PEMFC will supply the minimum power, and it will be turned off when SoC reaches 90%. For SoC < 50%, EMS will follow the power demand, provided that the maximum PEMFC power will be not exceeded. If SoC continues to fall below 30%, EMS will limit the engine power to 30% of its maximum power to avoid overheating the batteries.

All information about the FCHEV operation is delivered to the driver via a NI TP-12 touch screen. The front panel of LabVIEW has been designed to provide intuitive and easy to operate testing. It has nine tabs, the first six ones being for configuring, debugging, and viewing the most important FCHEV parameters and the last three tabs being real-time charts useful for viewing data across FCHEV.

As an example, the image of the front panel of the BMS in Fig. 6.28 and a real-time diagram of the HEV power balance in Fig. 6.29 are presented.

The tab in Fig. 6.28 offers valuable information about battery cells state of charge and thermal management. A window in the left corner of the tab shows the battery SoC versus cell average open voltage. Three slider indicators are providing intuitively information about instantaneous values of the battery pack charge/discharge current, voltage, and SoC. Several text indicators are providing information about voltage and temperature at cells level and also about the battery pack capacity, depth of discharge (DoD) and state of health (SoH), the charge current limit (CCL), and discharge current limit (DCL). A set of LEDs is signaling the BMS about the operating status and eventual errors.

The tab in Fig. 6.29 is showing a real-time diagram including power data across FCHEV: the load power, the PEM fuel cell stack power, the LiPo battery pack (high voltage) power, the auxiliary 24 V Li-ion battery pack, and 12 V Li-ion battery pack power.

It can be noticed that the PEMFC covers the load power demand and also charges the batteries (for batteries, the sign minus is assigned to the absorbed power during

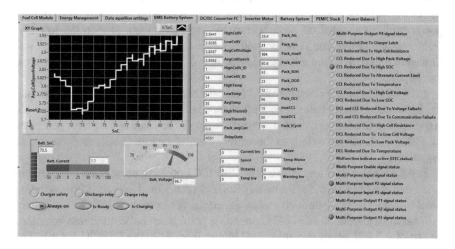

Fig. 6.28 Front panel of the software interface for the BMS battery system

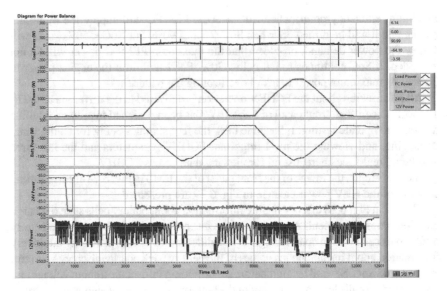

Fig. 6.29 Real-time diagram of FCHEV power balance

the charging process, and the sign plus is assigned to the power supplied to the system during discharging process).

Tests on Hy-DeMo fuel cells hybrid electric vehicle were carried out in order to estimate the hydrogen and electric power consumption and to assess the driving range.

To allow comparison, a standardized driving cycle, New European Driving Cycle (NEDC) was used. This driving cycle is composed of four repeated ECE-15 urban driving cycles followed by one Extra-Urban Driving Cycle (EUDC). Figure 6.30a shows the velocity profile of the driving cycle, and Fig. 6.30b, the corresponding power profile emulated on a programmable electronic load. Due to the low Hy-DeMo engine power, the velocity profile was halved and the maximum speed was limited al 60 km/h, so the mileage covered in one NEDC driving cycle became also halved, of 5.446 km.

In Hy-DeMo power train, the PEM fuel cells stack acts as main power source and the LiPo rechargeable battery, as a secondary power source to provide the electric engine power needs not covered by the main source. The value of the power generated by PEMFCs is set according to the battery state of charge (SoC), as presented in Fig. 6.31.

For the battery SoC lower than 50%, PEMFC operating point is set at maximum generated power of 3 kW, for the battery SoC between 50 and 80%, PEMFC operating point is set at the maximum efficiency generated power of 1.17 kW, for the battery SoC between 80 and 90%, PEMFC operating point is set at the minimum generated power of 0.7 kW, and for battery SoC higher than 90%, the PEMFC is OFF. The

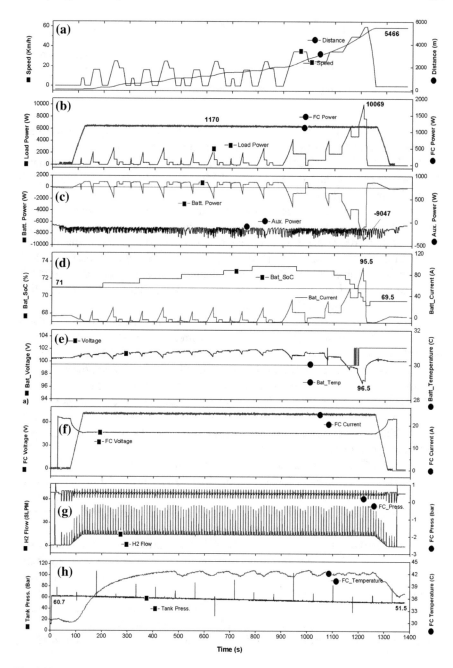

Fig. 6.30 Experimental results for FCHEV power train main parameters on NEDC

Batt. SoC [%]	0		50		80	90	100
FC power		$P_{FC\ max}$		$P_{Eff\ max}$	P_{FCmin}	0	

Fig. 6.31 PEMFC operating mode according battery SoC

LiPo battery power sizing was thought to allow PEMFC to operating at the maximum efficiency point, which is associated with the minimum hydrogen consumption.

On START/STOP, the PEMFC current load was controlled to avoid fast transients, its slope being maintained at 0.5 A/s. The PEMFC current load over the driving cycle was maintained constant at 25 A.

In Fig. 6.30c, the LiPo main battery power is shown. The positive values are assigned to the charging power received from PEMFC, and the negative ones, to the power delivered to the electric engine to supplement the power demand for the acceleration phases of the ECE 15 driving profile and for the aggressive EUDC, where the total power amount delivered by the high voltage battery is of—9047 W. It can be noted that the 12/24 V Li-ion batteries power delivered to the ancillary systems varied within a fairly narrow range.

Figure 6.30d, e are showing the LiPo main battery SoC and current, respectively, voltage and temperature. At the test starting, the battery SoC was 71% and it decreased to 69.5% at the end, after an increase to 73% after the four ECE 15 driving cycles, which indicates a charging process based on the PEMFC available power. Because the last portion of the driving cycle, EUDC, is very demanding, causing the LiPo battery current to reach 96.5 A, its voltage was monitored in order to prevent the voltage falling under the critical value of 80 V, which is representing an overheating and crashing hazard. It can be noticed that all this time, the battery warmed up by only one degree, to 31 °C.

The PEMFC operating parameters are shown in Fig. 6.30f–h, respectively, the stack current and voltage, the fuel supply (hydrogen) pressure and flow, and the stack temperature. Figure 6.30h shows also the hydrogen tank pressure. The maximum PEMFC voltage (OCV) was 66 V. As it was previously mentioned, the stack current over the whole driving cycle was maintained constant at 25 A, and as a result, the voltage was also constant at 47 V. Throughout the test, the PEMFC input pressure was between 0.3 and 0.6 bar, and the temperature oscillated around 42 °C. The hydrogen flow was counted during the whole test in order to specify the hydrogen consumption. All this time, the gas pressure in the hydrogen tank decreased from 60.7 to 51.5 bar.

Considering the fact that during a single NEDC driving cycle the hydrogen cylinder pressure drop was of 8.2 bar and the LiPo battery SoC decrease was of 1.5%, one could estimate that, starting from a 400 bar pressure of hydrogen tank and admitting a battery SoC drop of 70% (from 90 to 20%), a total of 48 halves of NEDC cycle could be driven using a hybrid fuel cell power train, meaning a distance of 261.4 km. The resulted hydrogen consumption per 100 km was of 0.45 kg.

After that a similar test was carried out with the vehicle acting as a pure electric one, using the same velocity and load profile as before, shown in Fig. 6.30a, b.

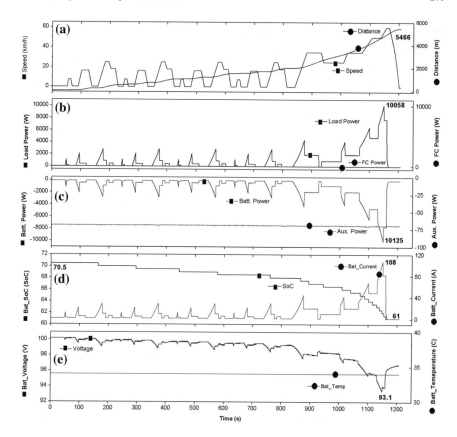

Fig. 6.32 Experimental results for FCHEV in pure electric mode

Figure 6.32c–e show the main parameters accomplished by the battery during this second test: state of charge, temperature, current, and voltage. The SoC initial value and final value were, respectively, 70.5 and 61%, with a slow decrease to 68% over the four ECE 15 cycles, followed by a more pronounced fall over the last part of the NEDC cycle, the aggressive EUDC cycle. The maximum current drawn from the battery, of 108 A, corresponding to the lowest voltage of 93.1 V, was reached at the end of this cycle. All over the EV test the battery temperature remained steady at 34 °C.

Starting from the 9.5% SoC drop over one NEDC cycle and allowing a total change in SoC of 70%, as in the case of HEV above, one could estimate a driving range of 7.3 halves of NEDC cycle, i.e., 40.2 km, for the pure electric drive train.

Comparing this result with that previously obtained in the case of the hybrid electric power train with PEM fuel cells, it can be noticed that hybridization improved the driving range by 6.5 times. Further improvements could be achieved by a more sophisticated energy strategy to be developed based on those analyzed in this chapter.

Table 6.1 Complexity versus performance for the strategies analyzed

Strategy	Performance indicators				
	η_{sys}	Fuel$_{eff}$	Fuel$_T$	Fuel$_{T(LC)}$	Complexity
sFF	Referential	Referential	Referential	Referential	Referential
S1	++	+	E + H +	E + H +	Medium
S2			E + H +	E +	Medium
S3	+		H ++	H ++	Medium
S4	++	++	E +++ H ++	E ++ H ++	Medium
S5	++		H +++	H +++	High
S6		++	E +++ H ++	E ++ H ++	High
S7	+	+	E +	E +	High

6.9   Conclusion

Complexity versus performance analysis is shown in Table 6.1 for the strategies considered in this chapter, compared to the sFF strategy (the referential). The performance indicators are as follows:

- FC electrical efficiency (η_{sys}): The comparison has been performed for the strategies that will improve this indicator across the range of load power; if not, the levels are not mentioned;
- Fuel consumption efficiency (Fuel$_{eff}$): The comparison has been performed for the strategies that will improve this indicator across the range of load power; if not, the levels are not mentioned;
- Fuel economy under constant load (Fuel$_T$): The comparison has been performed for the strategies that will improve this indicator in the entire load range (E) and/or in the range of high load (H);
- Fuel economy under load cycle (Fuel$_{T(LC)}$): The comparison has been performed for the strategies that will improve this indicator in the entire load range (E) and/or in the range of high load (H);

The levels used to compare increase of the performance are small (+), high (++), and very high (+++). The complexity of the strategies analyzed in this chapter is compared to the sFF strategy (which uses only the LF controller) considering the levels of medium complexity (one GES controller and the LF controller) and high complexity (two GES controllers and the LF controller);

Acknowledgements This work was supported by two grants of the Ministry of National Education and Scientific Research, National Agency of Scientific Research from Romania by the National Plan of R&D, CNCS/CCCDI-UEFISCDI within PNCDI III, project called "Experimental validation of a propulsion system with hydrogen fuel cell for a light vehicle—Mobility with Hydrogen Demonstrator," 53PED, ID: PN-III P2-2.1-PED-2016-1223, and project PN 19 11 02 02, PN-III-P-1.2-PCCDI-2017-0194/25 PCCDI, #117/2016, and within RDI Program for Space Technology and Advanced Research—STAR, project called "Concept Development of an Energy Storage Unit Using High Temperature Superconducting Coil for Spacecraft Power Systems (SMESinSpace)",#167/2017.

References

1. Luo X, Wang J-H, Dooner M, Clarke J (2015) Overview of current development in electrical energy storage technologies and the application potential in power system operation. Appl Energy 137:511–536
2. Meyar-Naimi H, Vaez-Zadeh S (2012) Sustainable development based energy policy making frameworks, a critical review. Energy Policy 43:351–361
3. Bizon N, Thounthong P (2018) Real-time strategies to optimize the fueling of the fuel cell hybrid power source: a review of issues, challenges and a new approach. Renew Sustain Energy Rev 91:1089–1102
4. Olatomiwa L, Mekhilef S, Ismail MS, Moghavvemi M (2016) Energy management strategies in hybrid renewable energy systems: a review. Renew Sustain Energy Rev 62:821–835
5. Ou K, Yuan W-W, Choi M, Yang S, Jung S, Young-Bae Kim Y-B (2018) Optimized power management based on adaptive-PMP algorithm for a stationary PEM fuel cell/battery hybrid system. Int J Hydrogen Energ 43:15433–15444
6. Bizon N, Lopez-Guede JM, Kurt E, Thounthong P, Mazare AG, Ionescu LM, Iana G (2019) Hydrogen economy of the fuel cell hybrid power system optimized by air flow control to mitigate the effect of the uncertainty about available renewable power and load dynamics. Energy Convers Manage 179:152–165
7. Guneya MS, Tepe Y (2017) Classification and assessment of energy storage systems. Renew Sust Energ Rev 75:1187–1197
8. Das V, Padmanaban S, Venkitusamy K, Selvamuthukumaran R, Blaabjerg F, Siano P (2017) Recent advances and challenges of fuel cell based power system architectures and control—a review. Renew Sust Energ Rev 73:10–18
9. Sikkabut S, Mungporn P, Ekkaravarodome C, Bizon N et al (2016) Control of high-energy high-power densities storage devices by Li-ion battery and supercapacitor for fuel cell/photovoltaic hybrid power plant for autonomous system applications. IEEE Trans Ind Appl 52(5):4395–4407. https://doi.org/10.1109/TIA.2016.2581138
10. Bizon N (2018) Effective mitigation of the load pulses by controlling the battery/SMES hybrid energy storage system. Appl Energ 229:459–473
11. Hedlund M, Lundin J, de Santiago J, Abrahamsson J, Bernhoff H (2015) Flywheel energy storage for automotive applications. Energies 8:10636–10663
12. Pires VP, Romero-Cadaval E, Vinnikov D, Roasto I, Martins JF (2014) Power converter interfaces for electrochemical energy storage systems—a review. Energy Convers Manage 86:453–475
13. Sharma A, Sharma S (2019) Review of power electronics in vehicle-to-grid systems. J Energ Storage 21:337–361
14. Ahmadi S, Bathaee SMT, Hosseinpour AH (2018) Improving fuel economy and performance of a fuel-cell hybrid electric vehicle (fuel-cell, battery, and ultra-capacitor) using optimized energy management strategy. Energy Convers Manage 180:74–84
15. Sulaiman N, Hannan MA, Mohamed A, Ker PJ, Majlan EJ, Daud WRW (2018) Optimization of energy management system for fuel-cell hybrid electric vehicles: issues and recommendations. Appl Energ 228:2061–2079

16. Sedaghati R, Shakarami MR (2019) A novel control strategy and power management of hybrid PV/FC/SC/battery renewable power system-based grid-connected microgrid. Sustain Cities Soc 44:830–843

17. Fernández RÁ, Caraballo SC, Cilleruelo FB, Lozano JA (2018) Fuel optimization strategy for hydrogen fuel cell range extender vehicles applying genetic algorithms. Renew Sust Energ Rev 81(1):655–668

18. Yuan J, Yang L, Chen Q (2018) Intelligent energy management strategy based on hierarchical approximate global optimization for plug-in fuel cell hybrid electric vehicles. Int J Hydrogen Energ 43(16):8063–8078

19. Chena S, Kumar A, Wong WC, Chiu M-S, Wang X (2019) Hydrogen value chain and fuel cells within hybrid renewable energy systems: Advanced operation and control strategies. Appl Energ 233–234:321–337

20. Haseli Y (2018) Maximum conversion efficiency of hydrogen fuel cells. Int J Hydrogen Energ 43(18):9015–9021

21. Weyers C, Bocklisch T (2018) Simulation-based investigation of energy management concepts for fuel cell—battery—hybrid energy storage systems in mobile applications. Energy Proc 155:295–308

22. Daud WRW, Rosli RE, Majlan EH, Hamid SAA, Mohamed R, Husaini T (2017) PEM fuel cell system control: a review. Renew Energ 113:620–638

23. Pukrushpan JT, Stefanopoulou AG, Peng H (2004) Control of fuel cell power systems. Springer, New York

24. Bizon N (2017) Energy optimization of fuel cell system by using global extremum seeking algorithm. Appl Energ 206:458–474. https://doi.org/10.1016/j.apenergy.2017.08.097

25. Zhao J, Ramadan HS, Becherif M (2018) Metaheuristic-based energy management strategies for fuel cell emergency power unit in electrical aircraft. Int J Hydrogen Energ. https://doi.org/10.1016/j.ijhydene.2018.07.131

26. Song K, Chen H, Wen P, Zhang T, Zhang B, Zhang T (2018) A comprehensive evaluation framework to evaluate energy management strategies of fuel cell electric vehicles. Electrochim Acta 292:960–973

27. Harrag A, Messalti S (2018) How fuzzy logic can improve PEM fuel cell MPPT performances? Int J Hydrogen Energ 43(1):537–550

28. Bizon N, Radut M, Oprescu M (2015) Energy control strategies for the fuel cell hybrid power source under unknown load profile. Energy 86:31–41. https://doi.org/10.1016/j.energy.2015.03.118

29. Li H, Ravey A, N'Diaye A, Djerdir A (2018) A novel equivalent consumption minimization strategy for hybrid electric vehicle powered by fuel cell, battery and supercapacitor. J Power Sources 395:262–270

30. Vergara-Dietrich JD, Morato MM, Mendes PRC, Cani AA, Normey-Rico JE, Bordons C (2017) Advanced chance-constrained predictive control for the efficient energy management of renewable power systems. J Process Control. https://doi.org/10.1016/j.jprocont.2017.11.003

31. Behdani A, Naseh MR (2017) Power management and nonlinear control of a fuel cell–supercapacitor hybrid automotive vehicle with working condition algorithm. Int J Hydrogen Energ 42(38):24347–24357

32. Bizon N (2011) Nonlinear control of fuel cell hybrid power sources: part II—current control. Appl Energ 88(7):2574–2591. https://doi.org/10.1016/j.apenergy.2011.01.044

33. Zhou W, Yang L, Cai Y, Ying T (2018) Dynamic programming for new energy vehicles based on their work modes part II: fuel cell electric vehicles. J Power Sources 407:92–104

34. Chauhan A, Saini RP (2014) A review on integrated renewable energy system based power generation for stand-alone applications: configurations, storage options, sizing methodologies and control. Renew Sustain Energy Rev 38:99–120

35. Bizon N (2018) Real-time optimization strategy for fuel cell hybrid power sources with load-following control of the fuel or air flow. Energy Convers Manage 157:13–27. https://doi.org/10.1016/j.enconman.2017.11.084

36. Han Y, Chen W, Li Q, Yang H, Zare F, Zheng Y (2018) Two-level energy management strategy for PV-Fuel cell-battery-based DC microgrid. Int J Hydrogen Energ. https://doi.org/10.1016/j.ijhydene.2018.04.013

37. Bizon N, Mazare AG, Ionescu LM, Enescu FM (2018) Optimization of the proton exchange membrane fuel cell hybrid power system for residential buildings. Energy Convers Manage 163:22–37

38. Han J, Park Y, Dongsuk K (2014) Optimal adaptation of equivalent factor of equivalent consumption minimization strategy for fuel cell hybrid electric vehicles under active state inequality constraints. J Power Sources 267:491–502

39. Zhang W, Li J, Xu L, Ouyang M (2017) Optimization for a fuel cell/battery/capacity tram with equivalent consumption minimization strategy. Energ Convers Manage 134:59–69

40. Geng C, Jin X, Zhang X (2019) Simulation research on a novel control strategy for fuel cell extended-range vehicles. Int J Hydrogen Energ 44:408–420

41. Hames Y, Kaya K, Baltacioglu E, Turksoy A (2018) Analysis of the control strategies for fuel saving in the hydrogen fuel cell vehicles. Int J Hydrogen Energ 43:10810–10821

42. Ahmadi S, Bathaee SMT (2015) Multi-objective genetic optimization of the fuel cell hybrid vehicle supervisory system: fuzzy logic and operating mode control strategies. Int J Hydrogen Energ 40(36):12512–12521

43. Çelik D, Meral ME (2019) Current control based power management strategy for distributed power generation system. Control Eng Pract 82:72–85

44. Turkmen AC, Solmaz S, Celik C (2017) Analysis of fuel cell vehicles with advisor software. Renew Sust Energ Rev 70:1066–1071

45. Zhao D, Zheng Q, Gao F, Bouquain D, Dou M, Miraoui A (2014) Disturbance decoupling control of an ultra-high speed centrifugal compressor for the air management of fuel cell systems. Int J Hydrogen Energ 39(4):1788–1798

46. Odeim F, Roes J, Wülbeck L, Heinzel A (2014) Power management optimization of fuel cell/battery hybrid vehicles with experimental validation. J Power Sources 252:333–343

47. Bizon N (2019) Real-time optimization strategies of FC hybrid power systems based on Load-following control: a new strategy, and a comparative study of topologies and fuel economy obtained. Appl Energ 241C:444–460

48. Bizon N, Hoarcă CI (2019) Hydrogen saving through optimized control of both fueling flows of the fuel cell hybrid power system under a variable load demand and an unknown renewable power profile. Energ Convers Manage 184:1–14

49. Ishaque K, Salam Z (2013) A review of maximum power point tracking techniques of PV system for uniform insolation and partial shading condition. Renew Sustain Energy Rev 19:475–488

50. Zhang P, Yan F, Du C (2015) A comprehensive analysis of energy management strategies for hybrid electric vehicles based on bibliometrics. Renew Sustain Energy Rev 48:88–104

51. Peng J, He H, Xiong R (2017) Rule based energy management strategy for a series–parallel plug-in hybrid electric bus optimized by dynamic programming. Appl Energ 185(2):1633–1643

52. Das V, Padmanaban S, Venkitusamy K, Selvamuthukumaran R, Siano P (2017) Recent advances and challenges of fuel cell based power system architectures and control—a review. Renew Sustain Energy Rev 73:10–18

53. Liu Z-H, Chen J-H, Huang J-W (2015) A review of maximum power point tracking techniques for use in partially shaded conditions. Renew Sustain Energy Rev 41:436–453

54. Carignano MG, Costa-Castelló R, Roda V, Nigro NM, Junco S, Feroldi D (2017) Energy management strategy for fuel cell-supercapacitor hybrid vehicles based on prediction of energy demand. J Power Sources 360:419–433

55. Meyer Q, Himeur A, Ashton S, Curnick O, Clague R et al (2015) System-level electrothermal optimisation of aircooled open-cathode polymer electrolyte fuel cells: air blower parasitic load and schemes for dynamic operation. Int J Hydrogen Energ 40:16760–16766

56. Bizon N, Oproescu M (2018) Experimental comparison of three real-time optimization strategies applied to renewable/FC-based hybrid power systems based on load-following control. Energies 11(12):3537–3569. https://doi.org/10.3390/en11123537

57. Bizon N, Culcer M, Iliescu M, Mazare AG, Ionescu LM, Beloiu R (2017) Real-time strategy to optimize the fuel flow rate of fuel cell hybrid power source under variable load cycle. In: ECAI—9th edition of international conference on electronics, computers and artificial intelligence. Targoviste, ROMÂNIA. https://doi.org/10.1109/ecai.2017.8166513
58. Bizon N, Iana G, Kurt E, Thounthong P, Oproescu M, Culcer M, Iliescu M (2018) Air flow real-time optimization strategy for fuel cell hybrid power sources with fuel flow based on load-following. Fuel Cell 18(6):809–823. https://doi.org/10.1002/fuce.201700197
59. Bizon N, Stan VA, Cormos AC (2019) Optimization of the fuel cell renewable hybrid power system using the control mode of the required load power on the DC bus. Energies 12(10):1889–1904. https://doi.org/10.3390/en12101889
60. Bizon N (2019) Fuel saving strategy using real-time switching of the fueling regulators in the proton exchange membrane fuel cell system. Appl Energy 252:113449–113453. https://doi.org/10.1016/j.apenergy.2019.113449
61. SimPowerSystems TM Reference (2010) Hydro-Québec and the MathWorks, Inc. Natick, MA
62. Bizon N, Thounthong P (2018) Fuel economy using the global optimization of the fuel cell hybrid power systems. Energ Convers Manage 173:665–678
63. Nikiforow K, Koski P, Ihonen J (2017) Discrete ejector control solution design, characterization, and verification in a 5 kW PEMFC system. Int J Hydrogen Energy 42:16760–16772
64. Ettihir K, Cano MH, Boulon L, Agbossou K (2017) Design of an adaptive EMS for fuel cell vehicles. Int J Hydrogen Energ 42(2):1481–1489
65. Ariyur KB, Krstic M (2003) Real-time optimization by extremum-seeking control. Wiley-Interscience, Hoboken
66. Ramos-Paja CA, Spagnuolo G, Petrone G, Emilio Mamarelis M (2014) A perturbation strategy for fuel consumption minimization in polymer electrolyte membrane fuel cells: analysis, design and FPGA implementation. Appl Energ 119:21–32
67. Zhou D, Ravey A, Al-Durra A, Gao F (2017) A comparative study of extremum seeking methods applied to online energy management strategy of fuel cell hybrid electric vehicles. Energ Convers Manage 151:778–790
68. Bizon N (2014) Tracking the maximum efficiency point for the FC system based on extremum seeking scheme to control the air flow. Appl Energ 129:147–157. https://doi.org/10.1016/j.apenergy.2014.05.002
69. Kumar J, Agarwal A, Agarwal V (2019) A review on overall control of DC microgrids. J Energ Storage 21:113–138
70. Bizon N, Thounthong P, Raducu M, Constantinescu LM (2017) Designing and modelling of the asymptotic perturbed extremum seeking control scheme for tracking the global extreme. Int J Hydrogen Energ 42(28):17632–17644. https://doi.org/10.1016/j.ijhydene.2017.01.086
71. Bizon N (2019) Efficient fuel economy strategies for the fuel cell hybrid power systems under variable renewable/load power profile. Appl Energ 251:113,400–113,518. https://doi.org/10.1016/j.apenergy.2019.113400

Chapter 7
Energy Harvesting from the Partially Shaded Photovoltaic Systems

7.1 Introduction

The share of renewables in worldwide energy consumption will increase by 12.4% from 2018 to 2023, ensuring about 30% (16%—Hydropower, 6%—wind, 4%—solar PV), and 3%—bioenergy) of power demand [1, 2]. Energy from solar photovoltaic (PV) systems has a huge potential to be exploited compared to other renewable energy sources (RESs) due to the many advantages in its use, such as sustainability, local availability, pollution-free, noise-free, and a relatively simple technology [3, 4]. Thus, a 98 GW PV capacity has been installed worldwide in 2018 that resulted in a cumulative installed capacity of about 500 GW [5].

As it is known, the PV power characteristic depends on the level of irradiance and temperature. The level of irradiance changes quicker than the temperature and sometimes partial shading conditions (PSCs) may appear for large PV array. The PV power characteristic of the PV array under uniform irradiance has a unique maximum power point (MPP), but the PV array under PSCs has many local MPPs (LMPPs) due to the action of bypass diodes. It is worth mentioning that the harvested PV power increases up to 45% if the global MPP (GMPP) will be tracked instead of a LMPP [6–8].

Consequently, a GMPP tracking (GMPPT) algorithm that always tracks the GMPP on the multimodal PV power characteristic obtained under partial shading conditions (PSCs) must be used [9–15]. The recent reviews [9–15] define the features of an ideal GMPPT algorithm as follows: ensuring efficient operation of the PV array under uniform irradiance and PSCs; fast finding and accurate tracking of the MPP for a PV array under uniform irradiance; fast localization and accurate tracking of the GMPP for a PV array under PSCs; 100% hit count in localization of the GMPP; robust search.

N. Bizon, *Optimization of the Fuel Cell Renewable Hybrid Power Systems*, Green Energy and Technology, https://doi.org/10.1007/978-3-030-40241-9_7

In the aforementioned reviews researches, the MPPT algorithms have been classified into conventional (the well-known techniques such as perturb and observe
(P&O), incremental conductance (IC), hill climbing (HC), and constant voltage
(CV), classical and advanced Extremum Seeking Control (ESC) [15–17], sweeping
techniques of the PV characteristics, etc.), soft computing (such as the evolutionary,
heuristic, and metaheuristic algorithms based on artificial intelligence or bio-inspired
concepts), and hybrid techniques.

The performance of the MPPT algorithms has been evaluated using five criteria
(dependency on the PV array's size, tracking speed, operation under PSCs, algorithm's complexity, and hardware implementation) [16], ten criteria (tracking speed,
dependency on the PV array's size, performance under PSCs, performance under
uniform irradiance, tracking efficiency, algorithm's complexity, hardware implementation, periodic adjustment of the tuning parameters, dependency on the starting
point (initial conditions) and PV power ripple during stationary regime) [17], five
criteria (type of the hardware implementation (analog or digital), number of used
sensors, tracking speed, stability of the search loop and periodic adjustment of the
tuning parameters) [18] and four criteria (tracking speed, algorithm's complexity,
performance under PSCs, and hardware implementation) [19].

An evaluation index of the MPPT algorithms has been proposed in [6, 8] based on
four and eight criteria selected from the aforementioned criteria, as it will be shown
in this chapter, where a new evaluation index will be proposed to efficiently compare
the GMPPT algorithms.

Thus, the structure of the paper is as follows. The MPPT algorithms selection
criteria are discussed in Sect. 7.2 based on performance indicators and evaluation
indexes proposed in the literature. The evaluation of the global asymptotic perturbed
extremum seeking control (GaPESC)-based MPPT algorithm using the evaluation
indexes based on four and eight criteria, and new criteria class proposed in this
chapter. Section 7.3 analyses the shape of the PV power characteristic of a PV array
under PSCs and propose a multimodal function as PV pattern for testing the GaPESC-
based MPPT algorithm performance for GMPP search. This evaluation is performed
in Sect. 7.4 using dynamic sequences of PV patterns without and with noise. The
last section concludes the chapter.

7.2 MPPT Algorithms Selection Criteria

7.2.1 Performance Indicators

The tracking speed represents the time to accurately find the maximum (GMPP) and
is measured by the number of the periods of the sinusoidal dither in case of the PESC
algorithms.

The tracking accuracy (T_{acc}) represents how close the GMPP can be found during
a stationary regime:

$$T_{acc} = \frac{y_{GMPP}}{y^*_{GMPP}} \cdot 100 \, [\%] \qquad (7.1)$$

where y_{GMPP} and y^*_{GMPP} is the GMPP and the founded value.

The tracking efficiency (T_{eff}) measures the tracking performance during the transitory regimes:

$$T_{eff} = \frac{\int_0^t y \, dt}{\int_0^t y^* \, dt} \cdot 100 \, [\%] \qquad (7.2)$$

where y is the PV or FC power, and y^* is the value of this power tracked by the GaPESC algorithm.

It is obvious that a better tracking efficiency will be obtained if the searching time will decrease and/or the transitory tracking accuracy will increase.

The search resolution (S_R) defines how close the GMPP and local MPP (LMPP) can be on a multimodal pattern:

$$S_R = \frac{\min_i |y_{GMPP} - y_{LMPPi}|}{y_{GMPP}} \cdot 100 \, [\%] \qquad (7.3)$$

where y_{LMPPi} is the value of the LMPP (number i).

The hit count represents the success rate to find the GMPP during repetitive tests: the ratio of the positive results to the total number of tests. It is obvious that the hit count is lower than 100% for a low search resolution. So, $S_{R(100\% \, hit)}$ is the search resolution to always find the GMPP (the hit count is 100%). It is important to evaluate the $S_{R(100\% \, hit)}$ value, which is the resolution for 100% hit count [10].

These indicators will be used to compare the performance of GaPESC algorithm to other GMPPP algorithms on the case studies of a PV array under PSCs.

Also, the GaPESC algorithm will be evaluated using the evaluation index based on four and eight criteria proposed in [6, 8], and briefly presented in next two sections.

7.2.2 Evaluation Index Based on Four Criteria

The four criteria proposed in [6] are as follows: algorithm's complexity; hardware implementation; tracking speed; tracking efficiency for uniform irradiance and tracking accuracy of GMPP for PSCs. All criteria are evaluated using three values (1, 2, and 3), but the score of the last two criteria is doubled in the evaluation index formula. So, the score will be from 6 (=1 + 1 + 2 + 2) to 18 (=3 + 3 + 6 + 6). In this chapter, the evaluation will be performed using a scale of appreciations and scores same as in [8] (bad = 1, medium = 2, good = 3) in order to compare the percentage score obtained for GaPESC algorithm in both evaluation indexes. For example, if the tracking speed is less than 100 ms or the tracking efficiency for uniform irradiance is higher than 97%, the score 3 will be awarded, not 1 as in [6].

7.2.3 Evaluation Index Based on Eight Criteria

The first two criteria set the implementation cost of the MPPT algorithm, so they will be also used in the formula of evaluation index based on eight criteria [8]. The other six criteria set the performance of the MPPT algorithm. The following criteria have been selected: tracking speed, convergence speed, tracking efficiency, initial parameters required, performance under uniform irradiance (without PSCs) and performance under PSCs.

All criteria are evaluated using five values: very worst = 1, bad = 2, medium = 3, good = 4, and very good = 5. So, the score will be from 8 (=1 × 8) to 40 (=5 × 8). Because the scale of appreciations in [6, 8] is different, the percentage score will be used to compare a MPPT algorithm using the evaluation index based on four and eight criteria.

It is worth mentioning that the tracking speed and tracking efficiency are also used in the formula of the valuation index based on four criteria [6] and they are the best criteria to evaluate the performance of the MPPT algorithms during stationary (uniform irradiance) and transitory (variable irradiance or PSCs) regimes that may appear in the search for MPP. The convergence speed is rated with the same score as tracking speed for the 17 MPPT algorithms analyzed in [8], so it will not be considered in the new criteria class proposed in next section. The last three criteria may be used for a subjective evaluation, but will be replaced with other criteria that can be easily used for an objective evaluation a MPPT algorithm.

7.2.4 Evaluation Index Based on New Criteria Class

The two criteria that determine the implementation cost of the MPPT (algorithm's complexity and hardware implementation) are used in both evaluation indexes [6, 8], so they must also be considered in the new criteria class. Hardware implementation must consider the required sensors, the resources needed for the microcontroller, the effort for preprocessing of the control variables, etc. Algorithm's complexity must consider the need for a periodic tuning of the control parameters, the PV array dependency, the dependency to the initial conditions, additional technique needed to reduce the oscillation around MPP, etc.

The tracking efficiency (T_{eff}) and tracking speed, which measures the tracking performance during the transitory regimes and the time to accurately find the maximum, respectively, are used in both evaluation indexes, so they must also be considered in the new criteria class.

The tracking accuracy (T_{acc}) represents how close the MPP can be found during a stationary regime that appears under uniform irradiance conditions (UICs) (being named stationary tracking accuracy), but it can be a criterion for evaluating how close the GMPP can be found during a transitory regime that appears under partial shading conditions (PSCs) (being named transitory tracking accuracy). Thus, the tracking accuracy must be also considered in the new criteria class.

The search resolution (S_R) defines how close the GMPP and LMPP can be on a multimodal pattern under PSCs:

The two criteria called in [8] as performance without PSCs and performance with PSCs are not defined, the MPPT algorithms being subjectively evaluated on a scale from 1 to 5. So, the new index class will consider the tracking accuracy and the search resolution instead of them. But the search resolution is closely related to the ability to track true maxima and it must be evaluated considering the initial conditions (the starting points) and hit count, where the last parameter represents the success rate to find the GMPP during repetitive tests: the ratio of the positive results to the total number of tests [20, 21]. It is obvious that the hit count is lower than 100% for a low search resolution and rapid changes in the irradiance conditions involving PSCs. So, the performance criterion $S_{R(100\% \text{ hit})}$ is proposed in order to avoid a subjective evaluation. Thus, $S_{R(100\% \text{ hit})}$ is the search resolution for 100% hit count of the GMPP search on a multimodal function from a set that may represent the real PV patterns or generated PV patterns using a PV simulator.

The robustness to the perturbations (noise), parametric disturbance and structural changes (called sensitivity in [8], but not considered in the eight-criteria class defining the evaluation index) is an important criterion to evaluate the PV system stability, so it must be considered in the new criteria class.

Therefore, the new evaluation index will consider the criteria as follows: (1) algorithm's complexity, (2) hardware implementation, (3) tracking efficiency, (4) tracking speed, (5) tracking accuracy, (6) search resolution for 100% hit count ($S_{R(100\% \text{ hit})}$), and (7) robustness.

The GaPESC-based MPPT algorithm proposed in [21] will be evaluated using the PV patterns defined in next section.

7.3 The PV Characteristics

The PV power characteristic under uniform irradiance has a unique maximum (MPP) [22], but under PSCs (due to shading of the PV panels by the clouds, buildings, or dust [23]) the PV power characteristic has a global maximum (GMPP) and many LMPPs [24].

To exemplify, the PV power characteristics for a 1Px7S PV array (with 7 SX60 panels in series defining one string) are presented in Fig. 7.1 for the mentioned irradiance sequences (G_1, G_2, G_3, G_4, G_5, G_6, G_7). It is worth mentioning that the GMPP may be located in left, center, and right of the PV power characteristics represented as function to the PV current and PV voltage in the top and bottom

Fig. 7.1 The PV power characteristics [21]

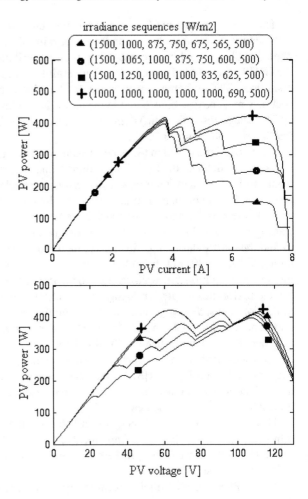

in Fig. 7.1 and the number of LMPPs depends on the defined irradiance sequence. Therefore, the function $y = h(p)$ given by (7.4) will be used as PV pattern:

$$y = \mathrm{sat}\left(M_l - (p - l)^2\right) + \mathrm{sat}\left(M_m - (p - m)^2\right) + \mathrm{sat}\left(M_r - (p - r)^2\right) \quad (7.4)$$

where the triplet $(l, m, r) = (20, 50, 80)$ defines the position of the GMPP and the highest two LMPPs and the triplet (M_l, M_m, M_r) defines the high of the GMPP and LMPPs on the PV pattern (7.4).

7.4 GaPESC-Based MPPT Algorithm Evaluation

The GaPESC-based MPPT algorithm proposed in [21] (see Fig. 7.2) has been briefly explained in Chapter 3. Based on the results presented in [22] and a scale of appreciations and scores same as in [8] (very worst = 1, bad = 2, medium = 3, good = 4 and very good = 5), the criteria can be evaluated as follows:

(1) The algorithm's complexity is low, so the obtained score is 4;
(2) The hardware implementation (see Fig. 7.2) is moderate and low for a digital and analog implementation, using voltage and current sensors in both implementations, so the obtained score is 3.5;
(3) The tracking efficiency is about 99.9 > 99.7%, so the obtained score is 5;
(4) The tracking speed is about 15 dither's periods (which means a tracking time of about 150 ms for 100 Hz sinusoidal dither or 15 ms for 1000 Hz sinusoidal dither), so the obtained score is 5;
(5) The tracking accuracy is higher than 99.95 > 99.8%, so the obtained score is 5;
(6) $S_{R(100\% \text{ hit})}$ is about 0.25% (20 times lower that 5% resolution reported for other GMPPT algorithms), so the obtained score is 5;
(7) The robustness to rapid change of the PV patterns with and without noise is very good (due to adaptive feature of the ES control), so the obtained score is 5.

Thus, the score obtained in the seven criteria is 4 + 3.5 + 5+5 + 5+5 + 5=32.5, which represents 92.87% from the maximum score of 35.

The best performance is obtained for bio-inspired-based MPPT algorithms and the percentage score is in range of 75–81.2% and 67.5–87.5% using the data available and maximum score of 18 (four criteria) and 40 (eight criteria) mentioned in [6, 8]. The best performance of 87.5% is obtained using the flower pollination algorithm and firefly algorithm [8]. Therefore, the performance of the GaPESC-based MPPT algorithm is comparable with that of the best MPPT algorithms proposed in the literature.

Fig. 7.2 GaPESC-based MPPT algorithm

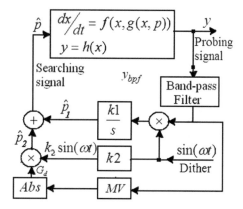

Some results from [21] are presented in next sections to highlight the aforementioned performance for the GaPESC-based MPPT algorithm.

7.4.1 Behavior and Performance of GMMP Searching and Tracking

GMPP tracking on the PV pattern (7.4) having a search resolution $S_R \cong 2.23\%$ ($=100 \cdot 10/430$) is presented in Fig. 7.3 in three GMPP positions on the PV pattern.

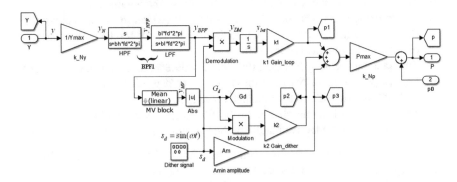

Diagram of the GaPESC-based MPPT algorithm

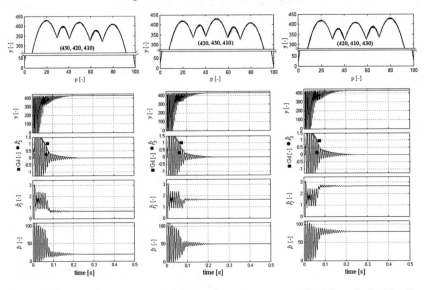

(a) GMPP is located in the left side (b) GMPP is located in the center (c) GMPP is locate in the right side

Fig. 7.3 GMPP tracking on the PV pattern (7.4)

The simulation diagram of the GaPESC-based MPPT scheme and the PV patterns are shown in the top of Fig. 7.3, and the plots' structure is as follows: the first plot presents the output (y); the second plot shows the localization signal (p_2) and its envelope (G_d); the third plot shows the tracking signal (p_1) from the classical ESC loop; the fourth plot shows the searching signal $p = p_1 + p_2 + p_3$, where p_3 is the dither component with minimum amplitude to start the search on static patterns.

The tracking speed is about 15 dither's periods, which means a tracking time of about 150 ms for 100 Hz sinusoidal dither. The 99.97% tracking efficiency can result using 1000 Hz sinusoidal dither because the transitory regime will be less than 15 ms. The tracking efficiency is about 99.9% for 100 Hz sinusoidal dither.

GMPP search behavior using the GaPESC scheme is presented in Fig. 7.4 using the GaPESC-based MPPT scheme and the PV pattern (7.4) having a search resolution $S_R \cong 0.223\%$ (=100 · 1/430). The GMPP search has been repeated using different starting points and GMPP has always been accurately found (see the first plot in Fig. 7.4). Therefore, the average value of the tracking accuracy is about 99.95% and $S_{R(100\% \text{ hit})}$ is about 0.25%. The tracking accuracy has been found in the range of 99.9–99.99% in different performed simulations reported in [25, 26] using the schemes GaPESCH, GaPESCH1, and GaPESDd. As it can be seen in Fig. 7.2, the GaPESCH scheme uses the same band-pass filter from the tracking loop to obtain the dither gain (G_d), by smoothing its output using a mean value (MV) filter. The schemes GaPESCH1 and GaPESDd obtain the dither gain (G_d) by using the first harmonic (H1) and derivative of the PV power (the probing signal in Fig. 7.2). It is obvious that the complexity increases using the fast Fourier transform for the GaPESCH1 scheme and the GaPESCd scheme is not recommended to be used due to high sensitivity to noise. The derivative operator may reduce the performance and stability of the PV system, as can be seen in Fig. 7.5 (where the tracking time increases to about 500 ms for 100 Hz sinusoidal dither).

The GMPP search behavior using the schemes GaPESCH1 and GaPESCd is presented in Fig. 7.5. It is worth mentioning that almost the same performance is obtained for the schemes GaPESC and GaPESCH1, except the algorithm's complexity and hardware implementation which is obviously higher for the GaPESCH1 scheme compared to the GaPESC scheme.

The robustness to rapid change of the PV pattern (with and without noise) will be highlighted in next section only for the GaPESC scheme.

7.4.2   The Search Robustness

The robustness analysis will be performed for closed loop gain $k_L < k_{L(max)}$ in order to obtain a stable PV system. For example, $k_{L(max)} \cong 9500$ for the PV system under study based on the roots locus presented in Fig. 7.6 [21]. The step responses for different values of the closed loop gain, $k_L = 500$ (▲), $k_L = 1000$, (●) and $k_L = 5000$ (■), are also shown in Fig. 7.6. It is worth mentioning that small oscillations appear

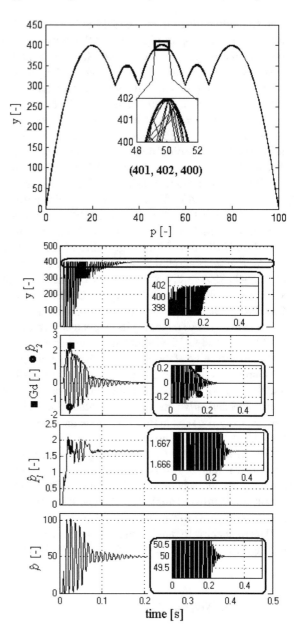

Fig. 7.4 GMPP search behavior using the GaPESC scheme [21]

at about $k_L \cong k_{L(max)}/2$, so the value of $k_L = 5000$ will be used in simulation in order to speed up the search loop.

The triplet (M_l, M_m, M_r) using the values (430, 300, 350), (300, 430, 350), and (300, 350, 430) will define three PV patterns based on (7.4) having the GMPP in the left, center, and right side of the PV pattern. All PV patterns have the GMPP

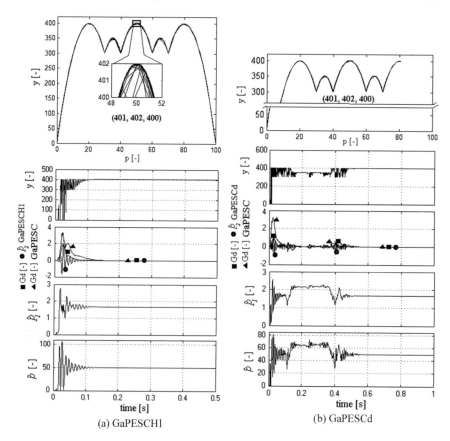

Fig. 7.5 GMPP search behavior using the schemes GaPESCH1 (left) and GaPESCd (right)

of 430 W. Two sequences of the aforementioned PV patterns, which are different ordered and changed every second, are used to test the robustness of the GaPESC scheme. The GMPP of 430 W is always tracked using the searching signal p for two sequences of the PV patterns, without or with noise (see the output y represented in Figs. 7.7, 7.8, and 7.9 presented in next sections).

7.4.2.1 Sequence of the PV Patterns Without Noise

Two sequences of PV patterns without noise are presented in Figs. 7.7 and 7.8. It is worth mentioning that the GPPP of 430 W is accurately found in less than 250 ms if 100 Hz sinusoidal dither is used. The search steps are presented on the PV patterns that are swept for GMPP search (the points 1–12 are mentioned on the PV patterns and the sequence).

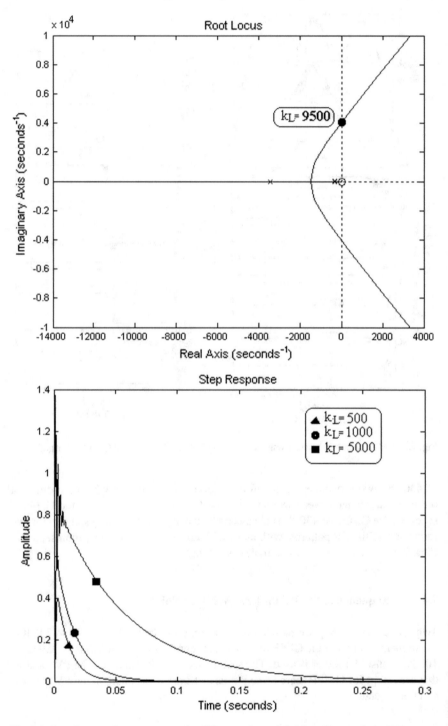

Fig. 7.6 Root locus and step responses for different values of the closed loop gain k_L [21]

Fig. 7.7 GMPP search
behavior for the first
sequence of PV patterns [21]

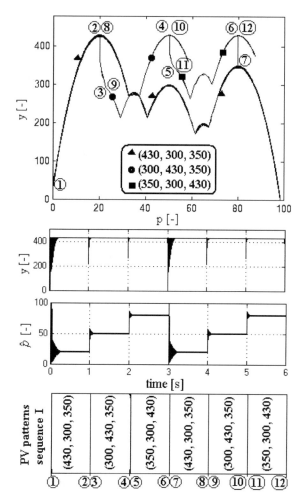

7.4.2.2 Sequence of the PV Patterns with Noise

If the sequence of the PV patterns has noise, then the position and high of the GMPPs
for any PV pattern will be different in time. The sweeping routes are presented in
Fig. 7.9 for random noise of $\pm 3_{p-p}$ (top) and $\pm 30_{p-p}$ (bottom) amplitude and the
sampling period of 10 ms. The GMPP may be in range of 427–433 W and 400–460 W
for $\pm 3_{p-p}$ and $\pm 30_{p-p}$ random noise. It is observed that the current GMPP in the
range of 427–433 W will be tracked, but this can be accurately done if the dither
frequency will be increased at 2000 Hz (so there will be 20 search periods during
the sampling period of 10 ms).

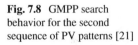

Fig. 7.8 GMPP search behavior for the second sequence of PV patterns [21]

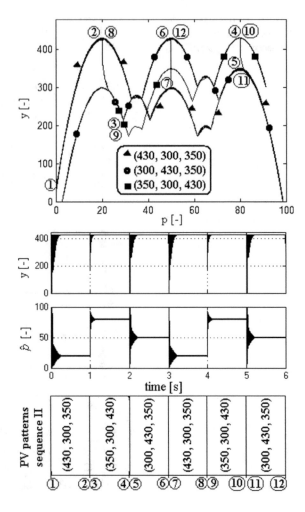

7.5   Conclusions

In this chapter, the evaluation indexes proposed in literature are analyzed and a new one is proposed.

The GaPESC-based MPPT algorithm has been evaluated using the aforementioned evaluation indexes and obtained the percentage score of 92.87% from the maximum score of 35. It is worth mentioning that the best performance has been reported for bio-inspired-based MPPT algorithms, with a percentage score in range of 75–81.2% and 67.5–87.5% from the maximum score of 18 (four criteria) and 40 (eight criteria), respectively. The best performance of 87.5% has been obtained for the flower pollination algorithm and firefly algorithm.

Fig. 7.9 Search of the
GMPP on noisy PV patterns
[21]

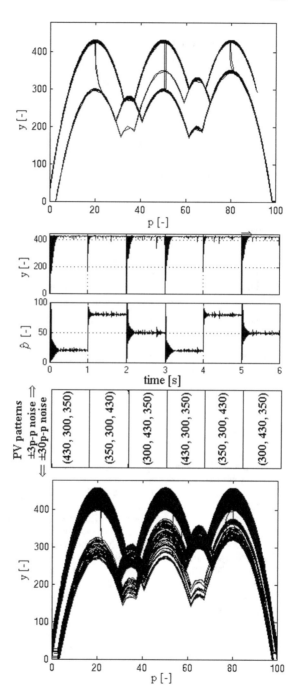

The behavior and performance of GMMP searching and tracking using the GaPESC-based MPPT algorithm have been highlighted for dynamic sequences of the PV patterns without and with noise as follows:

(1) The tracking efficiency is about 99.9%;
(2) The tracking speed is about 15 dither's periods (which means a tracking time of about 150 ms for 100 Hz sinusoidal dither or 15 ms for 1000 Hz sinusoidal dither);
(3) The tracking accuracy is higher than 99.95%;
(4) $S_{R(100\% \ hit)}$ is about 0.25% (20 times lower that 5% resolution reported for other GMPPT algorithms);
(5) The robustness to rapid change of the PV patterns with and without noise is very good (due to the adaptive feature of the ES control).

Also, other advantages of the GaPESC-based MPPT algorithm are as follows:

(6) Simple design of the tuning parameters to ensure the stability and performance of the search and localization loops of the GMPP;
(7) The design does not depend on the PV power (size of the PV array) due to use of the normalization gains for input and output;
(8) The design of the normalization gains and tuning parameters is not critical due to the adaptive feature of the ESC algorithm.

Furthermore, the algorithm's complexity is low and the hardware implementation is moderate, being easy to implement in cheap microcontrollers.

References

1. International Energy Agency (IEA) (2019) Renewables 2018. https://www.iea.org/renewables2018/. Accessed on 1 Aug 2019
2. International Energy Agency (IEA) (2019) World energy outlook 2018. https://webstore.iea.org/world-energy-outlook-2018. Accessed on 1 Aug 2019
3. International Energy Agency (IEA) (2019) Clean energy investment trends 2019. https://webstore.iea.org/clean-energy-investment-trends-2019. Accessed on 1 Aug
4. International Energy Agency For Photovoltaic Power Systems (IEA-PVPS) (2019) IEA-PVPS annual report 2018. http://www.iea-pvps.org/. Accessed on 1 Aug 2019
5. Bizon N (2017) Searching of the extreme points on photovoltaic patterns using a new asymptotic perturbed extremum seeking control scheme. Energ Convers Manage 144:286–302
6. Ahmad R, Murtaza AF, Sher AH (2019) Power tracking techniques for efficient operation of photovoltaic array in solar applications—a review. Renew Sustain Energy Rev 101:82–102
7. Belhachat F, Larbes C (2019) Comprehensive review on global maximum power point tracking techniques for PV systems subjected to partial shading conditions. Sol Energy 183:476–500
8. Eltamaly AM, Farh HMH, Othman MF (2018) A novel evaluation index for the photovoltaic maximum power point tracker techniques. Renew Sustain Energy Rev 174:940–956
9. Rezka H, Fathy H, Abdelaziz AY (2017) A comparison of different global MPPT techniques based on meta-heuristic algorithms for photovoltaic system subjected to partial shading conditions. Renew Sustain Energy Rev 74:377–386
10. Jiang LL, Srivatsan R, Maskell DL (2018) Computational intelligence techniques for maximum power point tracking in PV systems: a review 85:14–45

11. Belhachat F, Larbes C (2018) A review of global maximum power point tracking techniques of photovoltaic system under partial shading conditions. Renew Sustain Energy Rev 92:513–553
12. Danandeh MA, Mousavi GSM (2018) Comparative and comprehensive review of maximum power point tracking methods for PV cells. Renew Sustain Energy Rev 82:2743–2767
13. Batarseh MG, Za'ter ME (2018) Hybrid maximum power point tracking techniques: a comparative survey, suggested classification and uninvestigated combinations. Sol Energy 169:535–555
14. Husain MA, Tariq A, Hameed S, Saad Bin Arif M, Jain A (2017) Comparative assessment of maximum power point tracking procedures for photovoltaic systems. Green Energy Environ 2:5–17
15. Bizon N (2016) Global maximum power point tracking (GMPPT) of photovoltaic array using the extremum seeking control (ESC): a review and a new GMPPT ESC scheme. Renew Sustain Energy Rev 57:524–539
16. Salam Z, Ahmed J, Merugu BS (2013) The application of soft computing methods for MPPT of PV system: a technological and status review. Appl Energy 107:135–148
17. Seyedmahmoudian M, Horan B, Soon TK, Rahmani R, Muang Than Oo A, Mekhilef S, Stojcevski A (2016) State of the art artificial intelligence-based MPPT techniques for mitigating partial shading effects on PV systems–a review. Renew Sustain Energy Rev 64:435–455
18. Karami N, Moubayed N, Outbib R (2017) General review and classification of different MPPT techniques. Renew Sustain Energy Rev 68:1–18
19. Ram JP, Babu TS, Rajasekar N (2017) A comprehensive review on solar PV maximum power point tracking techniques. Renew Sustain Energy Rev 67:826–847
20. Bizon N (2016) Global maximum power point tracking based on new extremum seeking control scheme. Prog Photovoltaics Res Appl 24(5):600–622
21. Bizon N (2016) Global extremum seeking control of the power generated by a photovoltaic array under partially shaded conditions. Energy Convers Manage 109:71–85
22. Rahman MM, Hasanuzzaman M, Rahim NA (2015) Effects of various parameters on PV-module power and efficiency. Energy Convers Manage 103:348–358
23. Ismail MS, Moghavvemi M, Mahlia TMI (2013) Characterization of PV panel and global optimization of its model parameters using genetic algorithm. Energy Convers Manage 73:10–25
24. Adaramola MS, Vågnes EET (2015) Preliminary assessment of a small-scale rooftop PV-grid tied in Norwegian climatic conditions. Energy Convers Manage 90:458–465
25. Bizon N (2013) Energy harvesting from the PV hybrid power source. Energy 52:297–307
26. Moura SJ, Chang YA (2013) Lyapunov-based switched extremum seeking for photovoltaic power maximization. Control Eng Pract 21(7):971–980

Chapter 8
Mitigation of Energy Variability in Renewable/Fuel Cell Hybrid Power Systems

8.1 Introduction

The micro- and nano-grids may be sustainable architectural solutions to integrate the renewable energy sources (RESs) and the proton exchange membrane fuel cell (PEMFC) system into the hybrid power systems (HPSs) in order to support the increased energy demand, and environment and climate changes [1, 2].

The PEMFC system may be used as a non-polluting energy source for FC/ RES HPS to compensate the variability of power flows on the DC bus from the load and RESs [3, 4]. The advantages of the PEMFC technology compared to other fuel cell (FC) technologies make the PEMFC system to be widely used in stationary and mobile [5, 6], including aeronautics and space applications [7, 8].

Developing of new control methods [8, 9] and advanced fuel economy strategies [10–14] requires the use of accurate PEMFC models [15, 16] in the design phase in order to ensure safe operation under pulses [17] and increase the FC lifetime [18, 19].

The power-following (PFW) control is the modified variant of the load-following control proposed in [20] to compensate the power flow balance under dynamic load [21, 22] with minimum energy exchanged by the battery with the DC bus [23, 24]. The FC power estimated based on the power flow balance with battery operating in charge-sustained mode may be generated on the DC bus via the controlled boost DC-DC converter using the PFW control for the fueling regulators [25–27]. The global extremum seeking (GES)-based optimization strategies to improve the FC power and fuel economy were extensively presented in Chaps. 5 and 6.

In this chapter, the PFW control to mitigate the energy variability on the DC bus will be analyzed. Thus, the chapter is organized as follows. The design of PFW control of the PEMFC system in order to mitigate the energy variability on the DC bus with less energy support from the battery is presented in Sect. 8.2. The performance of the most efficient energy strategies proposed in Chap. 5 for a FC/RES HPS is analyzed in Sect. 8.3 under constant and variable load, without and with variable power from the RESs. Also, the performance of the best fuel economy strategies

N. Bizon, *Optimization of the Fuel Cell Renewable Hybrid Power Systems*,
Green Energy and Technology, https://doi.org/10.1007/978-3-030-40241-9_8

proposed in Chap. 6 for a FC/RES HPS is analyzed in Sect. 8.4, without and with
support of an electrolyzer (ELZ). Mitigation of energy variability in RES/FC/ELZ
hybrid power systems by appropriate control of the battery (Bat)/ultracapacitors (UC)
hybrid energy storage system (ESS) is also presented. The last section concludes the
chapter.

8.2   Power-Based Control to Mitigate the Energy Variability on the DC Bus

The architecture of a renewable fuel cell hybrid power system (FC/RES HPS) is
presented in Fig. 8.1.

If the power exchanged by the ESS with the DC bus is zero on average ($P_{ESS(AV)} \cong$
0), then the energy on the DC bus is supplied by the PEMFC system and RESs, and
consumed by the loads (DC and AC) and electrolyzer (only when $P_{RES} > P_{load}$, so
$P_{FC} = P_{FC(min)} \cong 0$). The surplus of energy ($P_{RES} - P_{load} > 0$) during the regime
when $P_{RES} > P_{Load}$ will supply the electrolyzer (ELZ) in order to produce hydrogen
that will be stored in pressurized tanks. The PEMFC system will operate at minimum
power ($P_{FC} = P_{FC(min)}$) in order to avoid complicated start-up procedures. The lack
of energy ($P_{RES} - P_{Load} < 0$) during the regime when $P_{RES} < P_{Load}$ will be supplied
by the PEMFC system. The RES and load power flows are variable on the DC bus
and the FC power requested to compensate the power flow balance on the DC bus
will be generated with a delay due to FC response time and slope limiters used by
the fueling regulators. Thus, the hybrid Bat/UC ESS will dynamically compensate

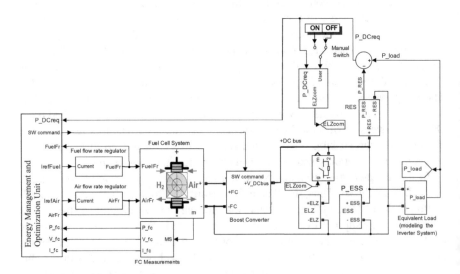

Fig. 8.1   FC/RES HPS using the power-based strategy

the power flow balance on the DC bus, but on average $P_{Batt(AV)} \cong 0$, which means that the battery will mainly operate in charge-sustained mode. The hybrid ESS will have the same role during the regime when $P_{RES} > P_{Load}$ and the electrolyzer is supplied with $P_{ELZ(AV)} = P_{RES(AV)} + \eta_{boost} P_{FC(min)} - P_{Load(AV)}$, compensating the power $P_{RES} + \eta_{boost} P_{FC(min)} - P_{Load} - P_{ELZ} \cong 0$. Thus, $P_{Batt(AV)} \cong 0$, which means that the battery will mainly operate in charge-sustained mode ($P_{ESS(AV)} \cong 0$) during the regime when $P_{RES} > P_{Load}$ and the electrolyzer is supplied with the energy excess on the DC bus in a controlled mode.

In the same manner, during the regime when $P_{RES} < P_{Load}$, the FC system must generate the requested power (8.1) in order to operate the battery in charge-sustained mode ($P_{ESS(AV)} \cong 0$):

$$P_{DCreq(AV)} = P_{Load(AV)} - P_{RES(AV)} \tag{8.1}$$

The FC net power $P_{FCnet(AV)}$ that is needed to be generated in a controlled mode will be obtained based on the power flow balance as follows:

$$0 = \eta_{boost(AV)} P_{FCnet(AV)} + P_{RES(AV)} - P_{Load(AV)} \Rightarrow P_{FCnet(AV)} = P_{DCreq(AV)}/\eta_{boost(AV)} \tag{8.2}$$

where the average (AV) value may be obtained using different filtering techniques.

Thus, in order to control the power generated by the FC system, the power-based strategy will generate based on (8.2) the power-following (PFW) reference ($I_{ref(PFW)}$) given by (8.3):

$$I_{ref(PFW)} = I_{FC(AV)} = P_{FCnet(AV)}/V_{FC(AV)} = P_{DCreq}/\left(V_{FC(AV)}\eta_{boost(AV)}\right) \tag{8.3}$$

The reference $I_{ref(PFW)}$ will be generated by the PFW controller (see Fig. 8.1), which has the diagram similar to that of the LPF controller shown in Fig. 3.3 (see Chap. 3), except that the input will be the power requested on the DC given by (8.1).

In the top and bottom of Fig. 8.2, the diagram of the FC/RES HPS and the Energy Management and Optimization (EMO) unit are presented. The diagram of the FC/RES HPS highlights the three ways to control the generated FC power based on the reference $I_{ref(PFW)}$ applied to one reference control input of the boost controller, air regulator, and fuel regulator, which are $I_{ref(Boost)}$, $I_{ref(Air)}$, and $I_{ref(Fuel)}$, respectively. The references $I_{ref(Boost)}$, $I_{ref(Air)}$, and $I_{ref(Fuel)}$ are generated by the EMO unit based on the input variable acquired from the FC/RES HPS, such as FC current and voltage (resulting the FC power), airflow rate (AirFr), and fuel flow rate (FuelFr), and RES power and load power.

The optimization function used in this chapter problem is given by (8.4):

$$k_{net} \cdot P_{FCnet} + k_{fuel} \cdot Fuel_{eff} = f(x, AirFr, FuelFr, P_{Load}) \tag{8.4}$$

where k_{net} and k_{fuel} are the weight coefficients, and x is the state vector.

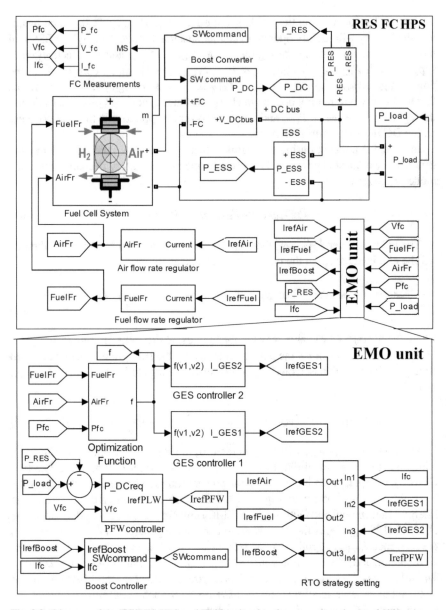

Fig. 8.2 Diagram of the FC/RES HPS and EMO unit using the power-based control [29]

This optimization function f will be computed by the respective block from the EMO unit and the output value f will be the input of two GES controllers. The GES controllers will search the global maximum of the optimization function f considering the dynamic model $\dot{x} = g(x, \text{AirFr}, \text{FuelFr}, P_{\text{Load}})$, $x \in X$, where g is a smoothing function that models the dynamics of the FC system [28]. The search for global maximum will be performed in real time on a surface or curve if two or one GES controllers will be used as optimization references (the references $I_{ref(GES2)}$ and $I_{ref(GES1)}$ in Fig. 8.2). The references $I_{ref(Boost)}$, $I_{ref(Air)}$ and $I_{ref(Fuel)}$ are generated by setting strategy considering the references $I_{ref(PFW)}$, $I_{ref(GES2)}$ and $I_{ref(GES1)}$ (see Figs. 8.3 and 8.4).

Thus, seven real-time optimization (RTO) strategies (S1–S7) can be obtained and these have been analyzed in Chaps. 3 and 6 to maximize the FC net power ($k_{net} \neq 0$ and $k_{fuel} = 0$) or the fuel economy ($k_{net} \neq 0$ and $k_{fuel} \neq 0$) using different set of weight parameters k_{net} and k_{fuel}.

The best strategies will be analyzed in this chapter in terms of the performance obtained when the RES power is considered variable, besides a constant or variable load. The reference strategy considered will be the static Feed-Forward (sFF) strategy proposed in [28].

8.3 Efficient Energy Strategies for Renewable/Fuel Cell Hybrid Power Systems Using the Power-Based Control

The strategies S3–6 to maximize the FC net power ($k_{net} \neq 0$ and $k_{fuel} = 0$) will be analyzed for a RES/6 kW FC HPS under constant and variable load. Note that all strategies S3–6 use the setting $I_{ref(Boost)} = I_{ref(GES1)}$ to optimize the FC net power, but different settings for the references of the fueling regulator. Strategies S3 and S5 use the setting $I_{ref(Air)} = I_{ref(PFW)}$, and strategies S4 and S6 use the setting $I_{ref(Fuel)} = I_{ref(PFW)}$. Also, it is worth mentioning that the strategies S3 and S4 use only one optimization reference ($I_{ref(GES1)}$), and the strategies S5 and S6 use two optimization references ($I_{ref(GES1)}$ and $I_{ref(GES2)}$).

8.3.1 Performance Under Constant Load Demand

The FC net power P_{FCnet} and total fuel consumption $Fuel_T$ are recorded in Table 8.1 for different levels of the load demand. It is worth mentioning that the maximum values for the FC net power have been obtained for the strategy S5 using the setting $I_{ref(Air)} = I_{ref(PFW)}$ and $I_{ref(Fuel)} = I_{ref(GES2)} + I_{FC}$ (see the bold values P_{FCnet} in the Table 8.1), and then for $P_{Load} < 4$ kW by the strategy S3 using the setting $I_{ref(Air)} = I_{ref(PFW)}$ and $I_{ref(Fuel)} = I_{FC}$, and for $P_{Load} > 4$ kW by the strategy S4 using the setting $I_{ref(Air)} = I_{FC}$ and $I_{ref(Fuel)} = I_{ref(PFW)}$ (see the values P_{FCnet} with italic font in Table 8.1).

Consequently, the strategies S5 and S3 (having the common setting $I_{ref(Air)} = I_{ref(PFW)}$) will maximize the FC net power.

$I_{ref(Boost)}$	$I_{ref(Air)}$	$I_{ref(Fuel)}$	Strategy	Class
$I_{ref(LF)}$	I_{FC}	I_{FC}	sFF	reference
$I_{ref(LF)}$	I_{FC}	$I_{GES1}+I_{FC}$	S1	C1
$I_{ref(LF)}$	$I_{ref(GES1)}+I_{FC}$	I_{FC}	S2	C1
$I_{ref(GES1)}$	$I_{ref(LF)}$	I_{FC}	S3	C2
$I_{ref(GES1)}$	I_{FC}	$I_{ref(LF)}$	S4	C3
$I_{ref(GES1)}$	$I_{ref(LF)}$	$I_{ref(GES2)}+I_{FC}$	S5	C2
$I_{ref(GES1)}$	$I_{ref(GES2)}+I_{FC}$	$I_{ref(LF)}$	S6	C3
$I_{ref(LF)}$	$I_{ref(GES1)}+I_{FC}$	$I_{ref(GES2)}+I_{FC}$	S7	C1

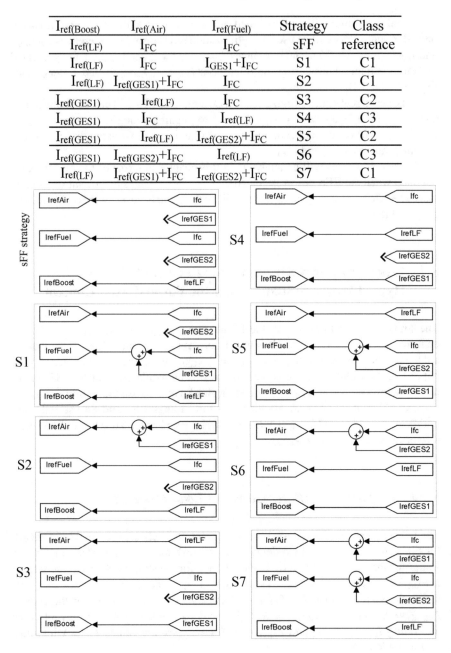

Fig. 8.3 The setting diagram of the fuel economy strategies [29]

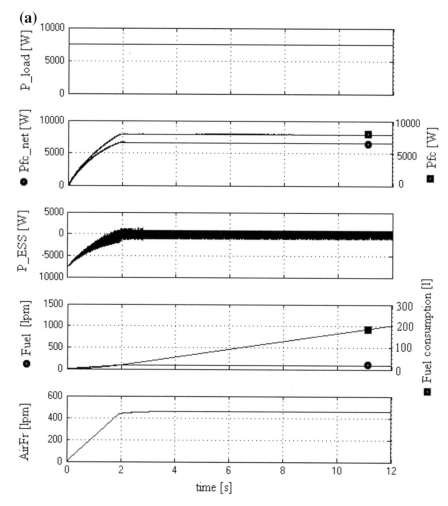

Fig. 8.4 a Strategy S3 **b** Strategy S4 **c** Strategy S5 **d** Strategy S6. Behavior of the HPS using strategies S3–S6 under constant load and $P_{RES} = 0$ [27]

Making the same analysis for the total fuel consumption $Fuel_T$, it is worth mentioning that minimum values for $Fuel_T$ have been obtained for the strategy S6 using the setting $I_{ref(Air)} = I_{ref(GES2)}$ and $I_{ref(Fuel)} = I_{ref(PFW)}$ (see the bold values $Fuel_T$ in Table 8.1), and then for $P_{Load} < 5.5$ kW by the strategy S4 using the setting $I_{ref(Air)} = I_{FC}$ and $I_{ref(Fuel)} = I_{ref(PFW)}$, and for $P_{Load} > 5.5$ kW by the strategy S5 using the setting $I_{ref(Air)} = I_{ref(PFW)}$ and $I_{ref(Fuel)} = I_{ref(GES2)} + I_{FC}$ (see the values $Fuel_T$ with italic font in Table 8.1).

Consequently, the strategies S6 and S4 (having the common setting $I_{ref(Fuel)} = I_{ref(PFW)}$) will minimize the total fuel consumption $Fuel_T$, so they will maximize the fuel economy during a load cycle. In addition, strategies S5 and S6 give the maximum

(b)

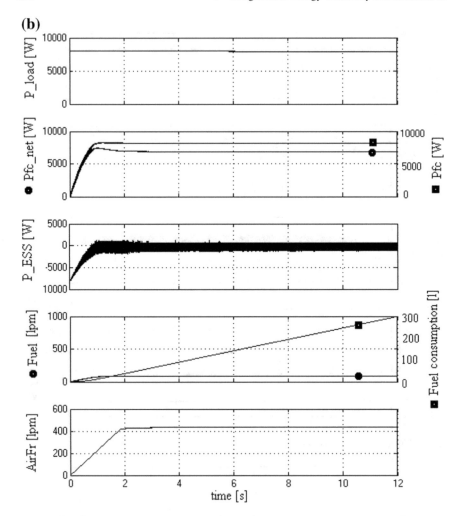

Fig. 8.4 (continued)

values for FC net power and the fuel economy, respectively, and both strategies use two optimization references ($I_{\text{ref(GES1)}}$ and $I_{\text{ref(GES2)}}$). These findings obtained for other values of the load power validate once again the conclusions of Chaps. 5 and 7.

The behavior of the HPS using strategies S3–S6 under constant load is presented in Fig. 8.4 for $P_{\text{Load}} = 7500$ W and $P_{\text{RES}} = 0$.

It is worth mentioning that $P_{\text{ESS}} \cong 0$ using any strategy S3–S6 (see the 4th plot in Fig. 8.4a–d), so the battery will operate in charge-sustained mode (with clear advantages related to its capacity (size) and lifetime).

Because the strategy S5 gives the maximum FC net power under different levels of constant load and strategy S4 gives good results in both P_{FCnet} and Fuel$_\text{T}$ indicators,

(c)

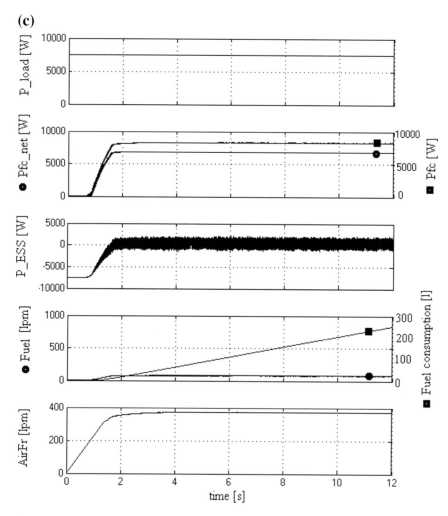

Fig. 8.4 (continued)

the performance of these strategies will be analyzed in next sections for variable load demand without or with variable power from renewable energy sources.

8.3.2 Performance Under Variable Load Demand Using Strategy S5

The variable load profile is of random type pulse using a 0.5 s sampling period (see the first plot in Fig. 8.5). Note that the FC system has 0.2 s constant time and 50 A/s slope

(d)

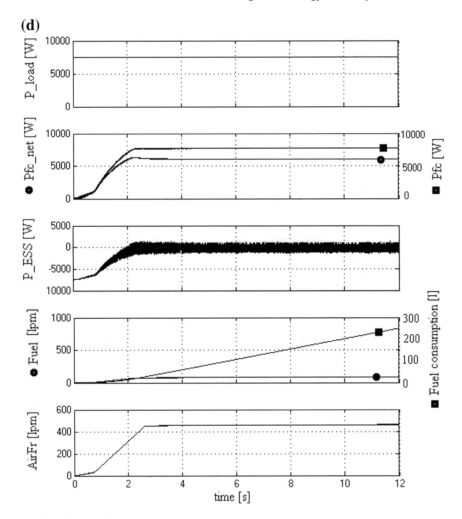

Fig. 8.4 (continued)

Table 8.1 FC net power P_{FCnet} and total fuel consumption $Fuel_T$

P_{Load} [W]	1250	2500	3750	5000	6250	7500	1250	2500	3750	5000	6250	7500
Strategy	\multicolumn{6}{} P_{FCnet} [W]						$Fuel_T$ [L]					
S3	*1588*	*2925*	*3989*	4874	5674	6639	39.2	75.8	105.4	130.8	156.7	203
S4	1306	2553	3699	*5442*	*5886*	*6836*	*25.5*	*54.7*	*83.9*	*117.4*	155.5	199
S5	**1613**	**3131**	**4309**	**5818**	**6131**	**6850**	32.5	70	99.7	125.3	*144.1*	*167.6*
S6	793.6	2003	3205	4152	5157	6064	**22.7**	**47.7**	**73**	**100.6**	**130.5**	**166.7**

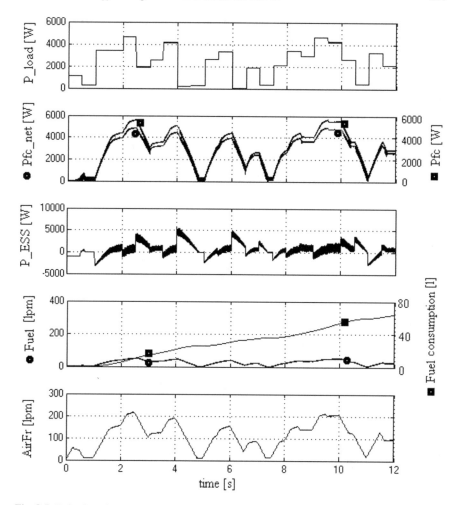

Fig. 8.5 Behavior of the HPS using strategy S5 under load pulses and $P_{RES} = 0$ [27]

limiters included in the fueling regulators to ensure safe operation. These constraints limit the FC response in generating the requested FC power on the DC bus based on the reference $I_{ref(PFW)}$ (see the variation of the AirFr and FC net power in the fifth and second plots in Fig. 8.5). Thus, the UCs' stack from semi-active ESS topology will dynamically compensate the power flow balance on the DC bus (see the third plot in Fig. 8.5).

8.3.3 Performance Under Variable Power from Renewable Energy Sources Using Strategy S5

The variable profiles of load and RES power are presented from 5 AM to 11 PM in Fig. 8.6 for a residential home (see the first plot in Fig. 8.6) without use of an electrolyzer as a consumer during the regimes when $P_{RES} > P_{Load}$. In this case, the surplus of energy ($P_{RES} - P_{load} > 0$) during these stages will charge the battery (see the profiles of the P_{RES} and P_{ESS} from 11 AM to 5 PM, when $P_{Load} = 0$).

A better solution is to use the surplus of energy ($P_{RES} - P_{load} > 0$) to supply an electrolyzer in order to produce hydrogen, as will be shown in Sect. 8.4 of this chapter.

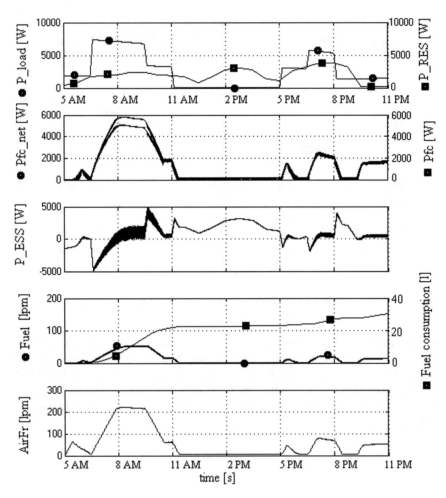

Fig. 8.6 Behavior of the HPS using strategy S5 under load pulses and filtered P_{RES} [27]

The load power and RES power have been filtered to better highlight the operation of the FC system (see the second plot in Fig. 8.6, where, e.g., the FC system operate in standby mode at minimum power $P_{FC} = P_{FC(min)} \cong 0$).

This RES power profile has been randomized with random power of 2 kW peak-to-peak (see first plot in Fig. 8.7). This profile of high RES power could appear during a sunny and windy day.

It is worth mentioning that only the smooth part of the requested power on the DC bus, $P_{DCreq(AV)} = P_{Load(AV)} - P_{RES(AV)}$, will be used to generate the reference $I_{ref(PFW)}$ given by (8.3) that will set the generated AirFr and PF net power (see the variation of the AirFr and FC net power in the fifth and second plots in Fig. 8.7).

Note that a 500 W load pulse appears around 2 PM, but the FC system remains in standby mode because $P_{RES} > P_{load}$.

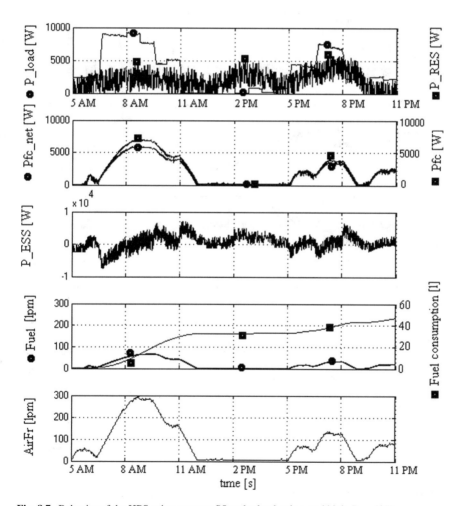

Fig. 8.7 Behavior of the HPS using strategy S5 under load pulses and high P_{RES} [27]

This is not the case when a low-RES power profile is also used to test the RES/FC HPS (see Fig. 8.8). This of RES power profile could appear during a cloudy day with no wind.

500 W load pulse appears around 2 PM, but the FC system passes from the standby mode to PFW-operating mode because $P_{RES} < P_{load}$ during the 500 W load pulse that appeared around 2 PM (could be the power consumed by a washing machine programmed at this hour).

These low energy load pulses may be supplied from the battery if the RES power is not available or $P_{RES} < P_{load}$, avoiding the starting of the FC system and increasing the energy efficiency of the HPS. A detection circuit of low energy load pulses can be designed using an appropriate power threshold.

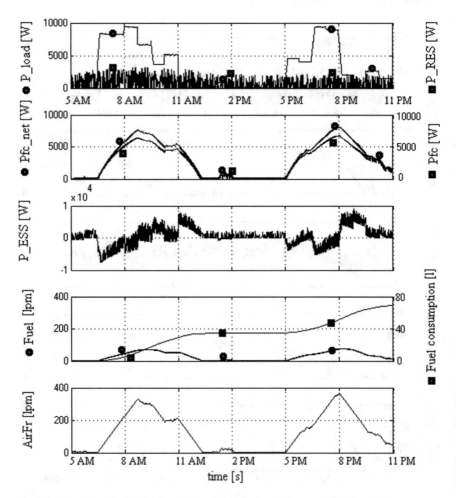

Fig. 8.8 Behavior of the HPS using strategy S5 under load pulses and high P_{RES} [27]

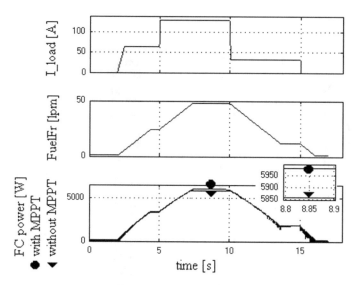

Fig. 8.9 Behavior of the HPS using strategy S4 under load pulses and $P_{RES} = 0$ [20]

8.3.4 Performance Under Variable Load Demand Using Strategy S4

The test performed without RES power ($P_{RES} = 0$) highlights an increase of the FC net power with about $(120/6000) \cdot 100 = 2\%$ if the GES-based MPP tracking (MPPT) control is used (see Fig. 8.9). The 25 A/s slope limiters have been used for safe operation of the FC system (see the second plot in Fig. 8.9).

8.3.5 Performance Under Variable Power from Renewable Energy Sources Using Strategy S4

Behavior of the HPS using strategy S4 under constant load and variable P_{RES} is presented in Fig. 8.10. The profiles of the 4-kW load and the RES power (1 kW$_{p-p}$ random power added on the levels of 2, 4, 2, 0 kW) are shown in the first plot in Fig. 8.10. The FC net power (see the second plot in Fig. 8.10) has been generated based on the FuelFr (see the seventh plot in Fig. 8.10) using the reference $I_{ref(PFW)}$ given by (8.3). Optimization of the FC net power is performed by the boost controller (see the switching command in the sixth plot in Fig. 8.10) using the GES-based MPPT algorithm (see the eighth plot in Fig. 8.10). The FC net power $P_{FCnet(AV)}$ that is needed to be generated on the DC bus via the boost DC-DC power converter is given by (8.2) and represented in the third plot in Fig. 8.10. The shapes of the 100 Ah/210 V battery voltage and 100 F/100 V ultracapacitors' stack voltage are shown in the plots fourth

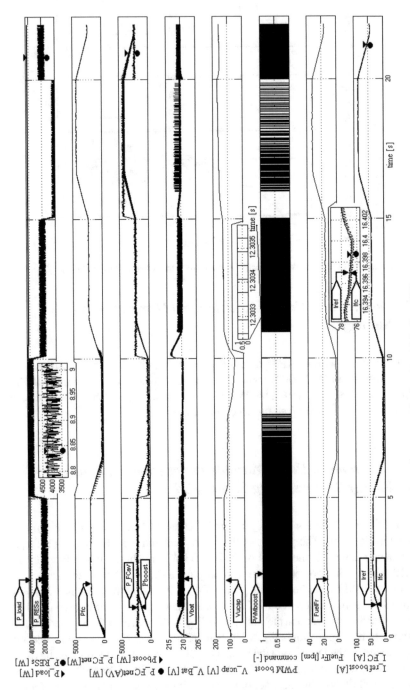

Fig. 8.10 Behavior of the HPS using strategy S4 under constant load and variable P_{RES} [20]

and fifth in Fig. 8.10. The design of the energy and power devices to ensure less than 10 V/210 V $\cong$ 5% ripple for the DC voltage and good dynamic compensation of the power flow balance on the DC bus has been performed in Chap. 3, but it is also available in [17, 30].

Behavior of the HPS using strategy S4 under constant load and filtered P_{RES} is presented in Fig. 8.11 to better highlight the PFW control of the fuel regulator (see the second plot in Fig. 8.10), good dynamic compensation of the power flow balance on the DC (see the third plot in Fig. 8.10), and DC voltage regulation (see the fourth plot in Fig. 8.10).

Behavior of the HPS using strategy S4 under load pulses and variable P_{RES} is represented in Fig. 8.12. The FC system will operate in standby mode when $P_{RES} > P_{load}$ and in PFW mode when $P_{RES} < P_{load}$ (see the second plot in Fig. 8.12). Without an electrolyzer, the surplus of energy ($P_{RES} - P_{load} > 0$) will charge the battery (see the third plot in Fig. 8.12). The better solution to supply an electrolyzer with this surplus of energy ($P_{RES} - P_{load} > 0$) will be shown in the next section.

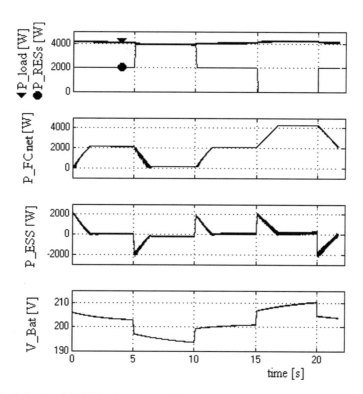

Fig. 8.11 Behavior of the HPS using strategy S4 under constant load and filtered P_{RES} [20]

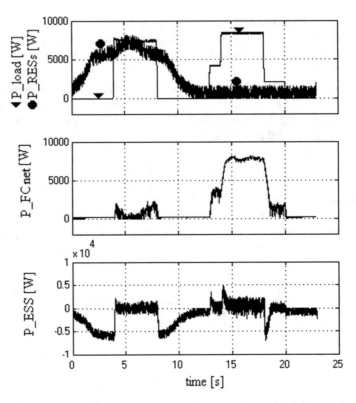

Fig. 8.12 Behavior of the HPS using strategy S4 under load pulses and variable P_{RES} [20]

8.3.6 Discussion

The following findings can be underlined based on the presented information:

- The strategies S5 and S3 are recommended to increase the FC net power generated on the DC bus; this operating mode will help a PV/FC vehicle to climb a hill (see Table 8.1);
- The fuel economy obtained using the strategy S5 is good, especially under high load (see Table 8.1);
- The best fuel economy is obtained in whole loading range using the strategy S6 (see FuelFr$_{S6}$ in Table 8.2);
- A good fuel economy is obtained using the switching strategy S4&S5 (see FuelFr$_{S4\& S5}$ in Table 8.2, which means the values of FuelFr$_{S4}$ and FuelFr$_{S5}$ for $P_{Load} < 6\,\text{kW}$ and $P_{Load} > 6\,\text{kW}$, respectively).
- About 2% increase in the FC net power generated on the DC bus may be obtained with the strategy S4 which gives good results in both P_{FCnet} and Fuel$_T$ indicators.

Table 8.2 Percentage of fuel economy [27]

	P_{Load} [W]					
	1250	2500	3750	5000	6250	7500
FuelFr$_{S6}$ [lpm]	113.6	238.3	365.0	502.8	652.4	832.8
FuelFr$_{S4\& S5}$ [lpm]	127.5	273.3	419.3	587.2	720.4	838.1
ΔFuelFr [lpm]	14.0	35.0	54.3	84.4	68.0	5.3
ΔuelFr/FuelFr$_{S6}$ [%]	12.3	14.7	14.9	16.8	10.4	0.6

The fuel consumption during 60 s has been recorded in liters per minute [lpm] for strategy S6 in row of the FuelFr$_{S6}$. The row of the fuel consumption FuelFr$_{S4\& S5}$ contains four values of FuelFr$_{S4}$ for $P_{\text{Load}} < 6\,\text{kW}$ and two values of FuelFr$_{S5}$ for $P_{\text{Load}} > 6\,\text{kW}$ (which are the next minimum values after the best values obtained using strategy S6). The fuel economy in liters per minute is given by ΔFuelFr = FuelFr$_{S4\& S5}$ − FuelFr$_{S6}$. Thus, the percentage of fuel economy (ΔFuelFr/FuelFr$_{S6}$) using strategy S6 compared with the switching strategy S4&S5 is presented in the last row of Table 8.2.

The findings reported in this chapter have been validated using a FC vehicle (see Chap. 6) or a FC emulator [31].

The advantages of a switching strategy have been highlighted in Chap. 3 and Ref. [13].

8.4 Efficient Fuel Economy Strategies for Renewable/Fuel Cell Electrolyzer Hybrid Power Systems Using the Power-Based Control

8.4.1 Performance Under Variable Load Demand Using Strategies S5 and S6

Considering the results presented in Chap. 6, it is worth mentioning that the best fuel economy has been obtained for $k_{\text{fuel}} = 25$. Thus, the optimization function (8.4) will use the weight coefficients $k_{\text{net}} = 0.5$ and $k_{\text{fuel}} = 25$.

The behavior of the HPS under a 5-kW load cycle is presented in Fig. 8.13a, b using strategies S5 and S6. The 5-kW load cycle uses the power levels of 3.75 kW, 6.25 kW, and 5 kW for each stage of 4 s (see the first plot in Fig. 8.13). Note that the battery operates in chargesustained mode in both strategies S5 and S6 (see the second plot in Fig. 8.13). The shapes of the performance indicators Fuel$_{\text{eff}}$, η_{sys}, and ΔFuel$_T$ presented in the last three plots in Fig. 8.13 are quite similar, and the differences $\Delta P_{\text{FCnet}} = P_{\text{FCnet6}} - P_{\text{FCnet5}}$, ΔFuel$_{\text{eff}}$ = Fuel$_{\text{eff6}}$ − Fuel$_{\text{eff5}}$, $\Delta\eta_{\text{sys}} = \eta_{\text{sys6}} - \eta_{\text{sys5}}$ represented in Fig. 8.14 are almost zero. But the difference ΔFuel$_T$ = Fuel$_{T6}$−Fuel$_{T5}$ represented in the fourth plot in Fig. 8.14 highlights that the fuel economy using the

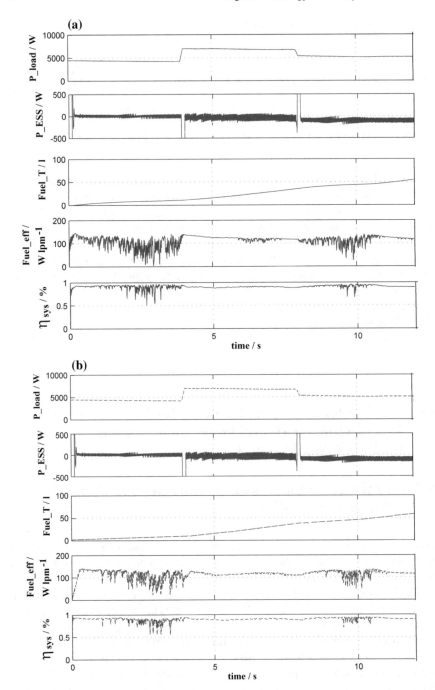

Fig. 8.13 **a** Strategy S5 **b** Strategy S6. Behavior of the HPS using strategies S5 and S6 under 5 kW load cycle [14]

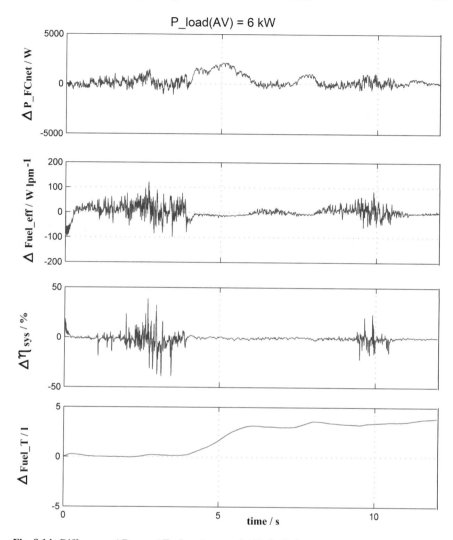

Fig. 8.14 Differences ΔP_{FCnet}, $\Delta Fuel_{eff}$, $\Delta \eta_{sys}$ and $\Delta Fuel_T$ [14]

strategy S5 is better than that obtained using the strategy S6. The fuel economy $\Delta Fuel_T = Fuel_{T6} - Fuel_{T5}$ is about 3.66 L for 12 s/5 kW load cycle and about 23 L for 12 s/6 kW load cycle (using the power levels of 4.5 kW, 7.5 kW, and 6 kW for each stage of 4 s). The economy in liters per minutes [lpm] will be about 18.3 lpm and 115 lpm for a load cycle of 5 kW and 6 kW using the strategy S5 instead of strategy S6.

8.4.2   Mitigation of Energy Variability in RES/FC/ELZ Hybrid Power Systems

The power from RESs such as photovoltaic (PV) and wind turbine (WT) systems (see p_{RES2} represented in the second plot in Fig. 8.15, which is the low-pass filtered shapes of the PV and WT power flows) is variable, and the power flow balance on the DC bus is difficult to be compensated by a battery with a reasonable capacity. The battery will operate in charge and discharge mode during the stage when $P_{Load} < P_{RES}$ and $P_{Load} > P_{RES}$. The frequent charge–discharge cycles reduce the battery's lifetime. In addition, the battery's state of charge (SoC) may vary in a large window if the battery's capacity is not appropriately designed.

In this section, the PFW control of the PEMFC system is proposed to operate the battery in charge-sustained mode (which means that the power exchanged by the battery with the DC bus is zero on average, $P_{ESS(AV)} \cong 0$) with help of an electrolyzer if $P_{RES} > P_{Load}$.

So, if $P_{ESS(AV)} \cong 0$ and $P_{RES} < P_{Load}$, then the lack of energy $(P_{Load} - P_{RES})$ on the DC bus will be supplied by the PEMFC system. The electrolyzer is off. If $P_{RES} > P_{Load}$ and $P_{ESS(AV)} \cong 0$ and $P_{RES} - P_{Load} < 0$, then surplus of energy

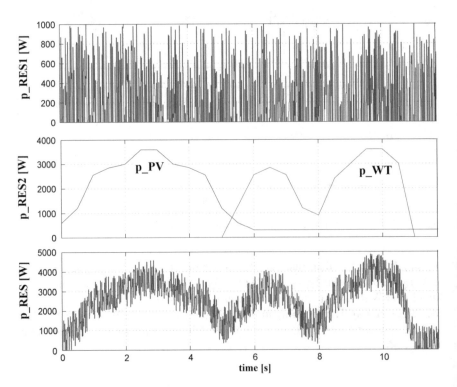

Fig. 8.15   The RES power and its components [14]

($P_{RES} - P_{load} > 0$) will supply the electrolyzer. The PEMFC system will be operated in stand-by mode (at $P_{FC(min)} \cong 0$).

The power flow balance on the DC is given by (8.5):

$$C_{DC}u_{dc}du_{dc}/dt \cong \eta_{boost}p_{FC} + p_{ESS} + p_{RES} - p_{ELZ} - p_{load} \qquad (8.5)$$

Neglecting the DC voltage ripple, the power deviation ΔP on the DC bus is given by (8.6):

$$p_{ESS} \cong \Delta P = p_{ELZ} + p_{load} - p_{DC} - p_{RES} \qquad (8.6)$$

where $p_{DC} = \eta_{boost}p_{FC}$.

ΔP must be exchanged by the battery with the DC bus, so the battery reference $I_{Bat(ref)}$ is the low-pass filtered (LPF) value of the $\Delta P/V_{DC}$:

$$I_{Bat(ref)} = LPF_{Bat}(\Delta P/V_{DC}) \qquad (8.7)$$

If the UCs' stack is used to mitigate the pulses on the DC bus, then the DC voltage regulation must be performed at battery side. So, the reference $I_{Bat(ref)}$ must be corrected with $V_{DC(correction)}$, obtaining the new battery reference $I'_{Bat(ref)}$:

$$I'_{Bat(ref)} = I_{Bat(ref)} - V_{DC(correction)} \qquad (8.8)$$

where $V_{DC(correction)}$ is the output of the proportional–integral (PI) controller based on the DC voltage error $e_{V_{dc}} = V_{DC} - V_{DC(ref)}$ (see Fig. 8.16). The switching command PWM_{Bat} is generated by a 0.1 A hysteresis controller using the error (8.9):

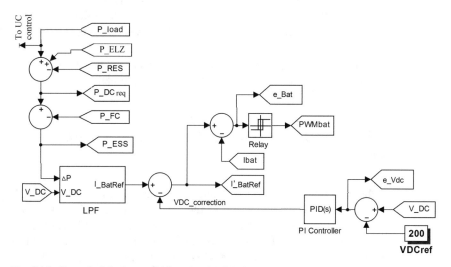

Fig. 8.16 Control of the battery bidirectional DC-DC converter [14]

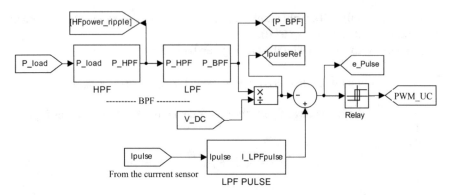

Fig. 8.17 Control of the ultracapacitors bidirectional DC-DC converter [14]

$$e_{\text{Bat}} = I'_{\text{Bat(ref)}} - I_{\text{Bat}} \tag{8.9}$$

where I_{Bat} is the battery current.

The UCs' stack will mitigate the pulses on the DC bus via the UCs bidirectional DC-DC converter using the band-pass filtered (BPF) reference, $I_{\text{Pulse(ref)}} = \text{BPF}(p_{\text{load}})/V_{\text{DC}}$ and the LPF signal $I_{\text{Pulse(LPF)}} = \text{LPF}_{\text{Pulse}}(I_{\text{Pulse}})$, where I_{Pulse} is the anti-pulse generated by the UCs bidirectional converter (see Fig. 8.17).

The switching command PWM_{UC} is generated by a 0.1 A hysteresis controller using the error (8.10):

$$e_{\text{Pulse}} = I_{\text{Pulse(ref)}} - I_{\text{Pulse(LPF)}} \tag{8.10}$$

The cutoff frequencies of the LPF and BPF are 1 kHz, and 0.1 Hz and 1 kHz, respectively. The cutoff frequency was set to 1 kHz in order to filter the noise of the acquired signals. The high-frequency noise on the DC bus will be filtered by the C_{DC} capacitor.

The variable RES power ($p_{\text{RES}} = p_{\text{RES1}} + p_{\text{RES2}}$) is presented in the third plot in Fig. 8.15, where p_{RES2} is a random power with 1 kW$_{\text{p-p}}$ amplitude.

8.4.3 Performance Under Variable Power from Renewable Energy Sources

Behavior of the HPS under a 4-kW load cycle is presented in Figs. 8.18 and 8.19 using strategy S7. The 4-kW load cycle uses the power levels of 3 kW, 5 kW, and 4 kW for each stage of 4 s (see the first plot in Figs. 8.18 and 8.19, where the RES power is also represented using the p_{RES} signal and p_{RES2} signal, respectively). The operation under PFW control of the PEMFC system and electrolyzer is shown in the second and third plots of both Figs. 8.18 and 8.19. The PEMFC system will

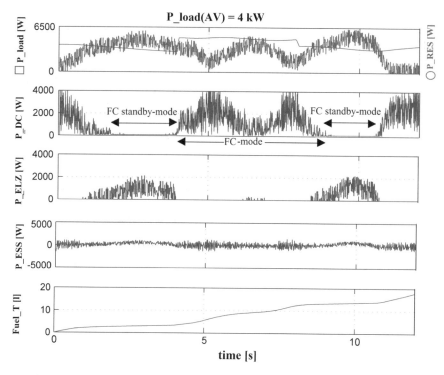

Fig. 8.18 The behavior of the HPS using the strategy S7 [14]

compensate the lack of energy ($P_{Load} - P_{RES} > 0$) on the DC bus if $P_{RES} < P_{Load}$ (when the electrolyzer is off) and the electrolyzer will consume the surplus of energy ($P_{RES} - P_{load} > 0$) on the DC bus if $P_{RES} > P_{Load}$ (when the PEMFC system will operate in stand-by mode at $P_{FC(min)} \cong 0$). In both stages, the battery will operate in charge-sustained mode (see the fourth plot in Fig. 8.18).

The DC voltage and UC voltage are represented in the fourth plot in Fig. 8.19. Note the good regulation of the DC voltage based on the $V_{DC(correction)}$ implemented at battery side using (8.8).

The search for the best fuel economy is performed around the value of the FC current using the references (8.11) for the air and fuel regulator:

$$I_{ref(Air)} = I_{ref(GES1)} + I_{FC} \text{ and } I_{ref(Fuel)} = I_{ref(GES2)} + I_{FC} \tag{8.11}$$

Thus, the shapes of the AirFr and FuelFr (see the fifth and sixth plots in Fig. 8.19) are quite similar with the shape of the FC current ($I_{FC} = P_{FC}/V_{FC}$). But minor differences in AirFr and FuelFr compared with the values obtained the sFF strategy (where the FC current is the reference for both air and fuel regulators) can significantly improve the performance of the strategy S7 (see Chaps. 5 and 6).

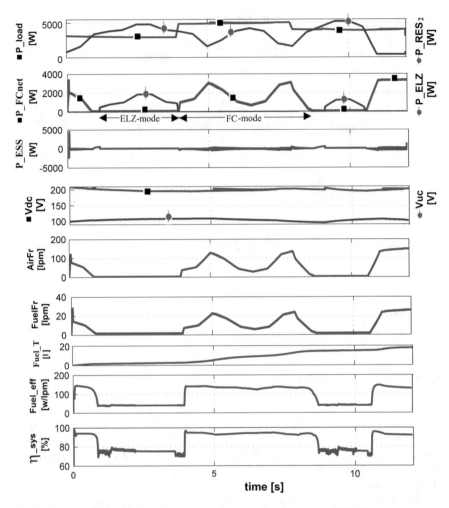

Fig. 8.19 The variation of the performance indicators using the strategy S7 [14]

The performance indicators ΔFuel$_T$, Fuel$_{eff}$ and η_{sys} are represented in the last three plots in Fig. 8.19. Note the almost same shape and final value of 18.2 L are obtained for the performance indicator ΔFuel$_T$ represented in Fig. 8.18 (the fifth plot) and Fig. 8.19 (the seventh plot) due to its definition relationship Fuel$_T$ = $\int$ FuelFr(t)dt.

The behavior of the HPS under 4 kW load cycle is presented in Fig. 8.20 using the strategy S6.

It is worth mentioning that the function of the PEMFC and electrolyzer is quite similar for strategy S6 with that using the strategy S7. Minor differences appear in the values of the performance indicators (see Chaps. 5 and 6), even if their shapes look similar.

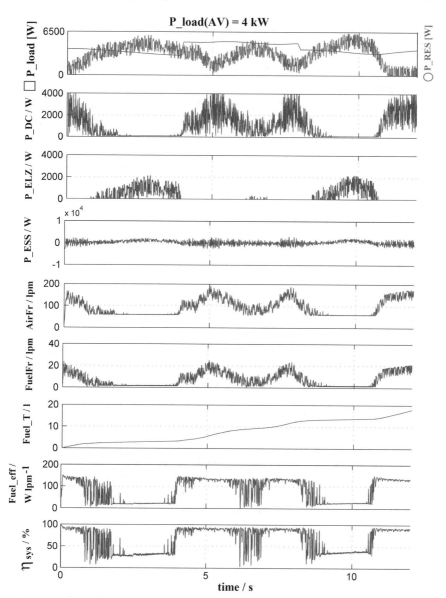

Fig. 8.20 The behavior of the HPS using the strategy S6

Note that these conclusions are valid for any strategy used to mitigate the power variability on the DC load produced by the RESs and loads.

8.4.4   Discussion

The results presented in this chapter have highlighted the following findings:

- The PFW control can mitigate the power variability on the DC load;
- The strategies using two optimization variables (strategies S6 and S5) have better performance compared with those using one optimization variable (strategies S4 and S3) due to higher flexibility in searching (a larger searching space for the optimum);
- The RES/FC/ELZ HPS architecture using the PFW control and two optimization variables may give the best performance compared with other strategies proposed in the literature.

8.5   Conclusion

The performance of the power-following control to mitigate the energy variability on the DC bus has been analyzed in this chapter under constant and variable load, without and with variable power from the renewable energy sources. Assuming that the battery operates in charge-sustained mode, the design of the power-following control has been made based on the power flows balance on the DC bus. Thus, the load exceeding the available RES power will be sustained by the proton exchange membrane fuel cell system based on the power-following control of the air or fuel regulator. If the available RES power exceeds the load demand, then the proton exchange membrane fuel cell system will operate in stand-by mode and the electrolyzer will be supplied with the excess power. The advantages of using an electrolyzer instead of a dump load to ensure the charge-sustained mode for the battery are fully highlighted.

In addition, the performance of the most efficient energy strategies and best fuel economy strategies proposed in Chaps. 5 and 6 for a renewable fuel cell hybrid power systems has been further highlighted in this chapter as follows:

- The strategies S5 and S3 are recommended to increase the FC net power generated on the DC bus, but the strategies S6 and S4 are recommended to increase the fuel economy; thus, a switching strategy based on the current objective of the EMO unit is best to use;
- The best fuel economy is obtained under high load using the strategies S5 and S3, but a fuel economy can be obtained in the entire loading range using a switching strategy;
- A good fuel economy is obtained using the switching strategy S4&S3 (both strategies S3 and S4 use one optimization variable) or the switching strategy S6&S5 (both strategies S5 and S6 use two optimization variables), but mixed variants, as the switching strategy S4&S5 analyzed in this chapter, are possible as well;
- About 2% increase in the FC net power generated on the DC bus may be obtained with the strategy S4 which gives good results in both P_{FCnet} and $Fuel_T$ indicators.

– The strategies S5 and S6 give better fuel economy compared to the strategies S3 and S4 due to a larger searching space for the optimum using two variables instead of one, but the switching strategy S6&S5 or the switching strategy S4&S3 are recommended in the entire loading range.

References

1. Gielen D, Boshell F, Saygin D, Bazilian MD, Wagner N, Gorini R (2019) The role of renewable energy in the global energy transformation. Energy Strateg Rev 24:38–50
2. Hesselink LXW, Chappin EJL (2019) Adoption of energy efficient technologies by households—barriers, policies and agent-based modelling studies. Renew Sustain Energy Rev 99:29–41
3. Sulaiman N, Hannan MA, Mohamed A, Ker PJ, Majlan EH, Wan Daud WR (2018) Optimization of energy management system for fuel-cell hybrid electric vehicles: issues and recommendations. Appl Energ 228:2061–2079
4. Sorrentino M, Cirillo V, Nappi L (2019) Development of flexible procedures for co-optimizing design and control of fuel cell hybrid vehicles. Energ Convers Manage 185:537–551
5. Wang F-C, Yi-Shao Hsiao Y-S, Yi-Zhe Yang Y-Z (2018) The optimization of hybrid power systems with renewable energy and hydrogen generation. Energies 11(8):1948. https://doi.org/10.3390/en11081948
6. Lawan Bukar AL, Wei Tan CW, A review on stand-alone photovoltaic-wind energy system with fuel cell: system optimization and energy management strategy. J Clean Prod 2019; in press: https://doi.org/10.1016/j.jclepro.2019.02.228
7. Bizon N (2019) Hybrid power sources (HPSs) for space applications: analysis of PEMFC/Battery/SMES HPS under unknown load containing pulses. Renew Sustain Energy Rev 105:14–37. https://doi.org/10.1016/j.rser.2019.01.044
8. Pan ZF, An L, Wen CY (2019) Recent advances in fuel cells based propulsion systems for unmanned aerial vehicles. Appl Energ 240:473–485
9. Bizon N (2018) Real-time optimization strategy for fuel cell hybrid power sources with load-following control of the fuel or air flow. Energ Convers Manage 157:13–27
10. Olatomiwa L, Mekhilef S, Ismail MS, Moghavvemi M (2016) Energy management strategies in hybrid renewable energy systems: A review. Renew Sustain Energy Rev 62:821–835
11. Bizon N (2018) Optimal operation of fuel cell/wind turbine hybrid power system under turbulent wind and variable load. Appl Energ 212:196–209
12. Bizon N, Stan VA, Cormos AC (2019) Optimization of the Fuel Cell Renewable Hybrid Power System using the control mode of the required load power on the DC bus. Energies 12(10):1889–1904. https://doi.org/10.3390/en12101889
13. Bizon N (2019) Fuel saving strategy using real-time switching of the fueling regulators in the Proton Exchange Membrane Fuel Cell System. Appl Energ 252:113449–113453. https://doi.org/10.1016/j.apenergy.2019.113449
14. Bizon N (2019) Efficient fuel economy strategies for the Fuel Cell Hybrid Power Systems under variable renewable/load power profile. Appl Energ 251:113400–113518. https://doi.org/10.1016/j.apenergy.2019.113400
15. Priya K, Sathishkumar K, Rajasekar N (2018) A comprehensive review on parameter estimation techniques for Proton Exchange Membrane fuel cell modelling. Renew Sustain Energy Rev 93:121–144
16. Yue M, Jemei S, Gouriveau R, Zerhouni N (2019) Review on health-conscious energy management strategies for fuel cell hybrid electric vehicles: degradation models and strategies. Int J Hydrogen Energy 44(13):6844–6861

17. Bizon N (2018) Effective mitigation of the load pulses by controlling the battery/SMES hybrid energy storage system. Appl Energ 229:459–473
18. Dafalla AM, Jiang J (2018) Stresses and their impacts on proton exchange membrane fuel cells: a review. Int J Hydrogen Energy 43(4):2327–2348
19. Chen H (2019) The reactant starvation of the proton exchange membrane fuel cells for vehicular applications: a review. Energ Convers Manage 182:282–298
20. Bizon N (2014) Load-following mode control of a standalone renewable/fuel cell hybrid power source. Energ Convers Manage 77:763–772
21. Ahmadi P, Torabi SH, Afsaneh H, Sadegheih Y, Ganjehsarabi H, Ashjaee M, The effects of driving patterns and PEM fuel cell degradation on the lifecycle assessment of hydrogen fuel cell vehicles. Int J Hydrogen Energy 2019; in press https://doi.org/10.1016/j.ijhydene.2019.01.165
22. Wang F-C, Lin K-M (2019) Impacts of load profiles on the optimization of power management of a green building employing fuel cells. Energies 12(1):57. https://doi.org/10.3390/en12010057
23. Daud WRW, Rosli RE, Majlan EH, Hamid SAA, Mohamed R, Husaini T (2017) PEM fuel cell system control: a review. Renew Energ 113:620–638
24. Bizon N, Radut M, Oproescu M (2015) Energy control strategies for the Fuel Cell Hybrid Power Source under unknown load profile. Energy 86:31–41
25. Bizon N, Lopez-Guede JM, Kurt E, Thounthong P, Mazare AG, Ionescu LM, Iana VG (2019) Hydrogen Economy of the Fuel Cell Hybrid Power System optimized by air flow control to mitigate the effect of the uncertainty about available renewable power and load dynamics. Energ Convers Manage 179:152–165. https://doi.org/10.1016/j.enconman.2018.10.058
26. Bizon N, Iana VG, Kurt E, Thounthong P, Oproescu M, Culcer M, Iliescu M (2018) Air flow real-time optimization strategy for fuel cell hybrid power sources with fuel flow based on load-following. Fuel Cell 18(6):809–823. https://doi.org/10.1002/fuce.201700197
27. Bizon N, Oproescu M, Raceanu M, Efficient energy control strategies for a standalone renewable/fuel cell hybrid power source. Energy Convers Manage 90 (15 January 2015), 93–110, Impact Factor:4.801. https://doi.org/10.1016/j.enconman.2014.11.002
28. Pukrushpan JT, Stefanopoulou AG, Peng H (2004) Control of fuel cell breathing. IEEE Control Syst Mag 24(2):30–46
29. Bizon N (2019) Real-time optimization strategies of FC Hybrid Power Systems based on Load-following control: a new strategy, and a comparative study of topologies and fuel economy obtained. Appl Energ 241C:444–460. https://doi.org/10.1016/j.apenergy.2019.03.026
30. Bizon N, Thounthong P (2018) Real-time strategies to optimize the fueling of the fuel cell hybrid power source: a review of issues, challenges and a new approach. Renew Sustain Energy Rev 91:1089–1102. https://doi.org/10.1016/j.rser.2018.04.045
31. Restrepo C, Ramos-Paja CA, Giral R, Calvente J, Romero A (2012) Fuel cell emulator for oxygen excess ratio estimation on power electronics applications. Comp Electr Eng 38:926–937

Index

© The Editor(s) (if applicable) and The Author(s), under exclusive license to Springer
Nature Switzerland AG 2020
N. Bizon, *Optimization of the Fuel Cell Renewable Hybrid Power Systems*,
Green Energy and Technology, https://doi.org/10.1007/978-3-030-40241-9

Printed in the United States
By Bookmasters